Lecture Notes in Earth Sciences

Edited by Somdev Bhattacharji, Gerald M. Friedman,
Horst J. Neugebauer and Adolf Seilacher

38

Werner Smykatz-Kloss
Slade St. J. Warne (Eds.)

Thermal Analysis
in the Geosciences

Springer-Verlag

Berlin Heidelberg New York London Paris
Tokyo Hong Kong Barcelona Budapest

Editors

Prof. Werner Smykatz-Kloss
Mineralogisches Institut, University of Karlsruhe
Kaiserstr. 12, D-7500 Karlsruhe

Prof. Slade St. J. Warne
Department of Geology, University of Newcastle
NSW, Australia

ISBN 3-540-54520-4 Springer-Verlag Berlin Heidelberg New York
ISBN 0-387-54520-4 Springer-Verlag New York Berlin Heidelberg

This work is subject to copyright. All rights are reserved, whether the whole or part of the material is concerned, specifically the rights of translation, reprinting, re-use of illustrations, recitation, broadcasting, reproduction on microfilms or in other ways, and storage in data banks. Duplication of this publication or parts thereof is only permitted under the provisions of the German Copyright Law of September 9, 1965, in its current version, and a copyright fee must always be paid. Violations fall under the prosecution act of the German Copyright Law.

© Springer-Verlag Berlin Heidelberg 1991
Printed in Germany

Typesetting: Camera ready by author
Printing and binding: Druckhaus Beltz, Hemsbach/Bergstr.
2132/3140-543210 – Printed on acid-free paper

Preface

The application of thermal analysis to earth science has a long history. This is evidenced by the extensive coverages by Mackenzie (1957, 1970), Langier-Kuzniarowa (1967), Schultze (1971), Liptay (1973), Smykatz-Kloss (1974), Todor (1976) and Heide (1982). The chief thermal method has been *differential thermal analysis* (DTA). Additionally, *thermogravimetry* (TG; Duval, 1963; Keattch, 1969; Earnest, 1988) and *thermodilatometry* (Schomburg & Strörr, 1984) have gained some importance. All these methods are still widely used. But recently several new techniques have gained attention, such as thermomagnetometry, thermomechanical analysis and thermosonimetry. Improved equipment made possible the application of thermal methods to problems in thermodynamics and kinetics (e.g. by means of differential scanning calorimetry, DSC). This progress in the construction of new instruments as well as the combination of existing methods to enable simultaneous determinations (e.g. TG/DTA; TG/IR spectroscopy; DTA/mass spectrometry; DTA/microscopy; high-pressure DTA) have led to a resurgence in the use and application of thermal analysis in the earth sciences. Here the applications cover such diverse areas as the examination of individual minerals, mineral mixtures, rocks, soils, ceramics, cements, raw materials as well as their industrial evaluation, performance assessment and quality control. In the field of solid fossil fuels thermal determinations range from proximate analysis of inorganic constituents and the measurement of calorific values to the assessment of the environmental aspects of fly ashes and mineral residues.

To support this tendency, the International Confederation for Thermal Analysis (ICTA) has recently founded a "Committee for Thermal Analysis in Geosciences". The aim of this committee shall be to discuss, improve and distribute the knowledge about the possibilities of solving geoscientific questions by means of thermal analytic methods.

On a joint scientific meeting with the German Society for Thermal Analysis (GEFTA) in Berghausen / Karlsruhe (October 2nd to 5th, 1990) the members of the committee met for a round-table discussion, to review the application of thermal analysis in the different fields of the geosciences. The session was introduced by the Doyen of the group, Robert C. Mackenzie from Aberdeen, Scotland, who illustrated the development of thermal analysis in earth sciences. It continued with papers on the use of DSC and DTA in measuring thermodynamic parameters of minerals and melts (J. V. Dubrawski, W. Eysel), in studying the kinetics of mineral and melting reactions (K. Heide, J. Rouquerol) , and in characterizing the different water species in minerals and rocks (M. Földvari, C. August, E. T. Stepkowska). A number of papers was concerned with clays and clay minerals, their exact determination and use (A. Langier-Kuzniarowa, Ch. Earnest) and their special importance in environmental studies (S. Yariv, W. Smykatz-Kloss). The lectures of L. Stoch, J. V. Dubrawski, J. Rouquerol and S. Warne included the basic principles of refined methods such as "variable atmosphere thermal analysis" (S. Warne) and "controlled transformation rate thermal analysis" (J. Rouquerol). A. M. Abdel Rehim, J. Schomburg and S. Starck reported the use of thermal methods in applied and technical mineralogy, e.g. for the study of raw materials and the control of technological processes.

Of course, the above-mentioned lectures may not cover all fields of geosciences in which thermal analytical methods can be profitably applied, but they help to envisage the scope of possible applications.

We have to thank many persons, institutions and companies for making the conference and this publication possible. Drs. H. Schubert, N. Eisenreich, W. Engel and H. Schmid from the "Institute for Chemical Technology (ICT)" of the Fraunhofer Society offered the place and the facilities for the conference in Berghausen. For financial support we are grateful to:

    Deutsche Forschungsgemeinschaft, Bonn
    Dorfner-Analysenzentrum, Hirschau
    Gesellschaft für Thermische Analyse (GEFTA)

Fa. Gödecke, Freiburg/Br.
Heraeus Feinchemikalien, Karlsruhe
Fa. Madaus AG, Köln
Fa. Marx-Bergbau, Ruppach-Goldhausen
Netzsch-Gerätebau, Selb
Oerlikon-Schweißtechnik, Eisenberg/Pfalz
Polymer Laboratories, Heidelberg
Fa. Riedhammer, Nürnberg
Ing.-Büro Roth, Karlsruhe
Chr. Seltmann GmbH, Weiden
Sonderabfall-Entsorgung Saar, Saarbrücken
Süd-Chemie AG, München

Also, we are grateful to our co-workers Anke Brannath, Lutz Kaeding, Jürgen Reichelt, Eva Roller, Sabine Wenigwieser and especially to Andreas Heil and Wolfgang Klinke for their assistance. Our colleague Egon Althaus gave an exciting invited lecture on the German continental deep drilling project. Last but not least: thanks to W. S.-K.'s daughter Bettina for acting as hostess and travel guide during the conference.

Karlsruhe and Newcastle/NSW,

July 1991   Werner Smykatz-Kloss, Slade St. J. Warne

References

DUVAL C. (1963) - Inorganic thermogravimetric analysis.- 2nd ed., Elsevier Publ. Co., Amsterdam.

KEATTCH C. (1969) - An introduction to thermogravimetry.- Heyden and Sons, London.

(for the other cited papers see the references of the special lectures)

Contents

I Introductory Lectures

MACKENZIE R. C., Aberdeen
Geosciences in Thermal Analysis Development 2

DUBRAWSKI J. V., Newcastle/NSW
Differential Scanning Calorimetry and its Applications to
Mineralogy and the Geosciences 16

II Petrography

WARNE S. ST. J., Newcastle/NSW
Variable Atmosphere Thermal Analysis - Methods, Gas
Atmospheres and Applications to Geoscience Materials 62

FÖLDVARI M., Budapest
Measurement of Different Water Species in Minerals by
Means of Thermal Derivatography 84

AUGUST C., Wrocław
The Determination of Hydrated Sulphates in the Weathered
Crystalline Rocks by Means of Thermal Analysis 102

III Physico-Chemical Mineralogy

STOCH L., Krakow
Internal Thermal Reactions of Minerals 118

ROUQUEROL J., BORDÈRE S. & ROUQUEROL F., Marseille
Kinetical Study of Mineral Reactions by Means of
"Controlled transformation Rate Thermal Analysis
(CRTA)" 134

EYSEL W., Heidelberg
Thermoanalytical Investigations of Binary Oxide
Systems   152

HEIDE K., Jena
Kinetic Analysis of the Crystallization of Silicate Melts by
Means of DSC, DTA and Thermal Optical Methods   172

IV Technical Mineralogy

ABDEL REHIM A. M., Alexandria
Application of Thermal Analysis in Mineral Technology . . .   188

SCHOMBURG J., Neubrandenburg
Thermal Investigations in Technical Mineralogy   224

STARCK S., Eisenberg/Pfalz
Application of Thermal Methods in Raw Material Control
and During the Production Process   234

V Clay Mineralogy and Applied Geology

STEPKOWSKA E. T., SUŁEK Z., PEREZ-RODRIGUEZ J. L.,
MAQUEDA C. & JUSTO A., Gdansk and Sevilla
A Study of the Thermal Behaviour and Geotechnical
Properties of a Marine Clay and Its Composites   246

EARNEST C. M., Rome/Georgia
Thermal Analysis of Selected Illite and Smectite Clay
Minerals.
Part I. Illite Clay Specimens   270

Part II. Smectite Clay Minerals   288

LANGIER-KUZNIAROWA A., Warszawa
Remarks on the Applicability of Thermal Analysis for the
Investigations of Clays and Related Materials   314

YARIV S., Jerusalem
Differential Thermal Analysis (DTA) of Organo-Clay
Complexes   328

SMYKATZ-KLOSS W., HEIL A., KAEDING L. & ROLLER E.,
Karlsruhe
Thermal Analysis in Environmental Studies   352

Subject Index   368

Lecturers

Abdel Rehim, Prof. Dr. Ali Mohammed, Head of the Department of Geology, Fac. of Science, Alexandria Univ., Egypt.

August, Docent Dr. Czesław, Inst. of Geological Sciences, University of Wrocław, Poland.

Dubrawski, Dr. Jules V., BHP Research Laboratories, Shortland NSW, Australia.

Earnest, Prof. Dr. Charles M., Berry College, Rome, Georgia/USA.
President of NATAS, North Amer. Thermal Anal. Society.

Eysel, Prof. Dr. Walter, Mineralogisch-Petrographisches Institut der Universität, Heidelberg, Germany.

Földvari, Dr. Maria, Geolog. Survey of Hungary, Budapest

Heide, Docent Dr. Klaus, Abt. für Glaschemie der Universität, Jena, Germany.

Langier-Kuzniarowa, Prof. Dr. Anna, Geolog. Inst. of the University, Warszawa, Poland.
Vice-Chairperson of the ICTA-Committee for "Thermal Analysis in Geosciences".

Mackenzie, Dr. Robert C., Retired Director of the Inst. for Soil Science, Aberdeen, U.K.

Rouquerol, Dr. Jean, Centre de Thermodynamique et de Microcalorimetrie du CNRS, Marseille, France.
Chairman of the ICTA-Committee for Science.

Schomburg, Docent Dr. Joachim, Durtec-Ges., Neubrandenburg, Germany.

Smykatz-Kloss, Prof. Dr. Werner, Mineralogisches Institut der Universität Karlsruhe, Germany.
Chairman of the ICTA-Committee for "Thermal Analysis in Geosciences".

Starck, Dipl. Mineral. Stefan, Oerlikon-Schweißtechnik, Eisenberg/Pfalz, Germany.

Stepkowska, Prof. Dr. Ewa T., Inst. of Hydroengineering, Polish Academy of Sciences, Gdansk, Poland.

Stoch, Prof. Dr. Leszek, Academy of Mining and Metallurgy, Krakow, Poland.

Warne, Prof. Dr. Slade St. J., Geological Department, University of Newcastle, New South Wales, Australia.
President of the ICTA (International Confederation for Thermal Analysis).

Yariv, Prof. Dr. Shmuel, Dept. of Inorganic and Analytical Chemistry, The Hebrew University of Jerusalem, Israel.

Introductory Lectures

GEOSCIENCES IN THERMAL ANALYSIS DEVELOPMENT

R. C. Mackenzie

3 Westholme Crescent South, Aberdeen AB2 6AF
Scotland, U.K.

Abstract

The interaction between geological materials and thermal analysis from the earliest times up to the mid 1950s is reviewed, with particular attention to the contribution the earth sciences have made to the devolopment of thermal analysis.

Introduction

Geological materials have played a prominent role in the development of thermal analysis - and, conversely, thermal analysis has contributed significantly to the earth sciences. It is, therefore, somewhat surprising and disappointing that in only a few countries do thermoanalytical techniques find their proper place in the geosciences at the present day. Consequently, the initiative of the International Confederation for Thermal Analysis (ICTA) in setting up a Geosciences Committee to encourage greater use of the methods is to be welcomed.

As most geoscientific thermal analysts (if one may use the term) will be familiar with the manner in which thermal analysis has been enriched by the study of geological materials over the past few decades, I intend here to draw attention to the connections between the two subjects from earliest times to about the 1950s. It must be admitted that history does not usually repeat itself, but it does hold lessons for those who come after and this review may show where we have gone wrong.

Thermal analysis resulted from a logical sequence of events that have occurred from the time man first possessed fire. This occurred about 500,000 years ago in China and enabled the distinction of combustible and non-combustible materials - surely the first conscious use of "thermal analysis". The ICTA definition, that thermal analysis "covers a group of techniques in which a physical property of a substance and/or its reaction products is measured as a function of temperature whilst the substance is subjected to a controlled temperature programme", means that thermal analysis, as we know it, was not possible before a generally acceptable temperature scale became available. Yet methods of temperature control predate that period and, in assessing development, it would be wrong to neglect such early information.

As the following account is based largely on material in an earlier publication (MACKENZIE, 1984), only references that did not appear therein are inserted.

Prehistory of Thermal Analysis

Archaeological studies have revealed how simple smelting furnaces developed from the ancient camp fire and how pottery kilns capable of reaching at least 1000°C were in use by about the fifth millenium BC in Iran, Palestine and China. By about the middle of the third millenium BC, the quality of pottery produced by the Indus Valley civilization evidences the fine temperature control possible in their kilns (Fig. 1). These developments in control of high temperatures are all connected with natural geological materials and illustrate how man gradually exploited his knowledge of the control of heat to improve his living standards. Advances like these took some time to penetrate to western Europe, but the vitrified forts found in northern France and north-east Scotland (Fig. 2) clearly reveal that, in the first millenium BC, man must have been aware that some rocks melted more readily than others.

Fig. 1: (a) Section through a pottery kiln of the fifth millenium BC in the Middle East; (b) section through a pottery kiln of the third millenium BC of the Indus Valley civilization.

The first written account of the use of heat or fire (often used interchangeably) to distinguish minerals or rocks (not then differentiated) appears to have been that of Theophrastus about 315 BC. Indeed his descriptions are so vivid that many can be identified in today's idiom. Somewhat later (about 27 BC), the Roman architect Vitruvius described that bugbear of thermal analyst, the temperature gradient in the sample, when he commented that sun-dried bricks "cannot dry throughout" in less than two years. He also presaged thermogravimetry (TG) and thermodilatometry when he noted that limestone lost about one-third of its mass on firing while retaining the same bulk.

Apart from some advances in mineralogical classification, dependent to some extent on the thermal behaviour of minerals (e.g. Christoph Entzel in Germany in 1557 used "fusibles" as one of his group names), and gradually increasing improvements in the attainment and control of high temperatures, there is little to be recorded for the next 1600 years or so. Then, in the latter part of the 17th century, thermoluminescence was first observed by Robert Boyle in England in 1664 for diamond and later, in 1676, by J. S. Elsholtz in Braunschweig for fluorite. It was not until the 1930s, however, that the use of controlled temperature programmes enabled this technique, which is finding increasing use in geological dating

(AITKEN, 1985), to be classified as thermoanalytical. At about the same time (1689), J. K. von Loewenstjern in Germany suggested that the blowpipe could be used as an analytical tool: its expert use by T. O. Bergman in Sweden in the following century establishes another link between mineralogy and what might loosely be called "thermal analysis".

Fig. 2: Wall of a vitrified fort in Aberdeenshire, Scotland, showing boulders cemented by a clinker of molten rock.

In 1727, Stephen Hales in England heated a wide variety of materials (mainly organic, but including some mineral materials) to high temperatures in an iron retort and collected the gases evolved over water. In one set of experiments, he merely checked whether gases were evolved and in another he measured the volumes of such gases, thus sowing the seeds of evolved gas detection (EGD) and evolved gas analysis (EGA). Later in that century, in 1782, Josiah Wedgwood in England used his observation of the shrinkage of ceramics on firing to construct what became known as the Wedgwood pyrometer. This consisted of a "piece" of china clay cut to size after firing to a dull red heat, which was placed in the furnace, taken out after firing, allowed to cool and had its shrinkage measured in a V-notch arrangement graduated in "degrees Wedgwood". This did not enable continuous measurement of temperature and the assumption that shrinkage was linear with temperature led to large errors when degrees Wedgwood were converted to degrees Fahrenheit (or Celsius) (Fig. 3) - but at least high temperatures could be compared for the first time. Wedgwood, in 1784, also attempted EGA in a more refined manner than Hales, but he failed to observe the water evolved from kaolinite, which had condensed in the bladder he used. He did, however, note that the decomposition of kaolinite occurred over a narrow temperature range - an observation of critical importance with regard to TG.

Thermal Analysis Proper - to 1920

The ICTA definition requires a thermoanalytical technique to satisfy three criteria: (a) a physical property (or variable of state) must be measured; (b) measurement must be as a function of temperature (or of time where the time/temperature relationship is known); (c) a controlled temperature programme must be used. Although criterion (b) could be satisfied once the Fahrenheit temperature scale became generally accepted in the early part of the 18th century (VAN DER STAR, 1983), there were limitations, as temperatures could not be accurately measured beyond the range of a mercury thermometer and the only temperature control available was passive - i.e. allowing the sample to heat or cool in, as near as possible, a uniform-temperature environment. It was not until the early 20th century that automatic dynamic temperature control became available.

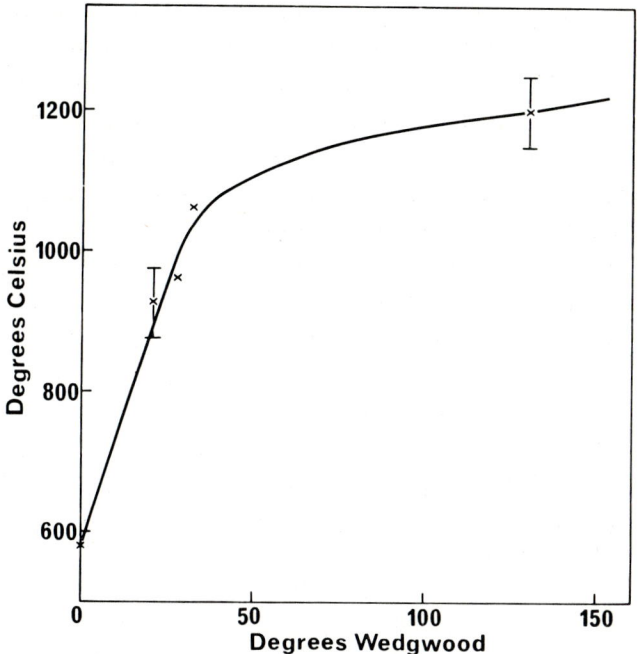

Fig. 3: Approximate relationship between degrees Wedgwood and degrees Celsius.

Nevertheless, several interesting studies, truly thermoanalytical within the above limitations, were performed, commencing with that described in 1739 by George Martine in Scotland, who first used the principle of DTA (MACKENZIE, 1989), although he did not pass through any phase transition points. On the basis of these experiments and some additional of his own, Joseph Black developed the concepts of heat capacity and latent heat. Unless water and mercury be regarded as geological materials, these experiments are outside the compass of this article. As indeed is the work of Fredrik Rudberg in Sweden in 1829, when he used inverse cooling-rate data to determine, with remarkable accuracy, the solidification points of metals and alloys. But the care with which he devised his cooling programme (Fig. 4) contrasts strongly with that used by M. L. Frankenheim in Germany in 1837 and 1854, when he obtained cooling-curve data for geological materials (sulphur and salpetre) by, apparently, simply allowing the hot samples to cool at the ambient temperature of the laboratory. Frankenheim observed both solidification and solid-state phase transition processes and the "Frankenheim method" was later (in the early 1900s) extensively used, with much better temperature control, by scientists in the Geophysical Laboratory in the USA to provide valuable data on phase transitions in feldspars, amphiboles and pyroxenes.

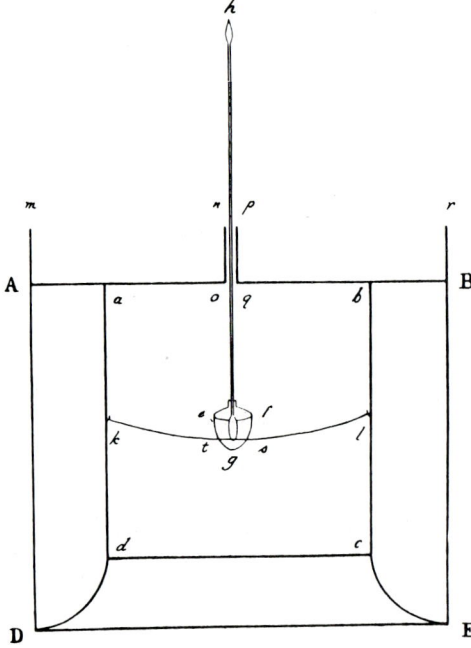

Fig. 4: Apparatus of Rudberg; the inner surfaces (abcd) of the double-walled container and the outside of the crucible (efg) were blackened to permit maximum heat transfer; the spaces between the two walls of the outer container and the top of the lid were filled with snow to ensure that the inner walls were always at 0°C.

Isothermal mass-change determination, as developed by J. B. Hannay in Scotland in 1877 is strictly not thermal analysis but is closely related thereto and a series of determinations at different temperatures (as Hannay performed) would satisfy the criteria. Hannay passed dried air over a thin layer of the sample in a U-tube immersed in a water- or oil-bath maintained at a constant temperature (Fig. 5), the water evolved being measured by weighing the absorption tube periodically. With this arrangement, the hemihydrate was shown to be a stable intermediate in the dehydration of gypsum and the existence of several hydrates of silica then supposed to exist was disproved. A few years later, in 1883, H. Le Chatelier in France, clearly unaware of Hannay's work, also studied the dehydration of gypsum by a heating-curve method. He positioned the bulb of a thermometer in the sample immersed in an oil-bath maintained at a constant 20°C above the temperature of the gypsum and plotted the temperature/time curve. Not only was this more accurate method of dynamic temperature control than any previous, but it must have approached the much more recent controlled transformation rate thermal analysis programme in characteristics.

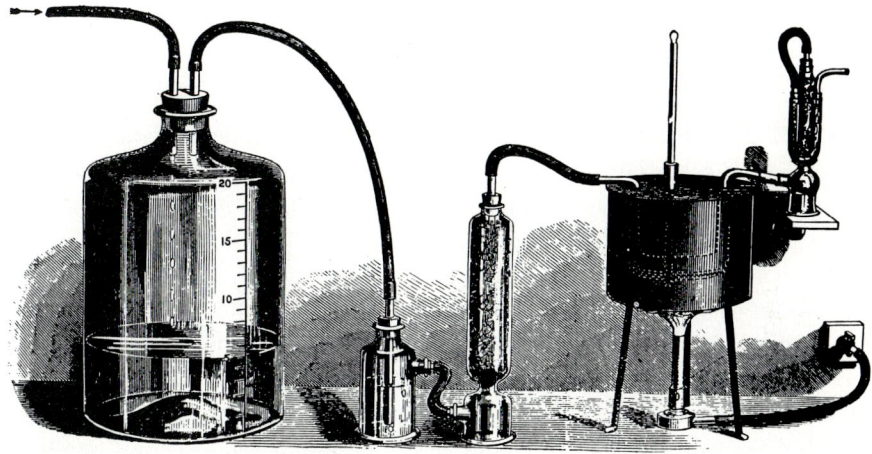

Fig. 5:   Apparatus used by Hannay to obtain isothermal mass-change curves.

Although most of the clay minerals we now know had been distinguished, mainly by the use of the five senses, by the 1880s, mistakes in identification often occurred. Well aware of the difficulties, Le Chatelier, who was interested in their industrial uses, considered they might be differentiated by their dehydroxylation temperatures. Consequently, in 1887, having in the previous year established the thermocouple as an accurate pyrometer, he heated samples of various clays, with thermocouples embedded in them, to about 1100°C at about 100°C/min and recorded on a photographic plate the reflections from the mirror of the galvanometer of the sparks from an induction coil set for two-second intervals. The spacings between the lines then indicated whether the sample heated more rapidly (open spacing) or less rapidly (close spacing) than its surroundings - i.e. the record represented heating-rate determination and not DTA as so often erroneously stated. By this technique, Le Chatelier distinguished five clay minerals (Fig. 6) and recognized several mixtures, thus for the first time putting clay mineralogy on a sound footing. This study, which revealed not only the value of thermal analysis in studying certain geological materials but also that the results could be recorded automatically is undoubtedly the fount of modern thermal analysis. More convenient recording systems were later developed by Le Chatelier himself, who refined the Saladin system (SALADIN, 1904), by W. C. Roberts-Austen in England, who used a moving photographic plate, and by N. S. Kurnakov in Russia, as it then was, who developed the more conventional recording drum. These developments were, incidentally, connected with metallurgy, which was the main home of thermal analysis during the late 19th and early 20th centuries.

The necessity for detecting transitions with small enthalpy changes in metallurgical studies led Roberts-Austen in 1899 to devise DTA. The first application to non-metals occurred in 1908, when W. Rosenhain in England applied it to detect quartz in some porcelains. Slavishly following Roberts-Austen, however, he used platinum as his reference material and obtained curves with so much base-line drift that he recommended the use of derivative DTA, the first mention of this variant. As later workers on non-metals used reference materials suited to their samples, Rosenhain's work may well have focussed attention on appropriate selection. The only other study on geological materials that contributed to the development of thermal analysis up to 1920 was that of H. Hollings and J. W. Cobb on coals in England in 1915, which showed the value of atmosphere control in DTA. All other DTA investigations, and the many heating-curve determinations, during this period contributed more to earth sciences than to thermal analysis. It is interesting to note, however, that K. Honda in Japan used calcite and gypsum in assessing the first thermobalance in 1915 (HONDA, 1915).

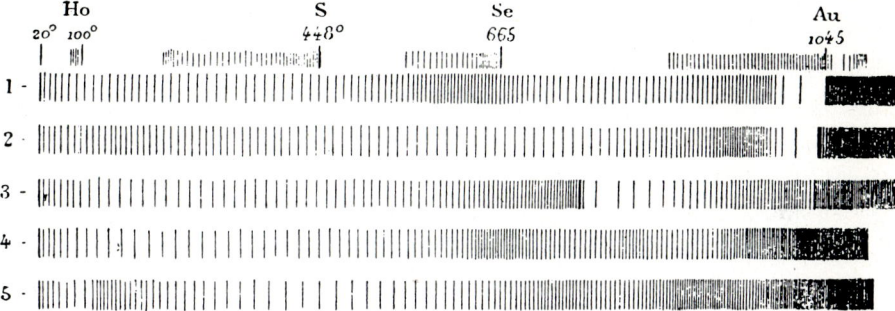

Fig. 6: Heating-rate traces obtained by Le Chatelier for: 1 - halloysite; 2 - allophane; 3 - kaolinite; 5 - montmorillonite. Also shown are the boiling points of water (Ho), sulphur (S) and selenium (Se) and the melting point of gold (Au).

Differential Thermal Analysis - from 1920 to 1955

Although thermoanalytical techniques such as TG and Thermodilatometry were used in mineralogical studies, particularly in France, during this period (and the value of the second has recently recieved detailed attention (SCHOMBURG & STÖRR, 1984)), attention here will be limited to DTA as the most common thermoanalytical techniques at that time.

During 1920-40, geological materials had considerable influence on the development of DTA and were indeed instrumental in ensuring its growth thereafter. Thus, in 1925, in their study of carnegieite, N. L. Bowen and J. L. Greig in the USA showed that DTA could be used quantitatively - a subject later intensively studied by their colleague F. C. Kracek in his investigations on polymorphic transitions. In 1927, the extensive and intensive DTA study of the chlorite minerals by J. Orcel in France revealed the value of the method in aiding the classification of complex mineral series - as did a similar study of the serpentine minerals by his pupil, S. Caillère, in 1936. In the same year, in the USSR, A. V. Kazakov and K. S. Andrianov, using geological materials, established the effects of many instrumental and experimental variables on the DTA curve. Unfortunately, this study seems to have been neglected outside the USSR: had its importance been appreciated elsewhere, much abortive work within the next 10 years might have been avoided. Despite the volume of thermoanalytical work emanating from Kurnakov's laboratory in the USSR from the early 1900s, DTA seems to have been first used there by E. Ya. Rode about 1930 and it was several years before its use became common. Yet its use in the USSR from that time was continuous and extensive and its value as a prospecting tool in connection with salt minerals and iron and manganese ores was well recognized in the late 1930s. This use was resuscitated by R. A. Nelson in the USA, quite independently, in 1942 for the assessment of bauxites (MACKENZIE, 1984) and led to the development and production of the first commercial, and only field-usable, DTA instrument in 1949. In contrast to the pattern in the USSR, DTA usage in the West was rather sporadic and the main impetus for its use came from the clay mineralogical investigations of F. H. Norton in the USA in 1939. Norton, in his enthusiasm, greatly overrated DTA by claiming in 1940 that the chemical composition of high-alumina clays could be deduced with an accuracy of 3-4% from DTA curves. Others, such as S. B. Hendricks and co-workers in the USA, also in 1939, were more realistic, merely recommending DTA as a useful complement to X-ray diffraction (XRD) and chemical analysis in the study of soil clay.

At that time interest in clay mineralogy was intense but few laboratories had the resources to equip themselves with XRD. Consequently, many clay mineralogists, probably infected by the enthusiasm of Norton but disregarding the precautions he carefully described, saw DTA as a fairly cheap alternative for, rather than complement to, XRD. Scientists in other disciplines were also interested but did not take the necessary precautions and had no idea of the amount of literature in languages other than English.

The result was disastrous, as DTA was applied in inappropriate circumstances to unsuitable materials and many concluded that DTA was of little value - an attitude that seems to have affected many earth scientists in the West to the present day.

Fortunately, some geoscientists appreciated the real value of DTA when properly applied. Foremost among these was R. E. Grim in the USA, who, with his collaborators, demonstrated, from 1942, the variety of information that DTA could yield and who employed it effectively in both academic and applied studies. Based on this work and the studies of others, such as J. A. Pask, I. Barshad and P. F. Kerr in the USA, A. L. Roberts in England, R. Norin in Sweden, R. Barta in Czechoslovakia and T. Sudo in Japan, centres of excellence in thermal analysis grew up, even although the majority of geoscience laboratories seemed to care little for thermoanalytical techniques. The impact of these centres spread gradually, although awareness has not yet reached the levels attained in the USSR and possibly Japan, where there was almost continuous use. Perhaps this explains why descriptions of new minerals from these countries usually carry DTA curves, whereas very few from other countries do.

To conclude this brief review in a more positive manner, however, let us note some significant contributions that the study of geological materials made to thermal analysis during the above, rather confusing, period. The introduction of the first commercial DTA instrument has been mentioned, but even further advances in instrumentation were made in 1951-52, when R. L. Stone in the USA designed and commercially produced an instrument for mineral studies capable of controlling the atmosphere within the sample over a wide pressure range - and later of enabling the study of milligram samples. Moreover, the work of MURRAY & WHITE (1949) in England on the dehydroxylation of kaolinite formed the basis for the derivation of kinetic data from DTA curves. Early theories, such as those of L. G. Berg and V. P. Anosov in the USSR in 1942, S. Speil in the USA in 1945, M. J. Vold in the USA in 1949 and P. L. Arens in Holland in 1951 were all derived as a result of work on minerals and the method of Berg in 1945 for resolving overlapping peaks was based on studies on dolomite - a mineral also used by B. Ya. Teitelbaum and Berg in initiating EGA in 1953. The theory of S. L. Boersma in Holland in 1955 that eventually led to the introduction of heat-flux differential scanning calorimetric (DSC) instruments was also based on work performed on clays.

Conclusion

The above account demonstrates that the close connection between geological materials and thermal analysis has, in the past, shown no respect for national boundaries and has persisted over a very long period of time. One would hope that similar close connection would persist into the future.

References

AITKEN M. J. (1985) - Thermoluminescence Dating.- Academic Press, London.

HONDA K. (1915) - Sci. Rep. Tohoku Univ., 4, 97-103

MACKENZIE R. C. (1984) - Thermochim. Acta, 73, 251-306, 307-367

MACKENZIE R. C. (1989) - J. Therm. Anal., 35, 1823-1836

MURRAY P. & WHITE J. (1949) - Clay Miner. Bull., 1, 84-86

SALADIN E. (1904) - Iron Steel Metallurg. Metallogr., 7, 237-252

SCHOMBURG J. & STÖRR M. (1984) - Dilatometerkurvenatlas der Tonmineralrohstoffe.- Akademie Verlag, Berlin.

VAN DER STAR P. (ed.) (1983) - Fahrenheit's Letters to Leibnitz and Boerhaave.- Rodopi, Amsterdam.

DIFFERENTIAL SCANNING CALORIMETRY AND ITS APPLICATIONS TO MINERALOGY AND THE GEOSCIENCES

J. V. Dubrawski

BHP Central Research Laboratories
Newcastle, NSW, Australia

Abstract

The review explores the contribution so far of DSC to the earth sciences. Outlined are the principles of power-compensated, heat-flux instruments and the measurement of enthalpies and specific heat. Areas covered in some detail include dehydration studies of minerals, mineral mixtures and the application of DSC to substitution effects in carbonates, particularly dolomite-ankerites. Also discussed is the use of DSC to the study of coal pyrolysis in relation to rank, hydrogenation as well as coal combustion. Specific heat and enthalpy measurements as an aid to oil shales and oil sands evaluation are also described.

Introduction

Differential scanning calorimetry (DSC) has been available for several decades and bears a resemblance to the older technique of differential thermal analysis (DTA). Thermally sensitive and easy to use DSC can provide quantitative thermodynamic information relatively quickly. However, because of the temperature restrictions of early instruments it has yet to fully establish itself in the field of the geosciences. This review discusses the method and its applications to mineralogy and materials relevant to earth sciences. General aspects of the technique have been described elsewhere (MCNAUGHTON & MORTIMER, 1975; MORTIMER, 1982; BROWN, 1988).

Instrumentation

The DSC instrument exists as two modifications employing different principles but yielding essentially the same result (WEBER-ANNELER & ARNDT, 1984). Perkin-Elmer developed the original concept and produced a system now referred to as power-compensated DSC (MCNAUGHTON & MORTIMER, 1975). Other manufacturers however, (e.g. DuPont and Setaram) have used a design superficially similar to DTA which has become known as heat-flux DSC (or sometimes as "quantitative DTA") (MORTIMER, 1982; BROWN, 1988).

In the power-compensated instrument (Fig. 1a) the individual sample and reference pans are provided with heaters which maintain a zero energy difference between both. During an endothermic or exothermic process electrical energy must be supplied to either the sample or reference so as to nullify the difference. The amount of this energy is equivalent to the enthalpy difference ΔH.

In the case of heat-flux DSC (Fig. 1b) the sample and reference are thermally connected through a conduction path that recieves heat from the surrounding furnace. As in DTA, difference-thermocouples are used to measure the temperature difference between sample and reference. The enthalpy ΔH is determined from the temperature difference which is directly proportional to the heat flux difference.

Several studies have reported on the performance of both types of instrument (HÖHNE, 1983; HÖHNE et al., 1983 and 1985; MARINI et al., 1984 and 1987; FLYNN et al., 1988). Currently the power-compensated design appears to yield slightly more accurate enthalpies and is thermally more stable (HÖHNE et al., 1985). Heat flux instruments are more susceptible to variations in sample characteristics and results can be affected by different heating rates (MARINI et al., 1984 and 1987). Such instruments require a reliable set of standards covering the entire range of investigation. The temperature response of both types of DSC has been compared and a dependence on heating rate observed for the heat-flux system. Both instrument types nonetheless are comparable with respect to their capabilities and have been shown to agree within an uncertainity of about 1% (HÖHNE et al., 1983).

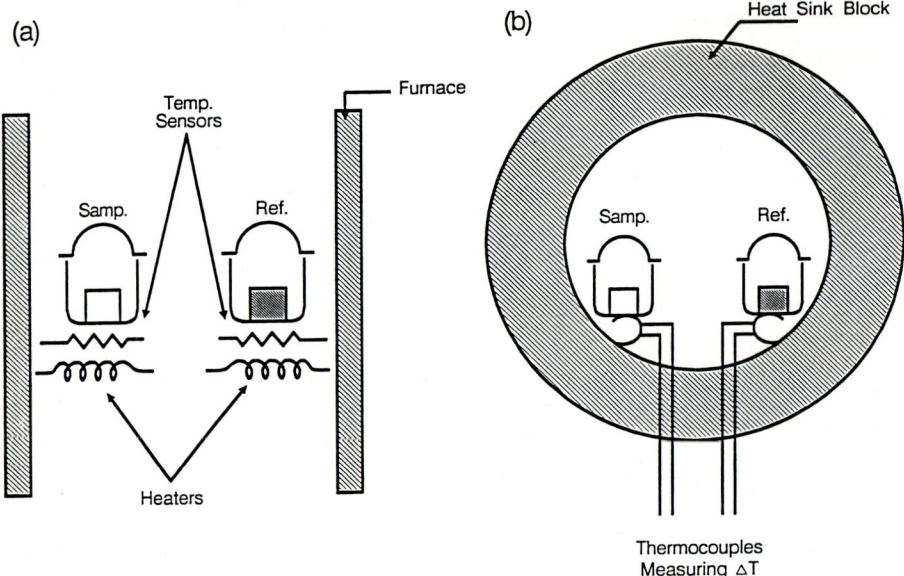

Fig. 1: Schematics of the principles of power-compensated DSC (a) and heat-flux DSC (b).

Power-compensated DSC however, is more restricted in its temperature range (max. ~800°C), whereas heat-flux units reaching 1500°C are commercially available (CHARLSLEY et al., 1984). Recently rapid heating (and cooling) DSC (200°C/min) of the power-compensated type has also been described (HIRAIRO et al., 1988).

Calibration and Enthalpy Measurements

The result obtained from a DSC scan is qualitatively similar to DTA although better thermal resolution can be achieved by the former. Heating a sample to the point where an endothermic or exothermic event takes place results in a peak whose intensity is a measure of the heat flow dH/dt, as a function of temperature (Fig. 2). The relationship between the enthalpy ΔH and the peak area A is given by,

$$\Delta H = K \cdot A / M \qquad [1]$$

where K is the "instrument" or "calibration" parameter and M is the sample mass (mg). As described below K is determined using a calibrant such as indium whose ΔH of fusion is accurately known.

The peak area (mm^2) must be converted into equivalent energy units, therefore both the instrument sensitivity (mJ/s) and chart speed (mm/min) are included in the enthalpy calculation, according to the expression,

$$\Delta H_s = (K \cdot S_c \cdot R_s \cdot A_s) / (S_s \cdot R_c \cdot M) \qquad [2]$$

where ΔH_s is the sample enthalpy, S_c and S_s are the recorder chart speed for the calibrant and sample runs respectively; R_s and R_c are the instrument sensitivity for the sample and calibrant runs respectively, and A_s is the sample peak area.

The parameter K is in fact temperature dependent and must be determined for each instrument over its temperature range. The extent of this dependence has been investigated by several authors. BARRALL & JOHNSON (1970) suggested that K varies by ~20% between 60°C to 330°C, whereas SCHWENKER & WHITWELL (1968) indicated a variation of about 4% for the range 156-420°C. Other workers (BREUER & EYSEL, 1982; DUBRAWSKI & WARNE, 1986a) have found K to vary substantially with temperature. DUBRAWSKI & WARNE (1986a) reported a 70% increase in K to a temperature of 820°C (Fig. 3). Such a variety of results arises presumably from the different behaviour of available instruments and reinforces the need to determine K over the entire temperature range of the experiment.

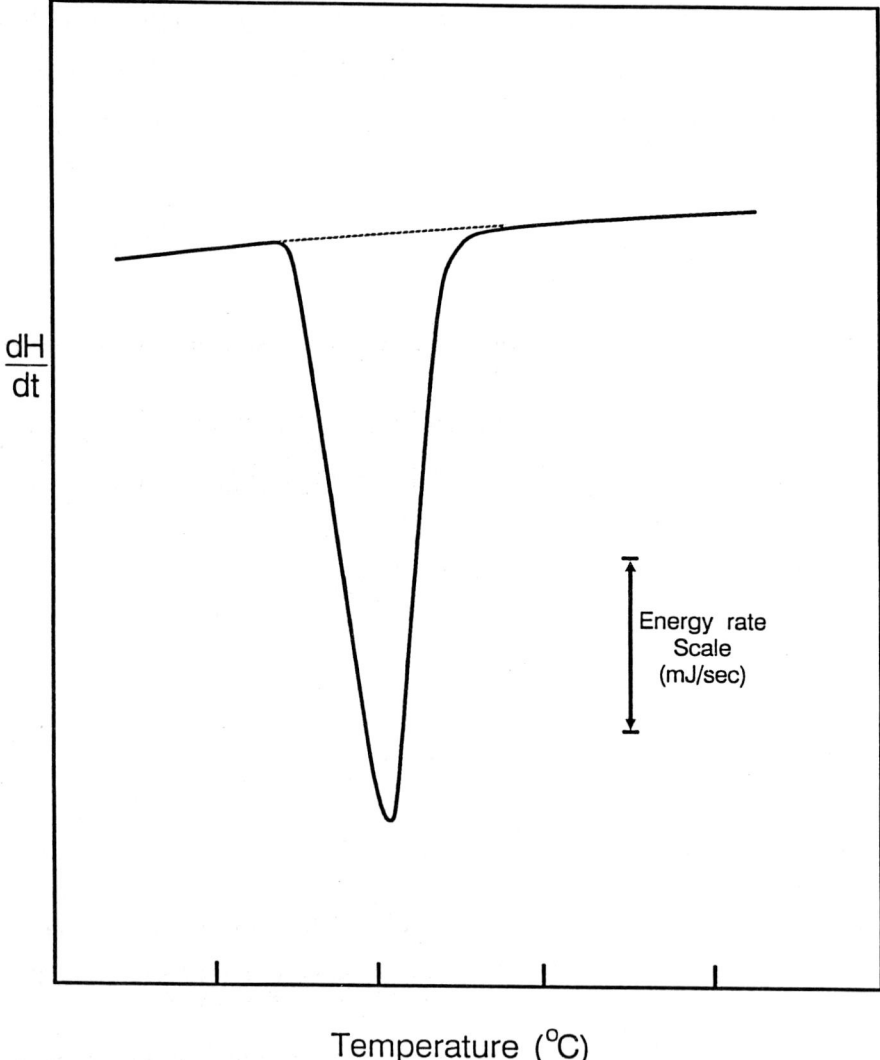

Fig. 2: Generalised DSC scan.

BROWN (1988) has remarked that power-compensated DSC yields K values less dependent upon temperature than heat-flux DSC. A theoretical description of the calibration of heat-flux instruments has been given by HÖHNE (1983).

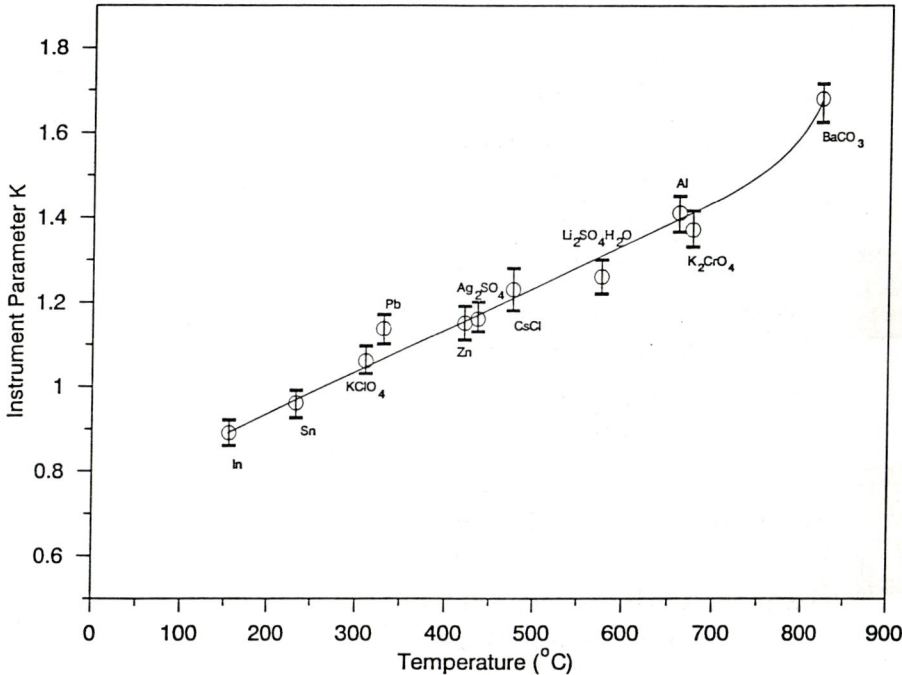

Fig. 3: Plot of instrument parameter K versus peak temperature (from DUBRAWSKI & WARNE, 1986a; reprinted with permission of Thermochimica Acta).

The accuracy of measured K depends upon the reliability of calibration materials. These have been investigated by several workers (GRAY, 1975; BREUER & EYSEL, 1982 and 1984; EYSEL & BREUER, 1984; DUBRAWSKI & WARNE, 1986a), in some cases to assist the ICTA committee with the recommendation of standards. BREUER & EYSEL (1982) examined 24 compounds providing melting or polymorphic transitions in the range 0-670°C, for which reliable ΔH values were available. They considered the effects of weighing errors, heating rate, size of ΔH and quality of background correction, upon enthalpy measurements. They concluded that the weighing error is underestimated and that ΔH did not vary with sample size. A major source of error, however, was delineation of the baseline since this affected the peak area. They found $Li_2SO_4 \cdot 2\,H_2O$, Sn, $AgNO_3$ and $RbNO_3$ to be good standards while Pb and K_2CrO_4 were acceptable. Quartz, however, was rejected since a reproducible baseline could not be established. They claimed that in many cases ΔH could be measured with a precision of <1% error. DUBRAWSKI & WARNE (1986a) also reported quartz and K_2SO_4 as unsuitable but found $KClO_4$, Pb, Ag_2SO_4 and $Li_2SO_4 \cdot 2\,H_2O$ yielded reliable results. SARGE & CAMMENGA (1985) studied different materials and compared K values for two heat-flux calorimeters in the temperature range of -100°C to 500°C. They reported the reproducibility of ΔH of fusion for both instruments to lie between 0.2% and 2%, but suggested that Pb be avoided as an enthalpy standard. The metals Zn, Sn and In have been used as standards during oil shales and oil sands analysis (RAJESHWAR et al., 1981; ROSENVOLD et al., 1982).

Measuring the DSC peak area is conventionally done by either counting squares, cutting out the peak and weighing, planimetry or digitisation and integration using a computer. Prior to this however, a reliable baseline must be drawn. When the heating rate is slow and specific heat changes small, the baseline can be accurately determined by merely joining the lines before and after a given thermal transition. Procedures for baseline extrapolation have been illustrated by BROWN (1988). A rigorous and more fundamental approach to baseline corrections has been described by several workers (BRENNAN et al., 1969; HEUVEL & LIND, 1970; GUTTMAN & FLYNN, 1973).

Sample Preparation and Experimental Techniques

Samples to be analysed are placed in cups fitted with lids that can be crimped, although loose fitting lids are also suitable. For dehydration studies lids with pin-holes can be used. A range of cups are available including aluminium for temperatures below 600°C, platinum, gold, silica and corundum. Samples can also be in various forms such as chips, granules or powders.

As with DTA however, the DSC curve is sensitive to sample configuration. Therefore to ensure optimum peak resolution and definition the thermal contact between pan and sample should be maximised. Lids should be polished and in good repair. Employment of a lid is essential to prevent radiation losses during heating. Enthalpies of carbonate minerals decomposed in nitrogen have been found to be reduced 20% when heated in uncovered cups (DUBRAWSKI & WARNE, 1987).

Particle size can also affect the DSC peak area as reported for quartz (DUBRAWSKI, 1987). The area of the α-β inversion and its reversible cooling β-α inversion, were observed to be unchanged within the size range 400 µm - 20 µm size, but decreased below 20 µm and markedly at 10 µm - 5 µm size (Fig. 4). Consequently discrepancies have resulted between XRD and DSC estimates for quartz. Such particle size effects have been reported in earlier DTA literature (DAWSON & WILBURN, 1970) and relate to the amorphous surface of fine quartz particles which effectively yield no thermal inversion. The influence of size has also been previously noted for carbonates (WEBB & KRUGER, 1970).

In dealing with complex materials such as coal and shale, additional factors must be considered as described by RAJESHWAR (1983). Of importance are sample geometry and heterogeneity. The poor thermal conductivity of coals and shales means that a thin uniform layer should be employed when the powder is available. Too small a sample however, leads to loss of representability.

The ambient atmosphere around a sample is almost important. During heating some samples generate their own atmosphere which can cause distortion of peak shapes or anomalous peak shifts for the pyrolysis reaction. It is possible that a reaction carried out in an oxidising atmosphere will produce overall endothermic rather than exothermic effects, should the supply of oxygen to the reacting particles be limited.

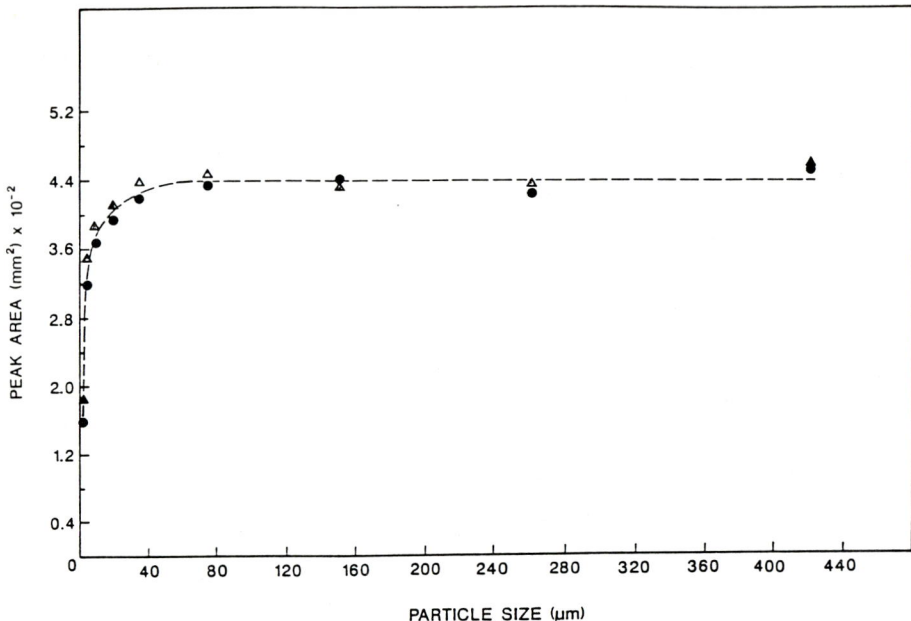

Fig. 4: The influence of quartz particle size upon DSC peak area; α-β transition (●); β-α transition (▲) (from DUBRAWSKI, 1987; reprinted with permission of Thermochimica Acta).

Therefore, the rate of products removal from the reaction zone is important. The residence time of products is determined by a combination of sample geometry, particle size, sweep gas flow rate and heating rate. In general, a high gas flow rate and a moderately high heating rate (>10°C/min) are recommended to minimise spurious effects.

Specific Heat

When a sample is subjected to a linear temperature rise heat flows into it at a rate proportional to the instantaneous specific heat, c_p. This is a measure of the heat needed to raise the temperature of unit mass of material through a degree of temperature, and is an important thermodynamic property. It is related to enthalpy change by,

$$\Delta H = \int_{T_1}^{T_2} c_p \, dT \qquad [3]$$

DSC allows c_p to be measured conveniently by the ratio method, described by O'NEILL (1966). A baseline is recorded within the temperature range of interest using empty cups. With a known mass of sample in the sample cup a heating curve is recorded as a function of temperature. An endotherm results from the sample absorbing heat. Thus,

$$\frac{dH}{dt} = m \cdot c_p \frac{dT}{dt} \qquad [4]$$

where dH/dt represents the heat flow, m the sample mass, c_p the specific heat ($Jg^{-1}K^{-1}$) and dT/dt the heating rate. The sample is replaced with a reference material of known c_p (e.g. sapphire, $\alpha\text{-}Al_2O_3$) (GINNINGS & FURUKAWA, 1953) which is heated in the same manner yielding a new curve (Fig. 5). Since the conditions are identical the following ratio can be derived.

$$\frac{c_p(\text{samp.})}{c_p(\text{ref.})} = \frac{x(\text{samp.})}{x(\text{ref.})} \cdot \frac{m(\text{ref})}{m(\text{samp.})} \qquad [5]$$

where x refers to the ordinate displacements of the sample and reference from the original baseline, and m is the mass of materials.

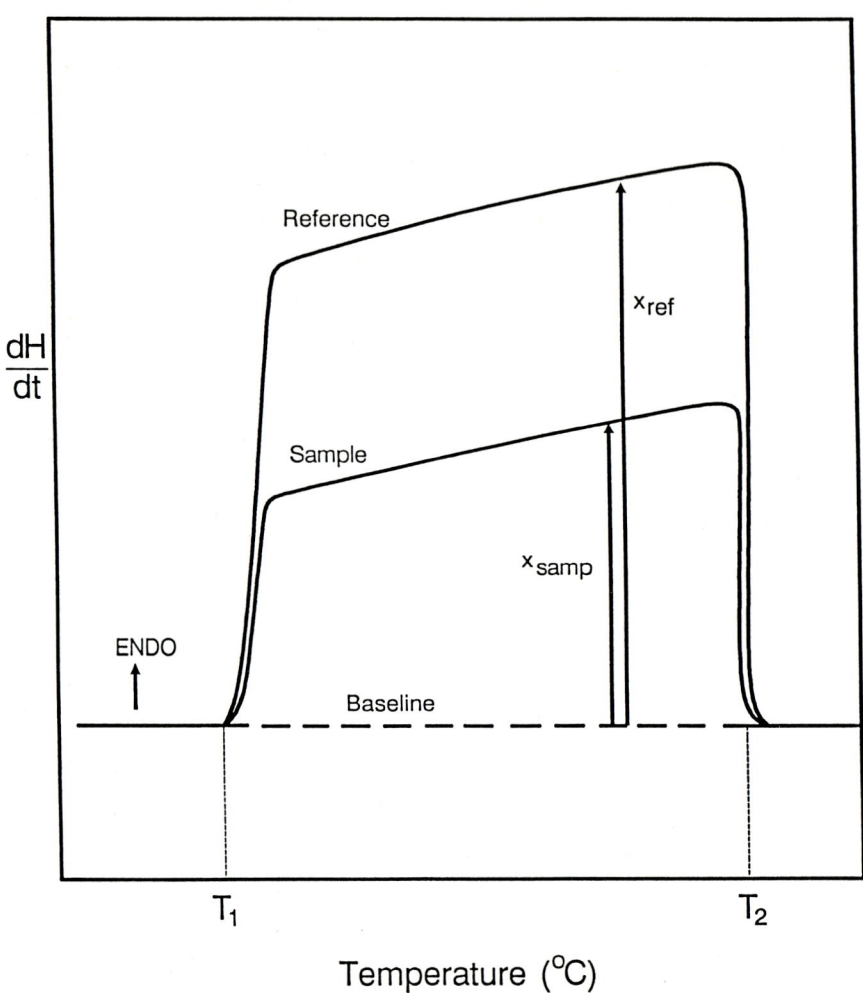

Fig. 5: Determination of specific heat c_p, using the ratio method.

Specific heats are small compared with enthalpies, therefore to maximise the displacements x instrumental sensitivity must be high (5-20 mJ/s) and scanning rates fast (~20°C/min). O'NEILL (1966) has reported a precision of 0.3% in the determination of c_p by the ratio method. He has also calculated that weight variations in the cups are not likely to cause significant errors.

Nevertheless other workers (BARRALL & JOHNSON, 1970; VARGHA-BUTLER et al., 1982) have taken this variation into account and derived the following expression for the correction in ordinate displacement,

$$\Delta x = (\Delta m \cdot c_p(M) \cdot S) / R \qquad [6]$$

where Δx is the ordinate displacement correction (mm) caused by the Δm difference in cup weights (mg); $c_p(M)$ is the specific heat of the cup material (e.g. M = Al, $Jg^{-1}K^{-1}$); S is the scan rate (Ks^{-1}) and R the setting of the range control divided by the chart span ($mJs^{-1}mm^{-1}$).

Despite the simplicity of the ratio method errors commonly arise from the following sources; (a) poor thermal contact between the base of the cup and holder; (b) errors in temperature linearity and heat read-out; (c) differences in the shapes of sample and reference materials. As SUZUKI & WUNDERLICH (1984) point out, a great deal of care is needed to obtain accurate and reproducible results.

Finally it should be noted that c_p is itself temperature dependent. It can be determined at each temperature within the measured range from equation [5] and is generally described in the form,

$$c_p = a + bT + cT^2 \qquad [7]$$

where a, b and c are constants.

Specific heat changes are large whenever a transitional change (e.g. solid to liquid) occurs (BROWN, 1988). In such cases a very marked shift in baseline occurs. Moreover the enthalpy of reaction measured by DSC should be corrected for the difference in c_p of reactants and products as described by MORTIMER (1982).

Dehydration Studies

Thermal methods are useful in characterising the behaviour of bound water in different mineral and geological systems.

The dehydration of gypsum has been studied because of its importance in plasters and cements (WEBB & KRUGER, 1972). A two-stage dehydration occurs,

$$CaSO_4 \cdot 2\,H_2O \rightarrow CaSO_4 \cdot 0.5\,H_2O \rightarrow CaSO_4$$

yielding hemihydrate ($CaSO_4 \cdot 0.5\,H_2O$) and anhydrite ($CaSO_4$). Using DSC (DUNN et al., 1987) samples of gypsum and cement mixtures have been heated in sealed Al cups from 80°C to 260°C. Resolution of both dehydration peaks was observed (Fig. 6) due to trapped water vapour, producing peaks at 150°C and 200°C. Air-tight sample cups have previously been used to investigate different hydrate phases (BUZAGH-GERE, 1980). SCHLICHENMEIER (1974 and 1975) has also employed sealed systems to study the dehydration of gypsum in plaster and determined the enthalpy for each dehydration step.

In the cement study a calibration curve was constructed from the area of the first dehydration step against the gypsum concentration to 8%. Analysis was reported to be rapid (20 min) and gypsum levels as low as 0.2% could be measured in cement. Moreover, cement samples milled at a range of temperatures were analysed and the amounts of gypsum and hemihydrate found to vary inversely with rising temperature.

DSC data for clays are rare. However, halloysite (10A) specimens formed under hydrothermal and weathering conditions have been investigated (MINATO, 1988). Tubular and spherical shaped halloysites were also included. Two dehydration steps between 30-120°C and 350-550°C were observed (Fig. 7). The low temperature endotherms produced ΔH values close to the latent heat of vaporization of water, suggesting that the water was loosely bound within the clay structure.

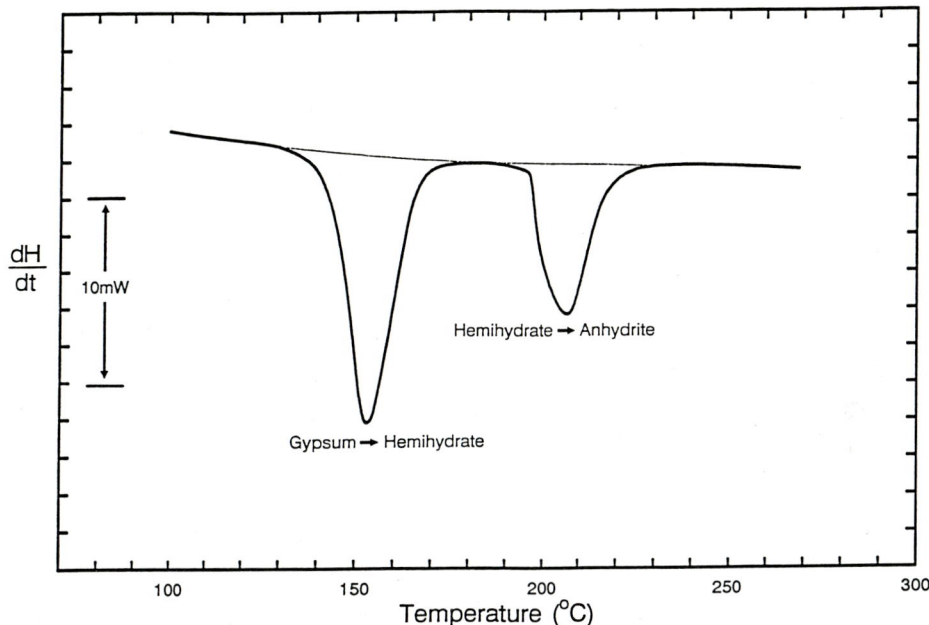

Fig. 6: DSC scan of gypsum heated in sealed cups (from DUNN et al., 1987; reprinted with permission of the Australian Chemical Society).

Dehydration enthalpies at 350-550°C varied for the different clays. Hydrothermal halloysites produced higher values than those formed under weathering conditions. Also, spherical halloysites possessed larger enthalpies than tubular halloysites formed under the same conditions. The degree of crystallinity was also significant.

Marine manganates selected from Pacific ocean nodules have been analysed using DSC/TG and XRD (OSTWALD & DUBRAWSKI, 1987a; DUBRAWSKI & OSTWALD, 1987). These tetravalent manganese oxides consisted of the minerals buserite, todorokite and vernadite. DSC revealed a two-step water loss process for buserite and broad ill-resolved steps for todorokite and vernadite (Fig. 8).

The first dehydration step for buserite at 100°C resulted in a loss of 2/3 of the bound water, corresponding to the formation of $Na_3MgMn_{14}O_{32} \cdot 7 H_2O$. This is close to the composition reported for synthetic birnessite (GIOVANOLI et al., 1970 and 1971). The product following the second dehydration peak was amorphous. Enthalpies of dehydration for marine manganates occurred in the range 250-400 Jg^{-1}. When converted to ΔH values per percentile of bound water buserite gave the lowest value with 20 Jg^{-1}, todorokite and vernadite between 26-30 Jg^{-1}. This was consistent with a layer structure already proposed from XRD analysis, and more rigid structures for todorokite and vernadite (OSTWALD & DUBRAWSKI, 1987b).

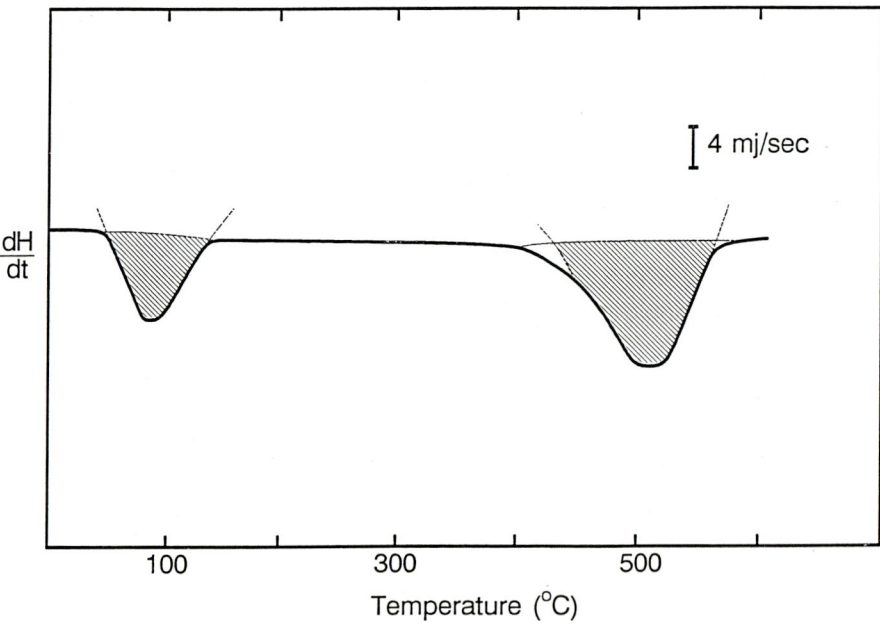

Fig. 7: Dehydration of halloysite (from MINATO, 1988; reprinted with permission of Thermochimica Acta).

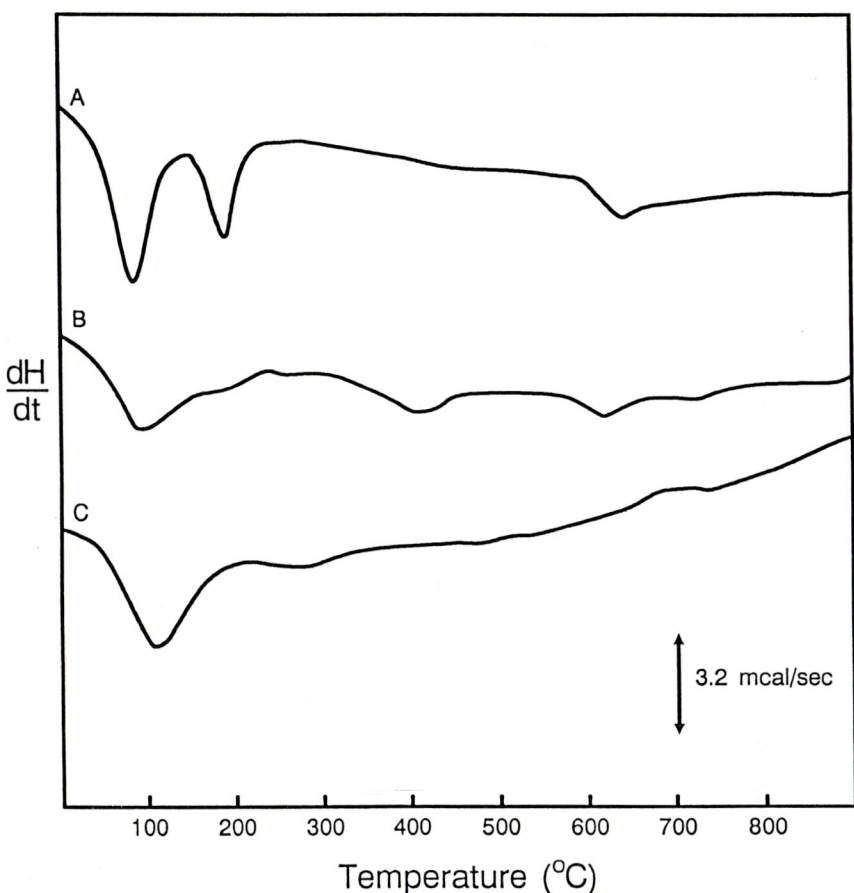

Fig. 8: DSC scans of marine manganates (a) buserite, (b) todorokite, (c) vernadite (from DUBRAWSKI & OSTWALD, 1987; reprinted with permission of N. Jb. Miner. Mh.).

Mineral Mixtures and Minerals in Coal

The applications of DTA to mineralogy (MACKENZIE, 1970 and 1972) and coal minerals (WARNE, 1979) are extensive but relatively few for DSC.

However the latter has been used (DUBRAWSKI & WARNE, 1986b) to quantitiatively analyse mixtures of kaolinite, gypsum and quartz (Fig. 9). Amounts of the component phases were determined from predetermined enthalpies of each mineral and equation [1]. For gypsum over the range 5-20% and kaolinite 30-80%, their amounts were measured within 3-6% of their true values using the respective enthalpies of dehydration.

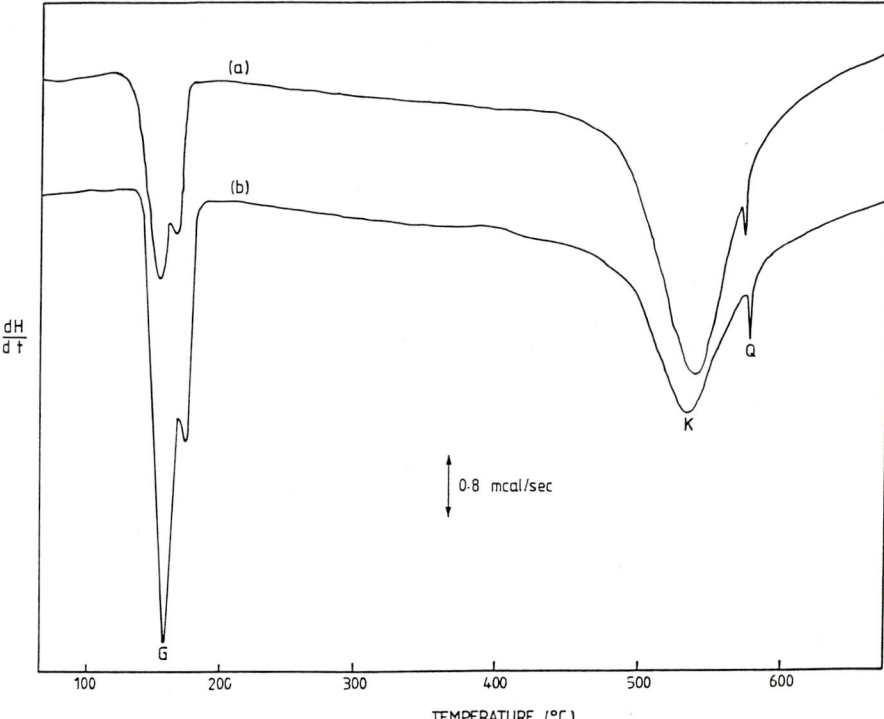

Fig. 9: DSC scans of mixtures of gypsum (G), kaolinite (K) and quartz (Q). Curve (a) G/K/Q (10/50/40); (b) G/K/Q (20/30/50) (from DUBRAWSKI & WARNE, 1986b; reprinted with permission of Thermochimica Acta).

Quartz, present in amounts between 5-50%, produced a weak α-β transition on the side of the kaolinite endotherm. Therefore it was determined using the reversible β-α exotherm upon cooling from 600°C. Relative errors were higher in this case, about 5-20%. This method was proposed as a general one for analysing quartz in the presence of thermally interfering species, although the effect of particle size must be considered. The mixtures of kaolinite, gypsum and quartz served as simple models for coal mineral matter.

It has been demonstrated that sulphur-bearing wastes can be quantitatively analysed by DSC (FICARA & MOREIRA, 1988). The presence of elemental sulphur was observed to produce two endotherms; a low temperature peak due to the rhombic-monoclinic transition and a higher temperature (~120°C) peak due to sulphur melting. The latter produced a linear relationship between peak area and percentage sulphur present.

Carbonate minerals occur commonly in coal and their effect upon the calorific value of coals has been considered (KAFRI et al., 1980; WARNE & DUBRAWSKI, 1988). It has been recognised that the endothermic breakdown of calcite lowers the calorific value of low-rank coals. The ΔH of decomposition of several carbonates has been determined using DTA (REDDICK, 1968). More recently DUBRAWSKI & WARNE (1987) have measured with DSC the decomposition enthalpies in N_2 of the carbonates siderite ($FeCO_3$), calcite ($CaCO_3$), magnesite ($MgCO_3$), dolomite ($CaMg(CO_3)_2$) and ankerite ($Ca(Mg,Fe)(CO_3)_2$).

Their results showed that, on a per gram basis, calcite possessed the largest ΔH followed in decreasing order by dolomite, magnesite, ankerite and siderite. In air siderite produced an exothermic effect due to oxidation of FeO to Fe_2O_3. It was concluded that coals containing mineral matter rich in calcite-dolomite would undergo a calorific reduction on heating, whereas the presence of siderite would provide a calorific enhancement. This enhancement however, would occur only for Fe-rich siderite and be reduced by Mg substitution because of the endothermic formation of magnesioferrite, $MgO \cdot Fe_2O_3$ (GALLAGHER & WARNE, 1981).

Substitution in Carbonate Minerals

The importance of the trigonal carbonate dolomite has resulted in various studies of its two-step decomposition mechanism (KULP et al., 1951; HAUL & HEYSTEK, 1952; BANDI & KRAPF, 1976; OTSUKA, 1986). The first endotherm is believed to originate from the $MgCO_3$ component of the structure, the second from the breakdown of the remaining $CaCO_3$. The effect of substitution by iron in the dolomite-ankerite isomorphous series, $CaMg(CO_3)_2$ - $Ca(Mg,Fe)(CO_3)_2$, has also been investigated by thermal methods (BECK, 1950; SMYKATZ-KLOSS, 1964; IWAFUCHI et al., 1983; OTSUKA, 1986). DSC has been applied to minerals in this series containing Fe in the molar ratio range of 0.082 to 0.49 (WARNE & DUBRAWSKI, 1987; DUBRAWSKI & WARNE, 1988a and 1988b).

DUBRAWSKI & WARNE observed three unresolved endothermic peaks, previously reported (BECK, 1950; KULP et al., 1951; SMYKATZ-KLOSS, 1964; WARNE et al., 1981; IWAFUCHI et al., 1983) by DTA, when these carbonates were heated in flowing N_2. The enthalpy of decomposition for each mineral was determined from the total peak area and found to decrease linearly with increasing Fe substitution. XRD of the products revealed dicalcium ferrite, $2\,CaO \cdot Fe_2O_3$, in amounts proportional to the iron content. MILODOWSKI & MORGAN (1981) previously observed the formation of this species.

This series was analysed further by DSC in flowing CO_2 where complete resolution of all endotherms was observed (DUBRAWSKI & WARNE, 1988a). Earlier DTA work (WARNE et al., 1981) had established the resolution of these peaks. The DSC scans produced two peaks for dolomite and three for the Fe-substituted carbonates (Fig. 10). Initial stages of decomposition for these substituted carbonates follow the scheme (KULP et al., 1951; MILODOWSKI & MORGAN, 1981; IWAFUCHI et al., 1983):

$$Ca(Mg_{1-x},Fe_x)(CO_3)_2 \rightarrow CaCO_3 + x\,FeO + (1-x)\,MgO + CO_2$$
$$2\,FeO + CO_2 \rightarrow Fe_2O_3 + CO$$
$$MgO + Fe_2O_3 \rightarrow MgO \cdot Fe_2O_3$$
$$FeO + Fe_2O_3 \rightarrow FeO \cdot Fe_2O_3$$

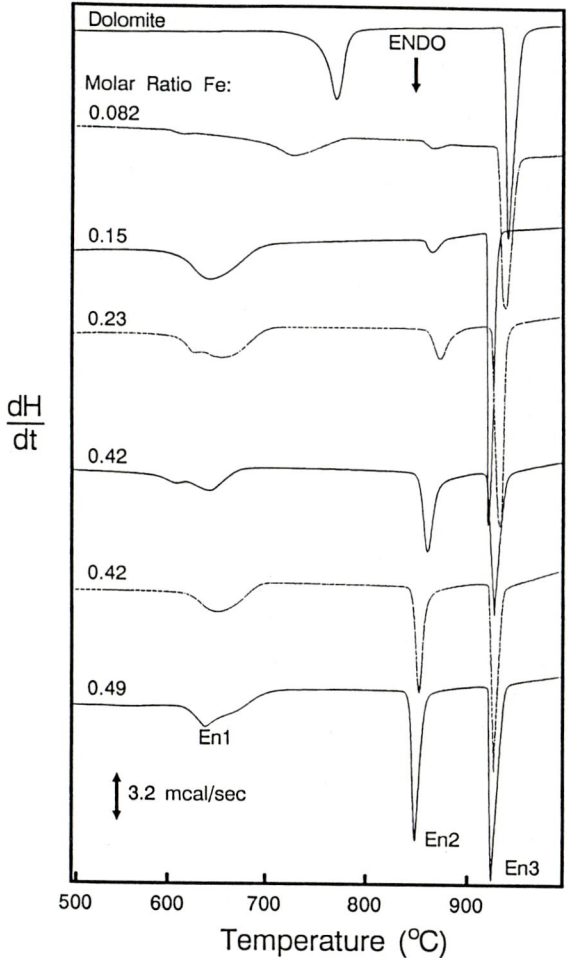

Fig. 10: DSC scans of dolomite - ferroan dolomite - ankerites in flowing CO_2 (from DUBRAWSKI & WARNE, 1988a; reprinted with permission of the Mineralogical Magazine).

This set of reactions produced the first endotherm, observed to be broad and sometimes split. The second endotherm resulted from the reaction,

$$2\,CaCO_3 + MgO \cdot Fe_2O_3 \rightarrow 2\,CaO \cdot Fe_2O_3 + MgO + 2\,CO_2$$

producing dicalcium ferrite. Finally, residual $CaCO_3$ decomposed to CaO resulting in the third endotherm.

Enthalpies ΔH_1, ΔH_2 and ΔH_3 were determined for these reaction steps. For dolomite ΔH_1 was about 100 kJmol^{-1}, close to the value for the decomposition of magnesite, indicating that the first endotherm resulted from the breakdown of the magnesium component of the mineral. Both ΔH_1 and ΔH_3 decreased with increasing Fe substitution. From the second endotherm a linear relationship was established between ΔH_2 and the degree of substitution (Fig. 11). ΔH_2 increased with Fe content but was influenced also by Mn substitution of the carbonates.

Finally the total enthalpy for each mineral carbonate ΔH_R, was calculated from the sum of ΔH_1, ΔH_2 and ΔH_3. It also decreased linearly with increasing substitution (Fig. 12). Extrapolation yielded a ΔH_R value of about 200 kJmol^{-1} for $CaFe(CO_3)_2$, the hypothetical end-member of the series. It was suggested that DSC is useful in distinguishing members of the series since the detection limit was estimated to be below 1 wt% FeO.

Decomposition in N_2 of members of the siderite-magnesite ($FeCO_3$-$MgCO_3$) series, has also shown a sharp decrease in enthalpy with Fe substitution (DUBRAWSKI, 1990). The influence of Mn has been clearly indicated and a linear relationship found between ΔH and mole fraction Fe+Mn. In most cases decomposition products consisted of substituted FeO and $(Fe,Mg)O \cdot Fe_2O_3$.

Finally, it has been reported that precursor carbonate solid solutions facilitate the formation of oxide solid solutions (FUBINI & STONE, 1985). The conversion of $Mn_xCa_{1-x}CO_3$ and $Sr_xCa_{1-x}CO_3$ to $Mn_xCa_{1-x}O$ and $Sr_xCa_{1-x}O$ respectively, were studied using DSC/TG and X-ray diffraction, which showed that an initial stage involves separation of the cations to form the oxide of one component and the carbonate of the other. The latter decomposes during a second stage and re-mixing of cations into a single oxide phase occurs.

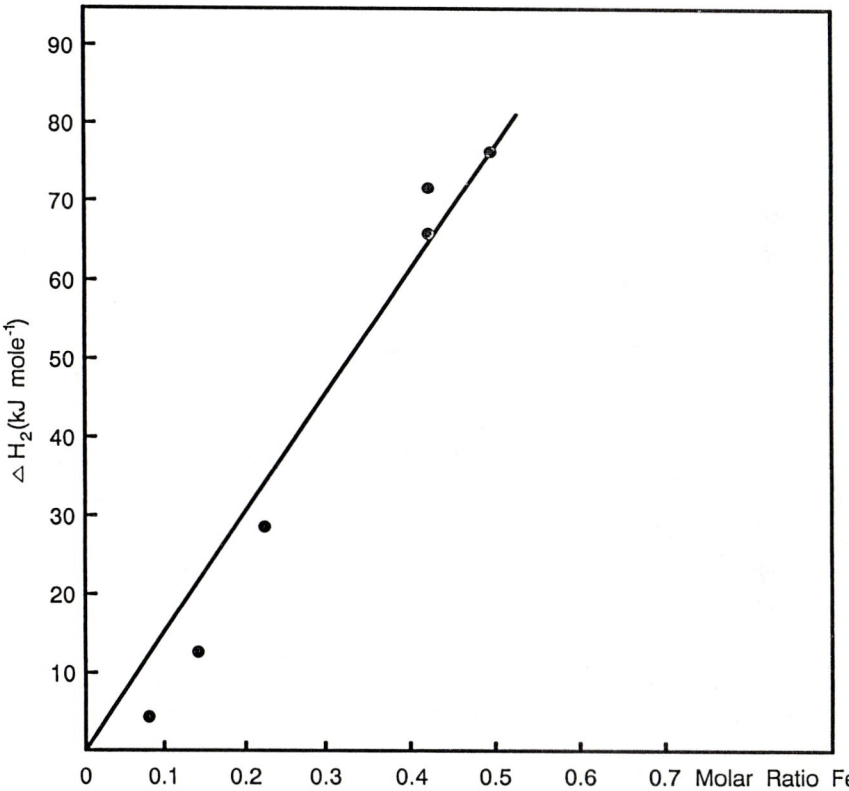

Fig. 11: Plot of ΔH_2 versus degree of Fe substitution for dolomite-ankerite minerals (from DUBRAWSKI & WARNE, 1988b; reprinted with permission of Thermochimica Acta).

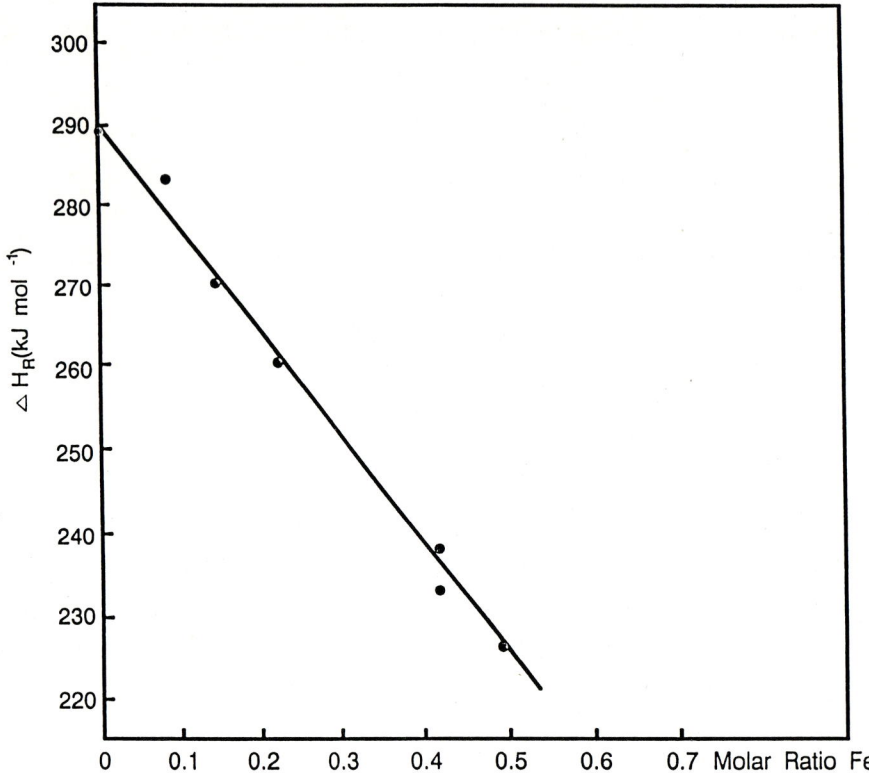

Fig. 12: Plot of enthalpy of decomposition ΔH_2 against degree of Fe substitution for dolomite-ankerite minerals (from DUBRAWSKI & WARNE, 1988b; reprinted with permission of Thermochimica Acta).

Coal

As reviewed by LAWSON (1970) many workers have employed DTA to study coal rank and elucidate the carbonisation process. Little quantitative data, however, is available. More recently reviews have appeared by RAJESHWAR (1983) on the TG and DSC of coal, shales and oil sands, and by WARNE & DUBRAWSKI (1989) on the DTA and DSC evaluation of coal and shales.

Coal Pyrolysis in Non-Oxidising Atmospheres

An early application of DSC involved pyrolysis to 580°C of 12 US coals in He atmosphere. MAHAJAN et al. (1976), applied pressures of 5.6 MPa in a DuPont pressure DSC cell (PDSC). During pyrolysis the weight loss was continually recorded and used to correct baseline variations in the original DSC curves. ELDER & HARRIS (1984) have since stressed the application of TG corrected DSC data to coal studies. This approach has been adopted in a degradation study of peats and lignites (OETER & STREMMLER, 1978), and the DSC of anthracite and coke (CARDILLO, 1980).

The thermal effects observed by MAHAJAN et al. (1976) were endothermic over the entire temperature range and for all coal ranks, anthracite to HVC bituminous. Low rank sub-bituminous and lignitic coals produced exothermic effects (Fig. 13).

Coal pyrolysis in N_2 near the plastic region (500°C) was carried out by GOLD (1980). Low-volatile bituminous to sub-bituminous coals produced exotherms between 400-500°C which did not correlate with rank. Peak intensity increased with heating rate and the exotherm shifted to higher temperature. Particle size also influenced the exotherm which decreased as the coal became finer. This was attributed to the tendency of fine particulate coal to oxidize, and the DSC curves of weathered coals revealed reduced exotherms. A relationship was proposed between the exothermic transition and production of volatile matter.

A suite of 21 Ohio bituminous coals analysed (ROSENVOLD et al., 1982) by DSC and TG in high purity N_2 to 575°C, produced completely endothermic peaks between 400-500°C, in agreement with MAHAJAN et al. (1976). It was suggested that DSC might be an effective indicator of coal rank. Three

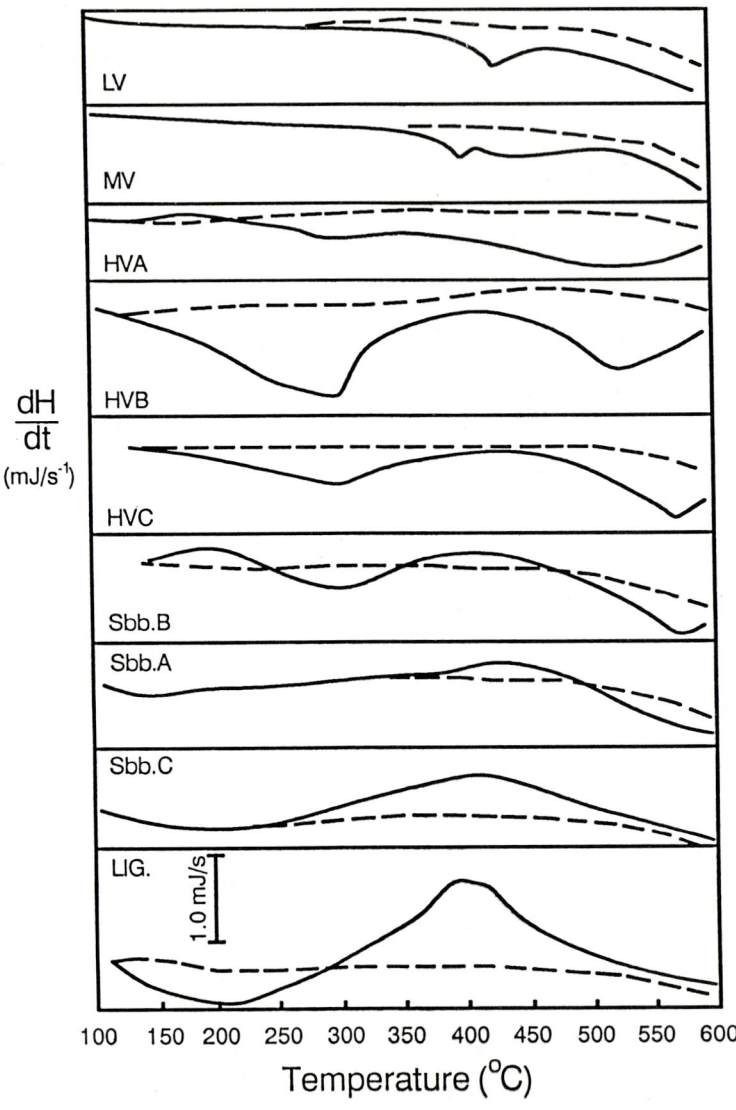

Fig. 13: Corrected DSC scans of US coals of different rank in He atmosphere. Baseline (– – –) (from MAHAJAN et al., 1976; reprinted with permission of Fuel).

regions of endothermic activity were identified. The first due to dehydration to ~150°C, the second at 400-500°C from devolatilization of organic matter, and a third partially resolved endotherm above 500°C.

Attempts to correlate the net area of the endotherm with volatile matter content were not very successful, and TG provided a more reliable estimate of this parameter. Evidence was presented for a catalytic effect induced by contact of the evolved gases with the pyrolysing sample. Autocatalytic effects have been reported for oil shales (RAJESHWAR et al., 1979). The extreme sensitivity of DSC peaks to atmosphere was demonstrated by introducing small amounts of O_2 whereupon a broad exotherm formed from the combustion of organic matter.

The inconsistencies between reported thermal effects using DSC (and DTA) has prompted RAJESHWAR (1983) to stress the sensitivity of peaks to local atmosphere, the possibility of autocatalysis, and the importance of choosing optimal experimental parameters.

The combination of DSC and TG has also been applied to Kentucky bituminous coals (ELDER & HARRIS, 1984). Kinetic analysis of the TG curves yielded activation energies for these curves in the range 198-220 kJmol^{-1} consistent with other work (WATERS, 1962). Heating rates up to 100°C/min were applied, to 600°C, and DSC curves were normalised by subtracting baselines. Particle size affected thermal results and reproducibility decreased with an increase in grain coarseness.

Following the endothermic loss of moisture the heat flow signal reflected the c_p of the dry coal. Several regions of thermal activity were recognised. From 200-300°C coal lost water from pyrolysis of phenolic moieties, and oxides of carbon from carbonyl and carboxylic groups (HOWARD, 1981). This enthalpy was measured to be 4.5±0.5 Jg^{-1}. At about 350°C, primary carbonisation occurred and CO_2 and H_2 were released (VAN KREVELEN, 1981). At higher temperatures CH_4, lower aliphatics, H_2, CO and alkyl aromatics were evolved and the entire effect was exothermic in agreement with GOLD (1980). However, the exotherm was less pronounced and the enthalpy determined as 22 ± 2 Jg^{-1}. Finally, prior to coke formation bituminous coals passed through a plastic state (WATERS, 1962).

Four of the Kentucky coals studied produced a final endotherm in 560-590°C region ascribed to pyrite/pyrrhotite transformation. This was consistent with the high pyritic sulphur character of these coals.

The hydrogenation of 20 US coals has also been investigated by MAHAJAN et al. (1977), using PDSC up to pressures of 5.6 MPa H_2 and to temperatures of 570°C. Although DSC output was corrected for weight loss enthalpies were conveniently expressed per gram of starting coal. Exothermic heats of hydrogenation for various coals ranged from 29 Jg^{-1} for anthracite to 640 Jg^{-1} for lignites.

Further when ΔH was plotted as a function of carbon content it decreased with increasing coal rank. The transition temperature marking the onset of the exothermic was also coal rank dependent.

The process of hydrogasification has been suggested to occur in the stages of devolatilisation, rapid methane formation and low rate gasification. A part of total heat evolved during hydrogenation was believed to result from the exothermic reaction between H_2 and surface carbon-oxygen complexes giving water and/or methane, (MAHAJAN et al., 1977).

The partial removal of pyrite from these coals by flotation decreased ΔH upon subsequent hydrogenation. This suggested that pyrite or its reduction products, possibly pyrrhotite $Fe_{1-x}S$, catalyse coal hydrogenation.

The heats of the pyrolysis of North Dakota lignite in H_2, Ar and CO have also been determined using DSC (HEFTA et al., 1986).

Coal Combustion

The combustion of solid fuels has been studied by DTA and relationships sought between the exothermic peak area and calorific value of the fuel

(LAWSON, 1970; MITCHELL & BIRNIE, 1970a and 1970b). Recently FYANS (1977) and EARNEST (1984a) employed DSC and reported good agreement between DSC calorific values and ASTM values for coal burned in O_2.

The heats of combustion or calorific values can be calculated from the usually two-stage exotherm of the coal. The total normalised peak areas are compared with a HV bituminous coal standard of known calorific value, according to the relationship,

$$\Delta H_s = \Delta H_r (A_s / A_r) \qquad [8]$$

where ΔH_s and ΔH_r are the calorific values of the sample and reference coal respectively; A_s and A_r are the peak areas of the sample and reference coal respectively.

Several empirical equations have been established yielding calorific values from elemental analysis data. According to EARNEST (1984b) the equation of MASON & GHANDHI (1980),

$$\Delta H(BTU/lb) = 198.11C + 620.31H + 80.93S + 44.95Ash - 5,153 \qquad [9]$$

appears to accord well with DSC determined heats of combustion. In this equation C, H, S represent weight percent of carbon, hydrogen and sulphur in the residues.

A number of workers, however (DOLLIMORE & MASON, 1981; VARHEGYI et al., 1986), have drawn attention to the problem of incomplete oxidation of organic materials, and indicated that DSC calorific values should be treated with caution. According to VARHEGYI et al. (1986) DSC and DTA oxidation start about 200°C and the majority of samples burn below 600°C. However, below this temperature catalysts are needed to ensure complete oxidation (SMOOT & PRATT, 1979; DE SOETE, 1982; MAZOR, 1983). Incomplete combustion is largely due to CO evolved during heating and is a primary product of carbonaceous materials.

VARGHEGYI et al. (1986) heated Hungarian sub-bituminous coals in pure O_2 and compared their results with those from bomb calorimetry. Values from DSC were low but could be increased by the presence of a catalyst such as $CuO/PbCrO_4$. Sulphur-bearing coals, however, poisoned a simple catalyst such as CuO and no enhancement was observed.

Oil Shales and Oil Sands

The thermal behaviour of oil shales and sands has been reviewed by several authors (RAJESHWAR et al., 1979; RAJESHWAR, 1983; WARNE & DUBRAWSKI, 1989). RAJESHWAR et al. (1981) demonstrated the effectiveness of DSC in characterising shale by relating the ΔH of decomposition of organic matter in shale to the oil yield determined by the traditional Fischer Assay.

The most important constituent of oil shale is the organic matter, kerogen and bitumen. In N_2 Green River Shales produced endotherms between 250-450°C corresponding to the decomposition of kerogen. RAJESHWAR et al. (1981) established a linear relationship between ΔH and oil yield with a precision of ±8 litres/metric ton of shale (Fig. 14). The relationship however, is valid only for the Green River formation and other deposits require their own calibration. Moreover the shales analysed did not contain any minerals decomposing below 500°C. The presence of species such as nahcolite ($NaHCO_3$), dawsonite ($NaAlCO_3(OH)_2$) and pyrite (FeS_2) would introduce thermal effects requiring correction.

The formation of the mixed carbonate $Na_2Ca(CO_3)_2$ in the Na_2CO_3 - $CaCO_3$ system is of significance in the utilisation of oil shales (SMITH et al., 1971). The extent of formation in the solid state has been determined by DSC and it has been proposed (GALLAGHER & JOHNSON, 1982) that structural disruption induced by formation of $Na_2Ca(CO_3)_2$ caused enhanced reactivity.

In his review RAJESHWAR (1983) mentioned the applicability of DSC to measurement of shale combustion enthalpies. Rotem, Kentucky and Green River shales have been combusted in O_2 and the latter found to have the highest enthalpy (LEVY & KRAMER, 1988). In this study variable atmosphere TG was applied to differentiate between volatile and non-volatile shale fractions. CRAWFORD et al. (1979) however, determined the enthalpies of Green River shale in a calorimeter by pelletising them with benzoic acid, thus providing an abundant local atmosphere of O_2 to promote complete combustion. They reported a precision of 21 Jg^{-1}.

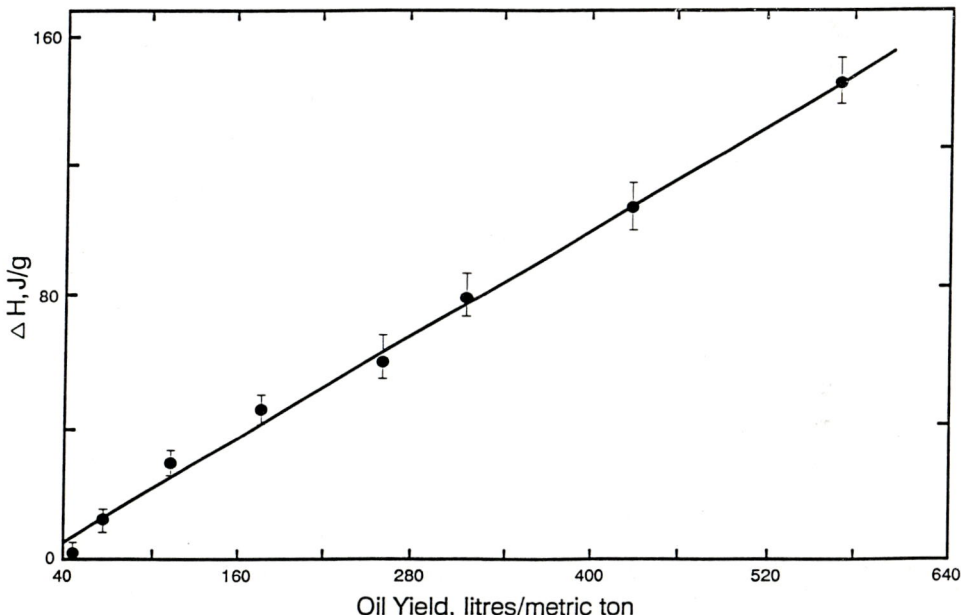

Fig. 14: Enthalpy of decomposition of kerogen versus oil yield for Green River oil shale (from RAJESHWAR et al., 1981; reprinted with permission of Analytical Chemistry).

Low temperature conversion of Mahogany Zone oil shales has been investigated with DSC/TG (WEN & KOBYLINSKI, 1983). Normal retorting involves heating to >500°C under air/gas mixtures. The breakdown of kerogen is complex producing pyrobitumen at first, followed by oil, gas and carbonaceous residue.

These shales produced a number of endotherms. The first at 50°C indicated the release of moisture and entrapped gas. A broad endotherm at 150-300°C resulted from the breakdown of kerogen to pyrobitumen, and some liberation of oil and gas. Further heating converted pyrobitumen to a polyoil. At 450°C a rapid weight loss occurred as the polyoil cracked to yield gas, liquid and coke. Finally, above 500°C decomposition of shale minerals took place.

The effect of temperature, pressure and residence time upon kerogen to pyrobitumen conversion was investigated. Increased H_2 pressure enhanced the low temperature DSC peak at 350-400°C and was beneficial to pyrobitumen formation. Reaction heats and kinetic parameters were determined. The low activation energies and high reaction orders in pyrobitumen formation and decomposition pointed to pyrobitumen being important in shale processing.

DSC curves of Green River shales have been compared with the electrical properties of the shales and the results of thermosonimetry (TS) (LØNVIK et al., 1980). Endotherms appeared between 100-300°C and 400-500°C, whose intensities were proportional to the amount of organic matter in the shale. A good correlation was obtained between TS, DSC and electrical measurements.

The combination of DSC/TG has been applied (ROSENVOLD et al., 1982) to the pyrolysis in N_2 of oil sands from four deposits in the US and Canada. Weight losses to 500°C were ascribed almost entirely to thermal changes to the indigenous bitumen. Extracted bitumen was also analysed thermally. In this case the major weight loss occurred below 500°C where a "semi-coke" was produced. The DSC curves indicated broad endotherms correlating with TG experiments. Peaks at about 100°C resulted from moisture loss, and between 400-500°C due to volatilisation of pyrolysis products. The net area of the two major endotherms gave ΔH, calculated to be between 400-672 Jg^{-1}, based on the starting weight of the material.

The pyrolysis and oxidation of Athabasca oil sands has been studied using PDSC/TG (PHILIPS et al., 1982). Samples were heated in N_2, air to 600°C in pressures up to 6.9 MPa. Enthalpies were calculated for the low temperature volatilisation reactions between 150-400°C, and for the high temperature cracking and volatilisation reactions at 400-500°C. Increases in pressure brought about increases in the enthalpies of endothermic and exothermic reactions.

Specific Heat of Coal and Oil Sands

The specific heats of bituminous and sub-bituminous Canadian coals have been determined by VARGHA-BUTLER et al. (1982). Both dried coals and coals with variable moisture content were analysed by the ratio method between 27-87°C. The c_p of dried coals increased almost linearly with temperature. Untreated coals however, showed a marked non-linear increase in c_p with temperature (Fig. 15). The authors could not explain this non-linear dependence and excluded the possibility of water loss since no weight change occurred during the experiment. The much smaller c_p of gaseous water, which could exist partially in the heated coal, might be a factor in this case. Generally the presence of water in coal is expected to have a marked effect upon c_p measurements since the c_p of water (4.19 Jg^{-1}K^{-1}) is about four times that of coal (KARR, 1978; BERKOWITZ, 1979). It was concluded that differences in c_p for bituminous and sub-bituminous coals resulted from their different water content rather than rank.

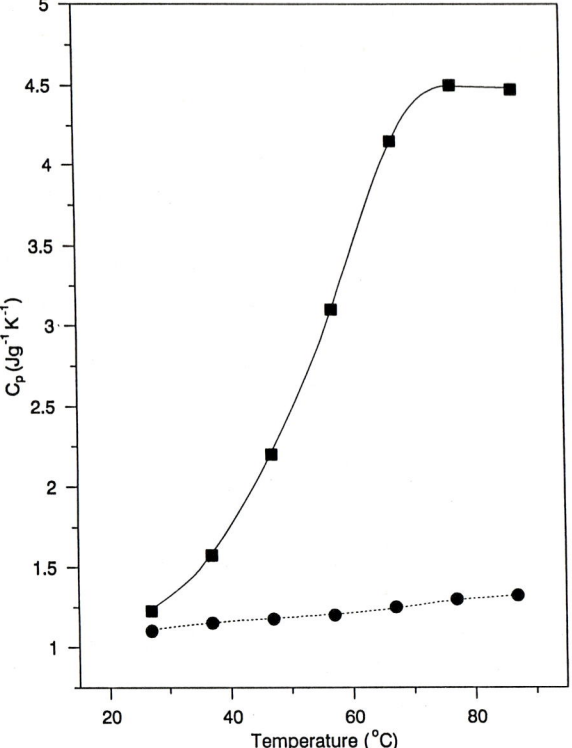

Fig. 15: Specific heat as a function of temperature for dried (●) and undried (■) Canadian coal (from VARGHA-BUTLER et al., 1982; reprinted with permission of Fuel).

VARGHA-BUTLER et al. (1982) also considered the effect of particle size upon c_p, which varied with mesh size. Since c_p is independent of particle size for chemically homogenous materials (VARGHA-BUTLER et al., 1982), the result observed with coal was attributed to its heterogeneous character. During grinding and screening minerals of lower hardness would be preferentially enriched in smaller particle size fractions. Pretreatment can also influence the composition of coal fractions. c_p values for given size fractions were different for wet and dry-screened coals.

Extrapolated to 100°C, the c_p values for dried coals obtained by these workers were 1.275-1.330 J/(gK), depending on size fraction. AGROSKIN & GONCHAROV (1965) obtained 1.35-1.40 $Jg^{-1}K^{-1}$ at 100°C for gas coal, and reported that the c_p of coal increases with temperature until decomposition

(350-550°C) occurs. ELDER & HARRIS (1984) determined c_p of dry bituminous Kentucky coals to be 1.21-1.47 Jg^{-1}K^{-1} in the range 100-300°C. This agreed with AGROSKIN et al. (1970), for coals of similar rank.

c_p measurements of coal have also been applied to probing the pore network of coals (MRAW & NAAS-O'ROURKE, 1979; MRAW & SILBERNAGEL, 1981). Differences in c_p versus temperature for bulk water and water bound within coal have been used to gain structural information.

Thermophysical characterisation of oil sands from US and Canadian deposits has included c_p measurements (RAJESHWAR et al., 1982). These were determined as 0.712-1.57 Jg^{-1}K^{-1} between 100-300°C. The temperature dependence of c_p fitted the form of equation [7] described earlier. Athabasca oil sand yielded a lower c_p value at 125°C (0.670 Jg^{-1}K^{-1}) relative to the other samples.

c_p values of raw oil sands were compared with those of extracted bitumens. Although a two to three-fold increase in c_p for bitumen was observed, no clear trend emerged linking c_p and bitumen content. This result was ascribed to the different origins and compositions of the sands.

The limited range of moisture in the sands investigated prevented a correlation of c_p and moisture content. However, as for coal, an increase in specific heat with water content is expected due to the high c_p of water relative to bitumen.

Conclusion

DSC has proved valuable in several areas pertaining to the earth sciences, especially when used in conjunction with other techniques such as TG. Nevertheless its contribution so far has been relatively minor. It can rapidly provide thermodynamic information, sometimes in situations where theoretical calculations are difficult, and with the advent of sophisticated high-temperature ranging instrumentation its widespread application is anticipated.

References

AGROSKIN A. A. & GONCHAROV E. I. (1965) - Determination of the specific heat of coals during carbonization.- Coke and Chem USSR, 11, 16-20

AGROSKIN A. A., GONCHAROV E. I., MAKEEV L. A. & YAKUNIN V. P. (1970) - Specific heat and heat of pyrolysis reaction for some Donbas coals.- Coke and Chem USSR, 5, 7-11

BANDI W. R. & KRAPF G. (1976) - The effect of CO_2 pressure and alkali salt on the mechanism of decomposition of dolomite.- Thermochim Acta, 14, 221-243

BARRALL II E. M. & JOHNSON J. F. (1970) - Instrumentation, techniques and applications of differential thermal analysis.- Tech Methods Polym Eval, 1, 1-42

BECK C. W. (1950) - Differential thermal analysis curves of carbonate minerals.- Am Mineral, 35, 985-1013

BERKOWITZ N. (1979) - An introduction to Coal technology.- Academic Press NY, p.80

BRENNAN W. P., MILLER B. & WHITWELL J. C. (1969) - Method of analysing curves in differential scanning calorimetry.- Ind Eng Chem Fundam, 8, 314-318

BREUER K.-H. & EYSEL W. (1982) - The calorimetric calibration of differential scanning calorimetry cells.- Thermochim Acta, 57, 317-329

BREUER K.-H. & EYSEL W. (1984) - Simultaneous temperature and calorimetric calibration of DSC cells.- ESTAC3 Interlaken Switzerland B19SC.

BROWN M. E. (1988) - Introduction to Thermal Analysis.- Chapman and Hall, London.

BUZAGH-GERE E. (1980) - Possibilities of detecting different hydrate phases by TG and DSC.- 6th ICTA, Bayreuth, Germany, 86CL.

CARDILLO P. (1980) - Use of combined thermoanalytical techniques for coal analysis.- Riv. Combust, 34, 129-137

CHARSLEY E. L., MARSHALL S. J., SWAN J. & PATEL R. (1984) - A new high temperature DSC apparatus.- ESTAC3 Interlaken Switzerland B18.

CRAWFORD P. C., ORNELLAS D. L., LUM R. C. & JOHNSON P. L. (1979) - Combustion calorimetry of oil shales.- Thermochim Acta, 34, 239-243

DAWSON J. B. & WILBURN F. W. (1970) - Silica minerals - Differential Thermal Analysis.- (MACKENZIE R. C. (ed.)), Academic Press, Vol. 1, p. 482-483

DE SOETE G. G. (1982) - Chemical aspects of the combustion of pulverized coal. Part 1.- Rev. Inst. Fr Pet, 37, 403-424

DOLLIMORE D. & MASON J. (1981) - Thermal decomposition of oxalates. Part 15. Effect of container material on DTA results for thermal decomposition of magnesium oxalate.- Thermochim Acta, 43, 183-187

DUBRAWSKI J. V. (1987) - The effect of particle size on the determination of quartz by differential scanning calorimetry.- Thermochim Acta, 120, 257-260

DUBRAWSKI J. V. (1990) - Thermal decomposition of some siderite - magnesite minerals using DSC.- Submitted to J Thermal Analysis.

DUBRAWSKI J. V. & OSTWALD J. (1987) - Thermal transformations in marine manganates.- N Jb Miner Mh, 9, 406-418

DUBRAWSKI J. V. & WARNE S. ST. J. (1986a) - Calibration of differential scanning calorimetry units for mineralogical studies.- Thermochim Acta, 104, 77-83

DUBRAWSKI J. V. & WARNE S. ST. J. (1986b) - The application of differential scanning calorimetry to mineralogical analysis.- Thermochim Acta, 107, 51-59

DUBRAWSKI J. V. & WARNE S. ST. J. (1987) - Use of differential scanning calorimetry in measuring the thermal decomposition of mineral carbonates occurring in coal.- Fuel, 66, 1733-1736

DUBRAWSKI J. V. & WARNE S. ST. J. (1988a) - Differential scanning calorimetry of minerals of the dolomite-ferroan dolomite-ankerite series in flowing carbon dioxide.- Min Mag, 52, 627-635

DUBRAWSKI J. V. & WARNE S. ST. J. (1988b) - Differential scanning calorimetry of the dolomite-ankerite mineral series in variable atmosphere.- Thermochim Acta, 135, 225-230

DUNN J., OLIVER K., NGUYEN G. & STILLS I. D. (1987) - Improved analysis of calcium sulphate hydrates in cement by thermal methods.- Proc 9th Aust Symp Anal Chem Sydney, 1, 88-91

EARNEST C. M. (1984a) - Characterization of coal liquefaction residues by thermal methods of analysis - Anal Calorimetry.- (JOHNSON, J. F. & GILL P. S. (eds)), Plenum Press NY, 5, 343-359

EARNEST C. M. (1984b) - Thermal Analysis of Clays, Minerals and Coal.- Perkin-Elmer Corp., Norwalk.

ELDER J. P. & HARRIS M. B. (1984) - Thermogravimetry and differential scanning calorimetry of Kentucky bituminous coals.- Fuel, 63, 262-267

EYSEL W. & BREUER K.-H. (1984) - in Analytical Calorimetry Proc Am Chem Soc Symp Anal Calorimetry.- (JOHNSON J. F. & GILL P. S. (eds)), Plenum Press NY, 5, 67

FICARA M. L. G. & MOREIRA E. (1988) - Utilization of differential scanning calorimetry technique for determination of high sulphur content.- Thermochim Acta, 134, 435-440

FLYNN R. M., FLYNN J. H. & BENT T. J. (1988) - Comparison of temperature response within and between power compensated and differential temperature DSC instruments.- Thermochim Acta, 134, 401-406

FUBINI B. & STONE F. S. (1985) - Two stage decomposition characteristics in the thermal decomposition of mixed carbonate systems.- Mater Sci Monogr 28A (React Solids Pt A), 85-92

FYANS R. L. (1977) - Thermal analysis application study, No. 21.- Perkin-Elmer Corp, Norwalk.

GALLAGHER P. K. & JOHNSON JR D. W. (1982) - Solid state reactivity in the system Na_2CO_3-$CaCO_3$.- J Phys Chem, 86, 295-297

GALLAGHER P. K. & WARNE S. ST. J. (1981) - Thermomagnetometry and thermal decomposition of siderite.- Thermochim Acta, 43, 253-267

GINNINGS D. C. & FURUKAWA G. T. (1953) - Heat capacity standards for the range 14 to 1200 K.- J Am Chem Soc, 75, 522-527

GIOVANOLI R., FEITKNECHT W. & FISCHER W. (1971) - Über Oxidhydroxide des vierwertigen Mangans mit Schichtengitter, 3. Mitteilung: Reduktion von Mangan (III) - Manganat (IV) mit Zimtalkohol.- Helv Chim Acta, 54, 1112-1124

GIOVANOLI R., STÄHLI E. & FEITKNECHT W. (1970) - Über Oxidhydroxide des vierwertigen Mangans mit Schichtengitter, 1. Mitteilung: Natrium Mangan (II, III) Manganat (IV).- Helv Chem Acta, 53, 209-220

GOLD P. I. (1980) - Thermal analysis of exothermic processes in coal pyrolysis.- Thermochim Acta, 42, 135-152

GRAY A. P. (1975) - The calorimetry of the NBS-ICTA temperature standards. Thermal Analysis.- Proc 4th ICTA Budapest, (BUZAS J. (ed)), Akademiai Kiado, 3, 991

GUTTMAN C. M. & FLYNN J. H. (1973) - On the drawing of the base line for differential scanning calorimetric calculation of heats of transition.- Anal chem, 45, 408-410

HAUL R. A. W. & HEYSTEK H. (1952) - Differential thermal analysis of the dolomite decomposition.- Am Mineral, 37, 166-179

HEFTA R., SCHOBERT H. & KUBE W. (1986) - Calorimetric pyrolysis of a North Dakota lignite.- Fuel, 65, 1196-1202

HEUVEL H. M. & LIND K. C. J. B. (1970) - Computerised analysis and correction of differential scanning calorimetric data for effects due to thermal lag and heat capacity changes.- Anal Chem, 42, 1044-1048

HIRAIRO K. I., OSHIMA T., OKAMATO H., KATO R. & MAESONO A. (1988) - A new DSC for high heating and cooling rates measurements.- Thermochim Acta, 134, 389-394

HÖHNE G. W. H. (1983) - Problems with the calibration of differential-temperature scanning calorimeters.- Thermochim Acta, 69, 175-197

HÖHNE G. W. H., BREUER K.-H. & EYSEL W. (1983) - Differential scanning calorimetry: comparison of power-compensated and heat flux instruments.- Thermochim Acta, 69, 145-151

HÖHNE G. W. H., EYSEL W. & BREUER K.-H. (1985) - Results of a round robin experiment on the calibration of differential scanning calorimeters.- Thermochim Acta, 94, 199-204

HOWARD J. B. (1981) - in Chemistry of Coal Utilisation, 2nd supplementary volume.- (ELLIOTT M. A. (ed)), Wiley-Interscience NY, chap. 12

IWAFUCHI K., WATANABE C. & OTSUKA R. (1983) - Thermal decomposition of ferromanganoan dolomite.- Thermochim Acta, 66, 105-125

KAFRI U., GERSCH S. & DOSORETZ C. (1980) - Corrections to calorific values of lignites of the Hula Basin, Israel, for contained $CaCO_3$.- Fuel, 59, 787-789

KARR JR C. (1978) - Analytical Methods for Coal and Coal Products NY, p. 248

KREVELEN D. W. VAN (1981) - Coal.- Elsevier NY, Chap 14, p. 265

KULP J. L., KENT P. & KERR P. F. (1951) - Am Mineral, 36, 643-670

LAWSON G. J. (1970) - Solid Fuels - Differential Thermal Analysis.- (MACKENZIE R. C. (ed)), Academic Press London, 1, 705-726

LEVY M. & KRAMER R. (1988) - Comparative TGA and DSC studies of oil shales.- Thermochim Acta, 134, 327-331

LØNVIK K., RAJESHWAR K. & DUBOW J. B. (1980) - New observations on chemical and structural transformations in Green River oil shales.- Thermochim Acta, 42, 11-19

MACKENZIE R. C. (ed) (1970) - Differential Thermal Analysis. Vol. 1.- Academic Press, London.

MACKENZIE R. C. (ed) (1972) - Differential Thermal Analysis. Vol. 2.- Academic Press, London.

MAHAJAN O. P., TOMITA A., NELSON J. R. & WALKER JR P. L. (1977) - Differential scanning calorimetry studies on coal 2. Hydrogenation of coals.- Fuel, 56, 33-39

MAHAJAN O. P., TOMITA A. & WALKER JR P. L. (1976) - Differential scanning calorimetry studies on coal 1. Pyrolysis in an inert atmosphere.- Fuel, 55, 63-69

MARINI A., BERBENI V., MASSAROTTI V., FLOR G. & CAMPARIVIGANO G. (1984) - Performances of a heat flux DSC cell.- ESTAC3 Interlaken Switzerland B32.

MARINI A., BERBENI V., FLOR G. & MASSAROTTI V. (1987) - On the quantitative reliability of heat flux DSC.- ESTAC4 Jena GDR, A15.

MASON D. M. & GHANDI K. (1980) - Formulas for calculating heating values of coals and chars.- Preprints ACS Fuels Division, 25, 325

MAZOR L. (1983) - Methods of Organic Analysis.- Elsevier Amsterdam.

MCNAUGHTON J. L. & MORTIMER C. T. (1975) - Differential scanning calorimetry.- Reprinted from Int Rev Sci Phys Chem Series 2, Vol 10 Butterworths.

MILODOWSKI A. E. & MORGAN D. J. (1981) - Thermal decomposition of minerals of the dolomite-ferroan dolomite-ankerite series in a carbon dioxide atmosphere.- Proc 2nd ESTAC Aberdeen, (DOLLIMORE D. (ed)), Heydon London, p. 468-471

MINATO H. (1988) - Dehydration energy of halloysite by means of DSC methods with the relationships of its mineralogy and modes of occurrence.- Thermochim Acta, 135, 279-283

MITCHELL B. D. & BIRNIE A. C. (1970a) - Organic Compounds - Differential Thermal Analysis.- (MACKENZIE R. C. (ed)), Academic Press London, 1, 611-637

MITCHELL B. D. & BIRNIE A. C. (1970b) - Biological Materials - Differential Thermal Analysis.- (MACKENZIE R. C. (ed)), Academic Press, London, 1, 673-702

MORTIMER C. T. (1982) - DSC in Thermochemistry and its Applications to Chemical and Biochemical Systems.- (RIBIERO DA SILVA M. A. V. (ed)), NATO ASI series, p. 47-60

MRAW S. C. & NAAS-O'ROURKE D. F. (1979) - Water in coal pores: low temperature heat capacity behaviour of the moisture in Wyodak coal.- Science, 205, 901-902

MRAW S. C. & SILBERNAGEL B. G. (1981) - in Chemistry and Physics of Coal Utilisation.- (COOPER B. R. & PETRAKIS L. (eds)), Am Inst Physic NY, p. 332

OETERT H. H. & STREMMLER J. (1978) - Thermal decomposition of younger fuels.- J Compend Dtsch Ges Mineraloelwiss Kohlechem, 78-79, 1107-1124

O'NEILL M. J. (1966) - Measurement of specific heat functions by differential scanning calorimetry.- Anal Chem, 38, 1331-1336

OSTWALD J. & DUBRAWSKI J. V. (1987a) - Buserite in a ferromanganoan crust from the south-west pacific ocean.- N Jb Miner Abh, 157, 19-34

OSTWALD J. & DUBRAWSKI J. V. (1987b) - An X-ray diffraction investigation of a marine 10A manganate.- Min Mag, 51, 463-466

OTSUKA R. (1986) - Recent studies in the decomposition of the dolomite group by thermal analysis.- Thermochim Acta, 100, 69-80

PHILLIPS C. R., LUYMES R. & HALAHEL T. M. (1982) - Enthalpies of pyrolysis and oxidation of Athabasca oil sands.- Fuel, 61, 639-649

RAJESHWAR K. (1983) - Thermal analysis of coals, oil shales and oil sands.- Thermochim Acta, 63, 97-112

RAJESHWAR K., JONES D. B. & DUBOW J. B. (1981) - Characterization of oil shales by differential scanning calorimetry.- Anal Chem, 53, 121-122

RAJESHWAR K., JONES D. B. & DUBOW J. B. (1982) - Thermophysical characterization of oil sands 1. specific heats.- Fuel, 61, 237-239

RAJESHWAR K., NOTTENBURG R. & DUBOW J. B. (1979) - Thermophysical properties of oil shales.- J Mater Sci, 14, 2025-2052

REDDICK K. L. (1968) - Heats of reaction for carbonate mineral decomposition.- Anal Calorim Proc Amer Chem Soc Symp, 155th, p. 297-303

ROSENVOLD R. J., DUBOW J. B. & RAJESHWAR K. (1982) - Thermal analysis of Ohio bituminous coals.- Thermochim Acta, 53, 321-332

ROSENVOLD R. J., DUBOW J. B. & RAJESHWAR K. (1982) - Thermophysical characterization of oil sands 4. Thermal analysis.- Thermochim Acta, 58, 325-331

SARGE S. & CAMMENGA H. K. (1985) - Calibration of differential scanning calorimeters using $\Delta H/T$ and c_p standards: extended application to two instruments.- Thermochim Acta, 94, 17-31

SCHLICHENMEIER V. (1974) - Determination of small amounts of $CaSO_4 \cdot 2H_2O$ in $CaSO_4 \cdot \frac{1}{2} H_2O$ by quantitative DTA.- 4th ICTA Budapest, p. 107

SCHLICHENMEIER V. (1975) - Detection and quantitative determination of $CaSO_4 \cdot 2H_2O$ in $CaSO_4 \cdot \frac{1}{2} H_2O$ using calorimetric DTA.- Thermochim Acta, 11, 334-338

SCHWENKER JR R. F. & WHITWELL J. C. (1968) - Differential enthalpic analysis as a calorimetric method: evaluation by a statistical design.- Proc Amer Chem Soc Symp Anal Calorim, (PORTER R. S. & JOHNSON J. F. (eds)), Plenum Press NY, p. 249-259

SMITH J. W., JOHNSON D. R. & ROBB W. A. (1971) - Thermal synthesis of sodium calcium carbonate - a potential thermal analysis standard.- Thermochim Acta, 2, 305-312

SMOOT L. D. & PRATT D. T. (ed) (1979) - Pulverised Coal Combustion.- Plenum Press NY.

SMYKATZ-KLOSS W. (1964) - Differential-Thermo-Analyse von einigen Karbonat-Mineralen.- Beitr. Mineral Petrogr, 9, 481-502

SUZUKI H. & WUNDERLICH B. (1984) - The measurement of high quality heat capacity by differential scanning calorimetry.- J. Thermal Anal, 29, 1369-1375

VARGHA-BUTLER E. I., SOULARD M. R., HAMZA H. A. & NEUMANN A. W. (1982) - Determination of specific heats of coal powders by differential scanning calorimetry.- Fuel, 61, 437-442

VARHEGYI G., SZABO P. & TILL F. (1986) - Problems in the DSC and DTA study of the burning properties of fuels and other organic materials.- Thermochim Acta, 106, 191-199

WARNE S. ST. J. (1979) - Differential thermal analysis of coal minerals - Analytical methods for coal and coal products.- (KARR C. (ed)), Academic Press, 3, 447-477

WARNE S. ST. J. & DUBRAWSKI J. V. (1987) - Differential scanning calorimetry of the dolomite-ankerite mineral series in flowing nitrogen.- Thermochim Acta, 121, 39-49

WARNE S. ST. J. & DUBRAWSKI J. V. (1988) - Potential for coal calorific value corrections, dependent on the carbonate mineral type present.- J Thermal Anal, 33, 435-440

WARNE S. ST. J. & DUBRAWSKI J. V. (1989) - Applications of DTA and DSC to coal and oil shale evaluation.- J Thermal Anal, 35, 219-242

WARNE S. ST. J., MORGAN D. J. & MILODOWSKI A. E. (1981) - Thermal analysis studies of the dolomite, ferroan dolomite, ankerite series I. Iron content recognition and determination by variable atmosphere DTA.- Thermochim Acta, 51, 105-111

WATERS P. L. (1962) - Rheological properties of coal during the early stage of thermal softening.- Fuel, 41, 3-14

WEBB T. L. & KRUGER J. E. (1970) - Carbonates - Differential Thermal Analysis.- (MACKENZIE R. C. (ed)), Chap 10, Academic Press, London, p. 306

WEBB T. L. & KRUGER J. E. (1972) - Building Materials - Differential Thermal Analysis.- (MACKENZIE R. C. (ed)), Vol 2, Chap 32, Academic Press, London, p. 187

WEBER-ANNELER H. & ARNDT R. W. (1984) - Thermal methods of analysis: differential scanning calorimetry in theory and application.- ESTAC3 Interlaken Switzerland A5.

WEN C. S. & KOBYLINSKI T. P. (1983) - Low-temperature oil shale conversion.- Fuel, 62, 1269-1273

Petrography

VARIABLE ATMOSPHERE THERMAL ANALYSIS - METHODS, GAS ATMOSPHERES AND APPLICATIONS TO GEOSCIENCE MATERIALS

S. St. J. Warne

Dept. of Geology, University of Newcastle
Newcastle, NSW 2308, Australia

Abstract

The application of thermal analysis (TA) in the geosciences is undergoing a resurgence due to the development of new equipment, methods and techniques. One of the latter, "variable atmosphere thermal analysis", involves the control, maintenance and or change of the atmosphere composition surrounding the sample during or between TA runs. It is advantageous in specifically affecting individual reactions, causing peak suppression or enhancement, the early or late appearance of reactions, increased peak definition, height and therefore detection limits, the resolution of superimposed peaks and improved mineral identifications. Applications are diverse as evidenced by the examples reviewed.

Introduction

The application of thermal analysis (TA) to earth science materials has a long and extensive history of use particularly in mineralogy. This is evidenced by the extensive coverages in the books and book chapters by MACKENZIE (1957, 1970 and 1972), SCHULTZE (1969), SMYKATZ-KLOSS (1974), TODOR (1976), WARNE (1967 and 1979), HEIDE (1982) and EARNEST (1984 and 1988). For much of this period the emphasis was almost entirely on the applications of differential thermal analysis (DTA).

This was subsequently complemented by thermogravimetry (TG) to give improved quantification of the thermal reactions recorded on DTA.

There followed a period where TA methods suffered overshadowing by a group of new mineralogical methods which ranged from atomic absorption and X-ray fluorescence spectroscopy to electron microprobe analysis. Latterly however, the simultaneous development of new methods e.g. thermomagnetometry (TM) and high temperature differential scanning calorimetry (DSC), up to approximately 1500°C and thermomechanical analysis (TMA), the coupling together of existing methods to give simultaneous determinations and the applications of the versatile technique of "*variable atmosphere thermal analysis*" has led to a resurgence in the use and applications of TA in the earth sciences. Here the applications cover such diverse areas as individual minerals, mineral mixtures, members of isomorphous substitution series, soils, ceramics, cements, natural geological materials characterization and their industrial evaluation, performance assessment and quality control. Further, in the field of solid fossil fuels determinations ranging from proximate analysis, inorganic contents/constituents and calorific values to the assessment of environmental aspects of fly ash, mineral residues, the release and extraction of evolved gases such as SO_2, together with their extraction from flue gases by various sorbents and their regenerative capacities have provided invaluable data and in many cases have become routine.

Variable Atmosphere Thermal Analysis

Of particular value has been the use of "variable atmosphere thermal analysis" as a technique for gaining additional information by its application to already well established methods (WARNE, 1986). This technique for TA studies is used where the furnace atmosphere conditions, surrounding the sample, may be preselected, controlled, maintained or changed as required, (WARNE, 1987), between or during individual TA runs (EARNEST, 1983). This may be achieved with individual or combinations of different purge gases in sequence or as mixtures at ambient and above ambient pressures or vacuum, under dynamic, isothermal or quasi-isothermal heating conditions. The required atmosphere conditions may be maintained by a positive gas flow through or over the sample under study, with the latter almost always used in order to minimize sample disturbance and loss from the sample holder.

With the application of this technique, involving the decrease or increase of partial pressures and the presence or absence of specific gases, individual reactions may be made to appear earlier or later, be suppressed or enhanced, produce greater sensitivity, peak height and therefore detection limits. This leads to the resolution of superimposed peaks and improved mineral identifications (WARNE, 1977; WARNE & FRENCH, 1984). In addition, organic contents, coal and coal chars and combustion characteristics (WARNE, 1985; CUMMING & MCLAUGHLIN, 1982; CUMMING, 1989), coal proximate analysis, pyrite contents and benefication products may be determined.

Variable atmosphere thermal analysis is of course equally applicable to single or simultaneous TA determinations. However, it is vital to note that for different purge gases the resultant DTA (and DSC) peak areas from duplicate samples determined under otherwise identical conditions will differ predictably. This is due to differences in their thermal conductivity which for air, N_2 and CO_2 is similar, while He is approximately 6 times greater and Ar 2/3 that of N_2 (WARNE, 1978).

Added advantages result from the latter case where synchronous determinations of complementary parameters (e.g. DTA/TG, DTA/TG/DTG or DTG/TM) may be obtained from a single sample under exactly the same conditions of thermal analysis. This obviates inconsistencies due to possible variations in, for example, the composition, grain size distribution and packing of "identical" duplicate samples, furnace atmospheres and heating rates.

General Applications

General examples of the types of applications involving variable atmosphere control are:

(1) the use of air or O_2 to promote reactions i.e. for oxidation, combustion, organic contents, mineral identification and coal mineral, ash and magnetic component determinations.

(2) for determinations under inert conditions e.g. using pure N_2, Ar or CO_2, to prohibit oxidation reactions, particularly where decomposition and oxidation reactions overlap, remove the effects of the partial pressure of other gases i.e. CO_2 in air in carbonate studies or O_2 in N_2 in sulphide studies, to provide improved individual peak and reaction definition, mineral identification, detection limits, coal mineral contents and volatile yields.

(3) purge gas partial pressure effects where increasing furnace atmosphere partial pressures of the same gas released during sample decomposition have to be overcome for this decomposition to take place. This results in increasing decomposition temperatures causing upscale movement of decomposition reactions and faster decomposition reaction speeds causing more vigorous reactions over a smaller temperature range. Hence detection limits are improved.

For example calcite ($CaCO_3$) decomposition temperatures respond to the different CO_2 partial pressures in air (low) and flowing pure CO_2 (high), see Fig. 1.

In addition to CO_2, other gases such as water vapour, sulphur dioxide and oxygen at different partial pressures may be used to preferentially move, suppress or enhance specific individual reactions. This assists in reaction suppression, definition, identification, superimposed peak resolution and the benefication/removal of iron minerals.

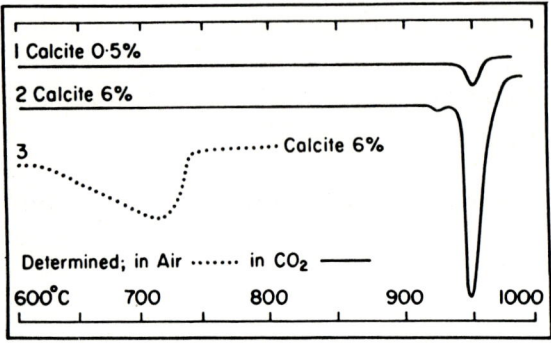

Fig. 1: DTA curves of calcite with Al_2O_3, in static air and flowing CO_2 to illustrate the marked effects of increased CO_2 partial pressure viz. improved peak resolution, definition, peak height, increased peak temperature and detection limits down to 0.25%.

Variations in Technique

Firstly the use of specific gas partial pressures, maintained at the same composition and flow rates over the sample for each complete TA run.

Secondly, a sequence of several flowing gases which replace one another at predetermined temperatures or positions during the progress of each specific TA run. The choice of which of these modes of gas atmosphere application depends on the type of TA determinations required, specific examples of which are described below.

These cannot aspire to be comprehensive but were specially selected to present a wide range of applied research examples and so illustrate the scope of employment of this technique in the earth sciences.

Detailed Applications

To date the majority of applications of variable atmosphere thermal analysis have utilized the single composition, flowing gas approach and it is only in the last few years that pre-selected sequences of purge gases have been used to any extent.

Single Purge Gas Conditions

The initial applications of single purge gases were directed towards the suppression of oxidation/burning reactions so that other associated reactions swamped or distorted by the complete or partial superposition of such large reactions could be detected, recorded in their entirety and evaluated, (see Fig. 2). For example, WARNE (1987) has described how in air the endothermic decomposition of siderite ($FeCO_3$) forms FeO which immediately oxidizes exothermically to Fe_2O_3. These reactions are virtually superimposed so that on the resultant DTA curves the endothermic decomposition peak is partly negated. This gives spuriously lower contents of siderite when based on the area of this endothermic peak and the application of the method of Probenmengen-Abhängigkeit (PA) curves of SMYKATZ-KLOSS (1974). The solution is to remove the oxidation effects by determinations in inert N_2 where the complete unmodified endothermic peak size is directly related to the siderite content (WARNE, 1987), see Fig. 2.

In a similar way, when determined in air, the combustion of the organic matter in coal and oil shale shows as a very large broad exothermic feature on the resultant DTA curves. This occurs within the temperature range 250-900°C and is typically characterized by two rounded exothermic maxima, (Fig. 3, curves 1 to 3).

On the one hand, although such very large peaks reduce progressively with decreasing organic contents, they are clearly recognizable down to contents of 0.25% (Fig. 3, curve 4) and considerably less if determined in flowing O_2 instead of air (WARNE, 1985). This provides a suitable method for the assessment, comparison and evaluation of carbonaceous materials ranging from very low quality solid fuels such as washery waste rejects or oil shale retorting residues, to small carbonaceous contents in power station fly ash (SWAINE, 1969), which may be used as an indicator of combustion efficiency, phytoclasts in sediments of different metamorphic grade (DIESSEL & OFFLER, 1975) and as detrimental contaminants in commercial geological materials, in fillers or in quality control (WARNE, 1985).

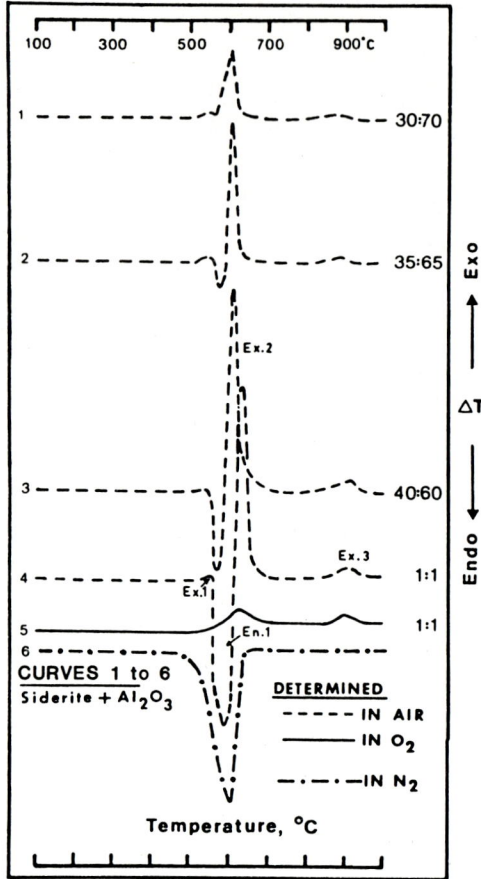

Fig. 2: DTA curves of siderite with Al_2O_3 in static air and flowing N_2 and O_2. These show the progressive superposition of the exothermic peak and negation of the endothermic peak with siderite dilution in air and O_2 (curves 1-5) and the suppression of exothermic effects in N_2 (curve 6), (after WARNE (1987), reprinted with permission of Thermochimica Acta).

On the other hand, these large exothermic combustion effects of coal may be suppressed by the use of an inert atmosphere of flowing N_2, (Fig. 3, cf. curve 1 with 5). Thus the much smaller, previously masked endothermic decomposition effects of the inorganic (mineral) constituents present may be identified (Fig. 3, cf. curve 5 with 6 and 7), down to contents as low as 1% in some cases (WARNE, 1979 and 1985).

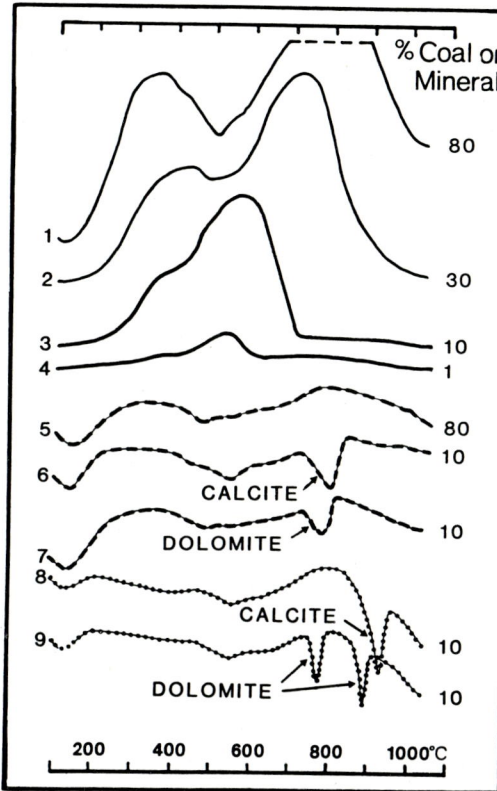

Fig. 3: DTA curves of coal with Al_2O_3 (curves 1-5) or 10% calcite or dolomite (curves 6-9), determined in static air (——), flowing N_2 (- - -) or flowing CO_2 (······), to show coal combustion or detection limits (curves 1-4). Also suppressed burning effects in N_2 (curve 5) and clear modifications, to the coal curve, due to calcite or dolomite in N_2 (curve 6-7) and their improved characterization in CO_2 (curves 8-9).

A further advance in this area was made with the establishment that the substitution of flowing CO_2 for N_2, also effectively prohibited coal combustion to produce a similar essentially featureless DTA curve. On this were equally well recorded the same mineral peaks with the exception of those of carbonates which were much improved (WARNE, 1979).

In particular the diagnostic two and three peak DTA curves of dolomite ($CaMg[CO_3]_2$) and ankerite ($Ca(Mg,Fe)[CO_3]_2$) respectively, become more widely separated (Fig. 4) and preserved right down to the limits of detection when determined in flowing CO_2 (WARNE et al., 1981). Conversely, they coalesce to a single wider more rounded peak in flowing N_2 as their content falls (Fig. 3 cf. curves 7 and 9). Complementary quantifiable reaction seperations are also shown on TG curves of these two minerals under the same conditions, (see Fig. 5) and have been discussed at length with regard to decomposition mechanisms (MILODOWSKI et al., 1989). These features for the DTA of 10% calcite and 10% dolomite (with coal) in flowing N_2 and CO_2 are illustrated in Fig. 3, curves 6 to 9.

Further, it has been shown that the size of the middle DTA peak of ankerite is directly relatable to Fe content isomorphously substituted in this mineral lattice and is clearly distinguishable from Fe which may be present due to admixtures with the mineral siderite ($FeCO_3$), (WARNE et al., 1981). This separation and peak definition in flowing CO_2 has been further improved by the application of DSC, including additional peak modifications, (DUBRAWSKI & WARNE, 1987 and 1988). In addition this variable atmosphere approach has led to the detailed elucidation of the complex decomposition reactions and products, of these two minerals, in papers by OTSUKA (1986), WARNE & DUBRAWSKI (1987), DUBRAWSKI & WARNE (1988), MILODOWSKI et al. (1989), ENGLER et al. (1989) and MCINTOSH et al. (1990).

As previously discussed, reactions caused by heating, which release specific gases, are delayed, to occur at higher temperatures dependent on the degree of the pre-selected increased partial pressure of this gas in the purge gas atmosphere immediately surrounding the sample. Different reactions liberating another gas or no gas at all are not affected i.e. the decomposition of calcite ($CaCO_3$) in flowing CO_2 compared to N_2 moves up scale while pyrite (FeS_2) remains at the same position (temperature) for both of these gases, but would be driven up scale by increased partial pressures of SO_2.

In this way, from duplicate runs, using the appropriate purge gas partial pressure increase (i.e. CO_2, SO_2 or H_2O), reactions liberating these gases may be identified by preferential up scale movements of individual decomposition peaks of carbonates, sulphides and minerals containing water respectively.

Thus many different DTA decomposition peaks may be specifically identified as to their reaction type while superimposed peaks of different gas producing types may be separated, as one will be affected, while another will not. For example, on DTA curves this leads to increased peak separation, resolution, peak height, detection limits and mineral identification, particularly of minerals present in mixtures.

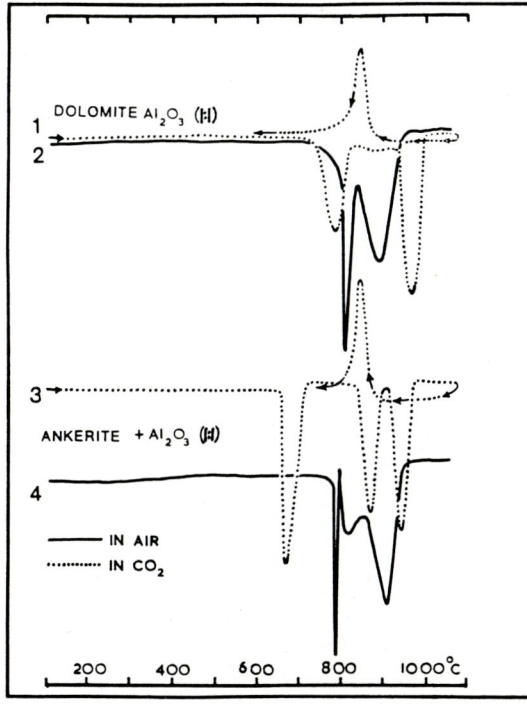

Fig. 4: DTA curves of 50% dolomite or ankerite with Al_2O_3 in static air or flowing CO_2 to show peak modifications and increased two and three peak separation respectively. The single exothermic peak on the cooling curves is due to recarbonation only of the CaO content.

An excellent example of this is shown by the DTA of oil shale containing pyrite, siderite and magnesite ($MgCO_3$) which all decompose at similar but not identical temperatures in flowing N_2 but are recorded as almost completely separated peaks in flowing CO_2 (WARNE & FRENCH, 1984). Similarly the endothermic peak of siderite, is separable from the similar peak of kaolinite in flowing CO_2 compared to determinations in N_2 where they are superimposed (BAYLISS & WARNE, 1972).

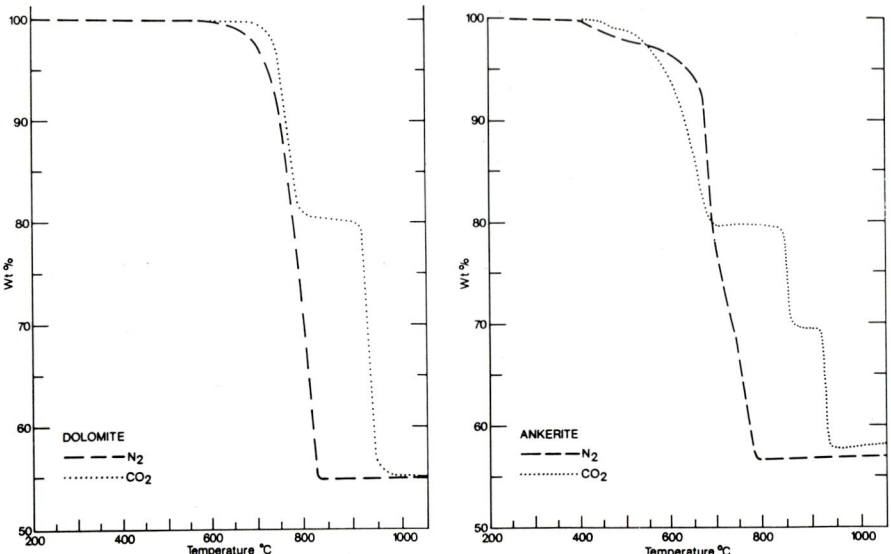

Fig. 5: TG curves in flowing N_2 and CO_2 of dolomite (above left) and ankerite (above right). These show the marked resolution of the weight loss reactions in CO_2 compared to N_2.

In contrast some minerals decompose to form products dependent on the partial pressure of a particular gas. A good example is skutterudite $((Co,Fe,Ni)As_{3-x})$, which on heating in the presence of O_2 dissociates and oxidizes to a mixture of the two cobalt arsenates, $2\ CoO \cdot As_2O_5$ and $3\ CoO \cdot As_2O_5$ and arsenic oxide vapour. The relative amounts of these depend on the partial pressure of O_2 during DTA and TG determinations (WILSON & MIKHAIL, 1989). The applications demonstrated here show that the removal of arsenic from skutterudite ores on roasting is not readily achievable but provides a model for further similar mineralogical studies.

In air or O_2 the decomposition product of pyrite is Fe_2O_3 (hematite) while in O_2 poor (but not O_2 free) conditions Fe_3O_4 (magnetite) is formed (WARNE & DUBRAWSKI, 1990). As the latter is magnetic this has considerable practical applications for the removal of pyrite (sulphur) from coal and coal liquefaction residues using a high intensity magnetic extraction process, provided that only the outer rim of the pyrite has reacted to form magnetite (JACOBS et al., 1978; MAXWELL et al., 1982). Similarly it has been shown that siderite decomposes to hematite or magnetite dependent on the O_2 availability (GALLAGHER & WARNE, 1981). Complementary work on pyrite decomposition in O_2 free N_2 and N_2 containing 3% H_2 has shown the formation of a series of somewhat different pyrrhotites (HURST et al., 1990). Another example is manganite (γ-MnO · OH) which has been shown to decompose to form different gases in O_2, air, CO_2, N_2 and Ar (MORGAN et al., 1988). This work is relevant to the quantitative determination of manganite or similar artificial analogues by TG and other thermal methods.

The recently developed technique of thermomagnetometry (TM) by which the magnetic susceptibility of a substance is measured as a function of temperature is also suitable for atmosphere control. The sample is subjected to a controlled temperature programme under suitable atmospheric conditions i.e. inert or oxidizing, usually with TG equipment where a static magnetic field gradient may be applied and removed at will. This is to detect the presence, absence, development or loss of magnetic phases during heating and or cooling programmes. To achieve this a magnet is placed above or below the sample to give additional "weight" losses or gains on the resultant TM curve. These are due to the added attraction of magnetic phases as they develop or suffer magnetic transitions (Curie points), (EARNEST, 1984).

The whole field of TM has recently been reviewed by WARNE & GALLAGHER (1987). A good example is the decomposition of siderite, mentioned above, which gives in air or O_2 *non-magnetic* Fe_2O_3, but in N_2 containing traces of O_2 gives *magnetic* Fe_3O_4. This very clearly shows from a comparison of the resultant TM curves of GALLAGHER & WARNE (1981), see Fig. 6. Further interesting work has been carried out on the stratigraphic recognition of red and green bed units and on the different contact metamorphic history of some Japanese volcanic tuffs, by SCHWARTZ (1971) and KAZUAKI & TAKUO (1972) respectively. In addition, the whole field of TM in mineralogy and geology has been reviewed in detail by WARNE et al. (1988).

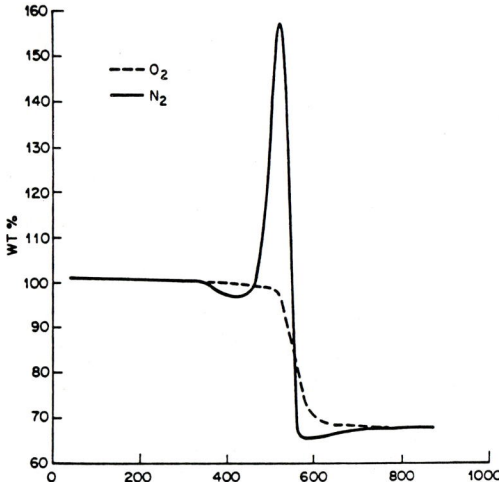

Fig. 6: TM curves of siderite showing the formation of a magnetic phase (magnetite Fe_3O_4) during the decomposition of siderite in O_2 poor conditions (in N_2 not oxygen free) but not in O_2, (after GALLAGHER & WARNE (1981), reprinted with permission of Thermochimica Acta).

Multiple Purge Gas Sequences

The advantages to be gained by the application of simultaneous thermal analysis where two or more methods are applied at the same time to a single sample under identical conditions of analysis are now well established i.e. DTA/TG, DTA/EGA or DTA/TMA (BROWN, 1988).

An additional advance has been made whereby a number of determinations may be made on one sample during the application of a single TA method, but using a *sequence of different purge gases*. To date this technique has

Variable Atmosphere Thermal Analysis

only been applied to individual TA methods, but its scope is such that it is only a matter of time before it is applied to simultaneous TA methods.

The pioneering research using flowing N_2 followed by flowing O_2 i.e. (N_2 - O_2), to provide a new method for the proximate analysis of coal using TG has been described by FYANS (1977). Also this has been shown to provide reliable results within the limits of the standard method (ELDER, 1983). The resultant TG curves give discrete weight losses for moisture, volatile matter, fixed carbon and a residual weight equal to the ash content. The success of this TG/gas sequence technique has led to a number of additional applications. CULMO & FYANS (1988) have used an extended form using N_2 followed by O_2 followed by CO_2 i.e. (N_2 - O_2 - CO_2) to study pre-heated coal, fly, bed and fluidized bed ash (see Fig. 7) and (N_2 - CO_2)

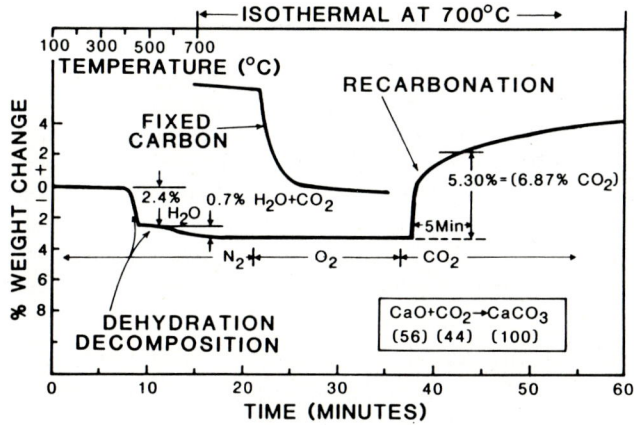

Fig. 7: Multi-gas TG of fluidized bed ash shows two initial weight losses in N_2 due to the dehydration of gypsum (2.4%) and the water loss from the hemihydrate + CO_2 from calcite (0.7%). Switching to O_2 gives a weight loss due to the burning of any fixed carbon present (upper curve), while the final weight gain in CO_2 is due to the recarbonation of CaO to form $CaCO_3$, (after CULMO & FYANS (1988) Figs. 7 and 8, copyright ASTM, reprinted with permission).

to study limestone quality, while DORSEY & BUECKER (1988) have analysed flue gas scrubber materials after utilization by using (flue gas - N_2). The ultimate in complexity (using air - air + SO_2 - N_2 - N_2 + H_2 - N_2 - air) has been very successfully applied to the comparison of sulphation, regenerative SO_2 and cyclic efficiency of sorbents (DUISTERWINKEL et al., 1989), see Fig. 8.

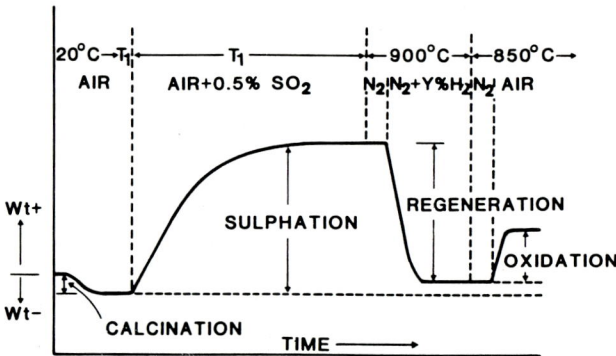

Fig. 8: Multi-gas TG for the sulphation, regeneration and oxidation (SRO) test for flue gas sorbents. The weight loss in air is due to calcination, followed in air+SO_2 by the large weight gain of sulphation, followed in N_2+H_2 by the weight loss caused by sorbent regeneration. After N_2 purging the final weight gain in air is due to oxidation to form $CaSO_4$, (after DUISTERWINKEL et al. (1989), reprinted with permission of Thermochimica Acta).

This contrasts with the simple measurment of pyritic sulphur contents of coal (using N_2 - O_2 - H_2) with the modified TG method of thermomagnetometry. This is used to determine the Fe content, by reduction in the H_2, of the iron oxide formed from pyrite during the oxidation of coal to give the ash component of proximate analyses (AYLMER & ROWE, 1984), see Fig. 9. A proviso is of course that no siderite ($FeCO_3$) is present in the coal under test, as this mineral will also contribute Fe, which in this case is not linked to sulphur (WARNE, 1985a).

The range of applications has been further extended to cover wood, pine cellulose, lignin, graphite and natural and synthetic diamond (WIEDEMANN et al., 1988), while a range of industrial applications has been described by CHARSLEY & WARRINGTON (1988). These cover brown coal, carbon black, charcoal, coke and diamond pastes and suspensions.

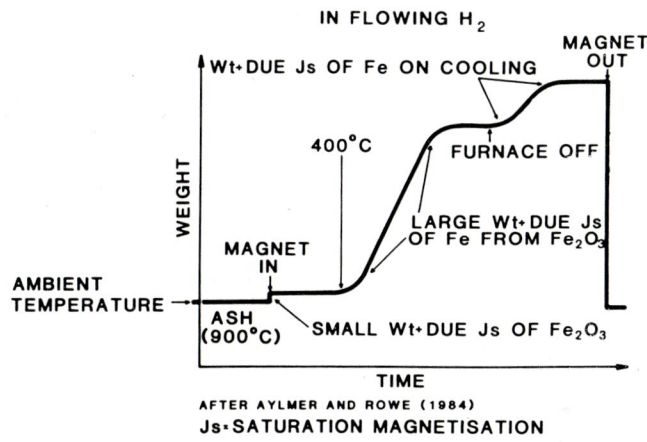

Fig. 9: Ash from coal proximate analysis subjected to TM in flowing H_2 causes reduction to metallic Fe of the iron oxide present which was originally in the coal as pyrite. The weight "loss" due to the magnetic attraction from the magnet above the sample, of this Fe, is thus a measure of the pyritic sulphur in the original coal, (after AYLMER & ROWE (1984), reprinted with permission of Thermochimica Acta).

Conclusion

The technique of variable atmosphere thermal analysis with its facility to affect individual reactions, be it directed at suppression, enhancement, identification, improved detection limits, preferential up or down temperature movements or the production of additional properties i.e. magnetism, has now been clearly established as a reliable technique. This has in part been due to:

1. the development of TA equipment suitable for accurate, reproducible and rapid control and change of the gas atmosphere surrounding the sample under analysis.

2. the establishment of the effects of gas partial pressure variations on thermal analysis results.

3. the realization of the scope of applications as evidenced by the wide range of applications reviewed above.

Because of the flexibility of this technique and its ability to contribute to the solution of specific reaction aspects of TA, it is expected to play an increasingly vital role in the resurgence of applications of a wide range of TA methods in the earth sciences. In this respect the development of a standard TG method, including variable atmosphere aspects (LARKIN, 1988) represents another important step in the right direction.

References

AYLMER D. M. & ROWE M. W. (1984) - A new method for simultaneous determination of pyrite content and proximate analysis of coal.- Thermochim Acta, 78, 81-92

BAYLISS P. & WARNE S. ST. J. (1972) - Differential thermal analysis of kaolinite-siderite mixtures.- Am Mineral, 57, 960-966

BROWN M. E. (1988) - Introduction to thermal analysis - techniques and applications.- Chapman and Hall, London.

CHARSLEY E. L. & WARRINGTON S. B. (1988) - Industrial applications of compositional analysis by thermogravimetry.- In EARNEST C. M. (ed) Compositional analysis by thermogravimetry. ASTM Spec Tech Publ, 997, 19-27

CULMO R. F. & FYANS R. L. (1988) - Thermogravimetry as a tool for determining combustion efficiency and calcium utilization of a fluidized bed combustion furnace.- In EARNEST C. M. (ed) Compositional analysis by thermogravimetry. ASTM Spec Tech Publ, 997, 245-253

CUMMING J. W. (1989) - A DTG combustion study on anthracite and other coal chars.- Thermochim Acta, 155, 151-161

CUMMING J. W. & MCLAUGHLIN J. (1982) - The thermogravimetric behaviour of coal.- Thermochim Acta, 57, 253-272

DIESSEL C. F. K. & OFFLER R. (1975) - Change in physical properties of coalified and graphitised phytoclasts with grade of metamorphism.- Neues Jb Miner Mh, 1, 11-26

DORSEY D. L. & BUECKER B. J. (1988) - Analysis of flue gas scrubber materials from a coal-fired power plant by thermogravimetry.- In EARNEST C. M. (ed) Compositional analysis by thermogravimetry. ASTM Spec Tech Pub, 997, 254-258

DUBRAWSKI J. V. & WARNE S. ST. J. (1987) - Use of differential scanning calorimetry in measuring the thermal decomposition of mineral carbonates occurring in coal.- Fuel, 66, 1733-1736

DUBRAWSKI J. V. & WARNE S. ST. J. (1988) - Differential scanning calorimetry of the dolomite-ferroan dolomite-ankerite series in flowing carbon dioxide.- Min Mag, 52, 627-635

DUISTERWINKEL A. E., DOESBURG E. B. M. & HAKVOORT G. (1989) - Comparing regenerative SO_2 sorbents using TG: the SRO test.- Thermochim Acta, 141, 51-59

EARNEST C. M. (1983) - The use of modern computerized instrumentation for rapid proximate and ultimate analysis of coals.- Proc 3rd Coal testing conf Lexington USA, 66-69

EARNEST C. M. (1984) - Thermal analysis of clays, minerals and coals.- Perkin Elmer Co, Norwalk USA.

EARNEST C. M. (1984) - Modern thermogravimetry.- Anal Chem, 56, 1471A

EARNEST C. M. (ed) (1988) - Compositional analysis by thermogravimetry.- ASTM Spec Tech Publ, 997, Philadelphia.

ELDER J. P. (1983) - Proximate analysis by automated thermogravimetry.- Fuel, 62, 580-584

ENGLER P., SANTANA M. W., MITTLEMAN M. L. & BALAZS D. (1989) - Nonisothermal, in situ XRD analysis of dolomite decomposition.- Thermochim Acta, 140, 67-76

FYANS R. L. (1977) - Rapid characterization of coal by thermogravimetric and scanning calorimetric analysis.- Perkin Elmer, Thermal analysis study, No. 21, 1-5

GALLAGHER P. K. & WARNE S. ST. J. (1981) - Thermomagnetometry and thermal decomposition of siderite.- Thermochim Acta, 43, 253-267

HEIDE K. (1982) - Dynamische thermische Analysenmethoden.- VEB Deutscher Verlag für Grundstoffindustrie, Leipzig.

HURST H. J., LEVY J. H. & WARNE S. ST. J. (1990) - The application of variable atmosphere thermogravimetry to the thermal decomposition of pyrite.- React of Solids, 8, 159-168

JACOBS I., LEVISON L. & HART H. (1978) - Magnetic and Mössbauer spectroscopic characterization of coal.- J Appl Phys, 49, 1775-1780

KAZUAKI M. & TAKUO Y. (1972) - Ferromagnetic minerals involved in volcanic ash-layers of the Osaka Group. Typical Plio-Pleistocene sediments in south west Japan.- Quat Res, 11, 270-280

LARKIN D. E. (1988) - Compositional analysis by thermogravimetry: The development of a standard method.- In EARNEST C. M. (ed) Compositional analysis by thermogravimetry. ASTM Spec Tech Publ, 997, 28-37

MACKENZIE R. C. (ed) (1957) - Differential thermal investigation of clays.- Min Soc Clay Min Gp London.

MACKENZIE R. C. (ed) (1970 and 1972) - Differential thermal analysis, Vol 1 Fundamental aspects and Vol 2 Applications.- Acad Press New York.

MAXWELL E., KELLAND D. R., JACOBS I. S. & LEVINSON L. M. (1982) - Magnetic separation and thermo-magnetic chemical properties of coal liquefaction mineral particles.- Fuel, 61, 369-376

MCINTOSH R. M., SHARP J. H. & WILBURN F. W. (1990) - The thermal decomposition of dolomite.- Thermochim Acta, 165, 281-296

MILODOWSKI A. E., MORGAN D. J. & WARNE S. ST. J. (1989) - Thermal analysis studies of the dolomite-ferroan dolomite-ankerite series II. Decomposition mechanism in flowing CO_2 atmosphere.- Thermochim Acta, 152, 279-297

MORGAN D. J., MILODOWSKI A. E., WARNE S. ST. J. & WARRINGTON S. B. (1988) - Atmosphere dependence of the thermal decomposition of manganite, γ-MnOOH.- Thermochim. Acta, 135, 273-277

OTSUKA R. (1986) - Recent studies on the decomposition of the dolomite group by thermal analysis.- Thermochim Acta, 100, 69-80

SCHULTZE D. (1969) - Differentialthermoanalyse.- Verlag Chemie, Weinheim/Bergstr.

SCHWARTZ E. J. (1971) - Magnetochemical aspects of the heating of red and green beds.- Geol Survey Canada paper 70-63, 31-34

SMYKATZ-KLOSS W. (1974) - Differential thermal analysis - Application and results in mineralogy.- Springer Verlag, Berlin, Heidelberg, New York.

SWAINE D. J. (1969) - The identification and estimation of carbonaceous materials by DTA.- Proc. 2nd ICTA conf Worcester USA, 2, 1377-1386

TODOR D. N. (1976) - Thermal analysis of minerals.- Abacus Press Tunbridge Wells UK.

WARNE S. ST. J. (1977) - Carbonate mineral detection by variable atmosphere differential thermal analysis.- Nature, 269, 678

WARNE S. ST. J. (1978) - Proben-abhängigkeit (PA) curves and simple anhydrous carbonate minerals.- J Thermal Anal, 14, 325-330

WARNE S. ST. J. (1979) - Differential thermal analysis of coal minerals.- In KARR C. JR. (ed) Analytical methods for coal and coal products. III, 447-477

WARNE S. ST. J. (1985) - The assessment of coal-organic matter contents of geological materials by differential thermal analysis.- Thermochim Acta, 86, 337-342

WARNE S. ST. J. (1985a) - Coal proximate analysis and pyrite contents by the TM/TG method. The problem of the iron bearing carbonates.- Thermochim Acta, 87, 353-356

WARNE S. ST. J. (1986) - Applications of variable atmosphere DTA (in CO_2) to improved detection and content evaluation of anhydrous carbonates in mixtures.- Thermochim Acta, 109, 243-252

WARNE S. ST. J. (1987) - Applications of thermal analysis to carbonate mineralogy.- Thermochim Acta, 110, 501-511

WARNE S. ST. J. & DUBRAWSKI J. V. (1987) - Differential scanning calorimetry of the dolomite-ankerite mineral series in flowing nitrogen.- Thermochim Acta, 121, 39-49

WARNE S. ST. J. & DUBRAWSKI J. V. (1990) - Applications of thermal analysis methods to iron minerals in coal - A synthesis.- Thermochim Acta, - in Press.

WARNE S. ST. J. & FRENCH D. H. (1984) - Siderite, pyrite and magnesite identification in oil shale by variable atmosphere DTA.- Thermochim Acta, 79, 131-137

WARNE S. ST. J. & GALLAGHER P. K. (1987) - Thermomagnetometry.- Thermochim Acta, 110, 269-279

WARNE S. ST. J., HURST H. J. & STUART W. I. (1988) - Applications of thermomagnetometry in mineralogy, metallurgy and geology.- Thermal Analysis Abs 17, 1-6

WARNE S. ST. J., MORGAN D. J. & MILODOWSKI A. E. (1981) - Thermal analysis studies of the dolomite, ferroan dolomite, ankerite series I. Iron content recognition and determination by variable atmosphere DTA.- Thermochim Acta, 51, 105-111

WIEDEMANN H. G., RIESEN R., BOLLER A. & BAYER G. (1988) - From wood to coal: a compositional thermogravimetric analysis.- In EARNEST C. M. (ed) Compositional analysis by thermogravimetry. ASTM Spec Tech Publ, 997, 227-244

WILSON L. J. & MIKHAIL S. A. (1989) - Investigation of the oxidation of skutterudite by thermal analysis.- Thermochim Acta, 156, 107-115

WOLF K. H., EASTON A. J. & WARNE S. ST. J. (1969) - Techniques of examining and analyzing carbonate skeletons, minerals and rocks.- In CHILINGAR G. V., BISSELL H. J. & FAIRBRIDGE R. W. (eds) Carbonate rocks Pt.B. Elsevier Amsterdam, 253-341

MEASUREMENT OF DIFFERENT WATER SPECIES IN MINERALS BY MEANS OF THERMAL DERIVATOGRAPHY

M. Földvári

Budapest, Hungaria

Abstract

Water can be bound in different ways in minerals. The most frequent types of occurrence of water in minerals have been tabulated to summarize the most important features. This clearly reveals that the different water types form a continous sequence.

Water escaping in the course of dehydration processes can be bound in minerals in two fundamentally different mechanisms, e.g. by adsorption forces on the different surfaces of the structure or by coordination forces around certain cations of the structure.

Adsorption Water

Minerals are capable of binding different amounts of water, both on their external and on their internal surfaces, namely on the surfaces of the internal channels or other cavities of the crystal. A typical example for water adsorbed on external and internal surfaces of minerals is provided by silicates, which contain polar AlO_4 and SiO_4 tetrahedra. The polar ions or atomic groups on the surface of a solid, bind the polar water molecules by van der Waals forces. The resulting monomolecular layer also creates a new polar surface, which permits the oriented adsorption of further molecule sequences. The most difficult type to remove is the water stuck directly to the solid surfaces of minerals.

Table 1

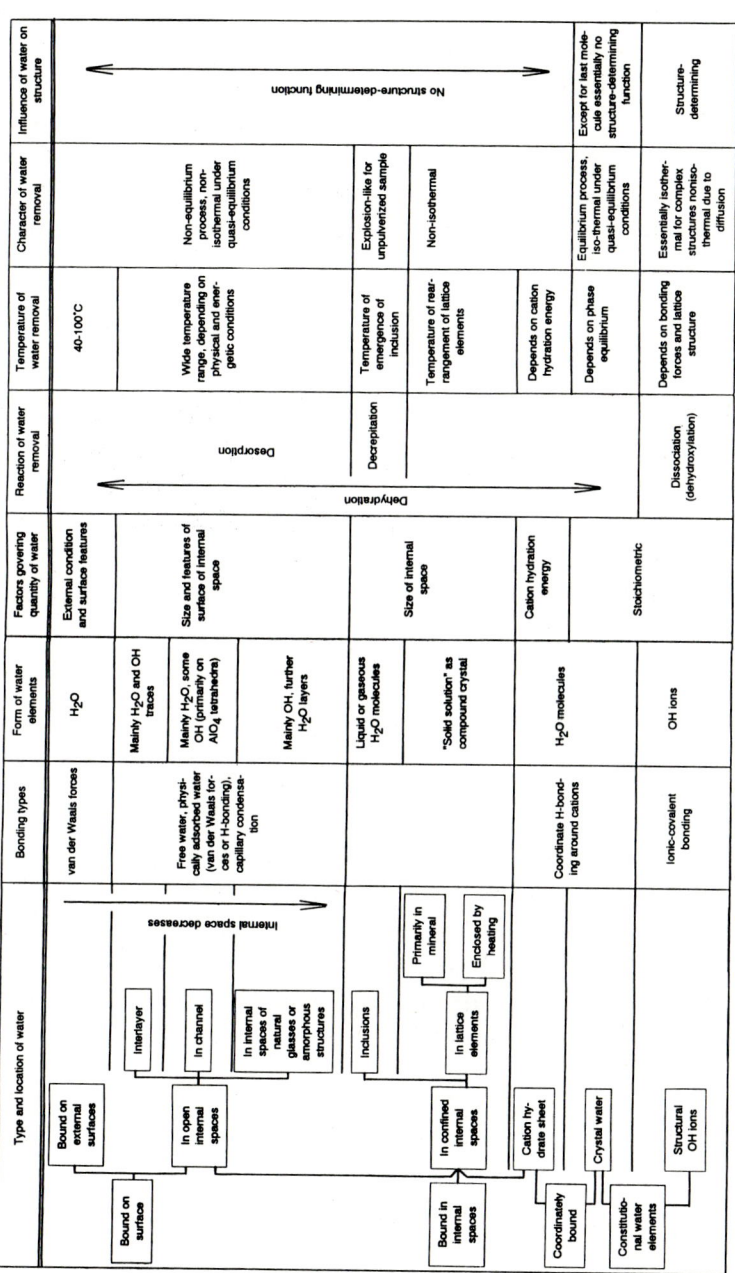

The linkage may be on the external surface of minerals (so-called sticking water) or can also appear on the surface of the internal spaces formed by the structure of minerals. Practically, binding of water on the internal surfaces is the same as described above, if the available internal space is large enough. These internal surfaces and cavities may be of different shapes and sizes, which can significantly influence the types of binding.

Interlayer Water Bound by Phyllosilicates

It may appear if there is an abundance of negative ions in the crystal-forming layer complex, that can be balanced by an interlayer, regularly exchangeable cation and by water. The interlayer space may contain free water molecules and water adsorbed to the active sites directly or in the second or third layer. Due to the large spaces and hence the comparatively weak binding, this water can escape at a relatively low temperature.

"Zeolitic Water"

In zeolites and in the structure of some other minerals (e.g. aluminite, palygorskite etc.) the so-called "zeolitic water" is found in the channels formed by the SiO_4 and AlO_4 tetrahedra, where it may move more or less freely. Relatively free movements are possible in the wide channels, whereas narrower channels or capillaries impede movement, and the water is adsorbed on the surface, e.g. mainly bound to the AlO_4 tetrahedra, partly in the form of OH groups. The temperature of escape of the latter is higher than that of molecular water.

The internal spaces may be capillaries with extensive internal surfaces, and in this case capillary condensation also plays a prominent role in water binding. These surfaces are covered by an H_2O monomolecular layer, which often shares neighbouring surfaces. Such internal spaces can generally be found with less ordered structure in materials of nearly amorphous type. Natural glassy rocks and amorphous formations often contain water bound similarly to zeolitic water. In glassy rocks, most of this water is bound in OH form. The network forming the glassy rock is composed primarily of SiO_4 and to a much lesser extent of AlO_4 tetrahedra to which the OH groups are bound freely or they bridge the SiO_4 tetrahedra by hydrogen-bonds. Further water molecules can be hydrogen-bound to the OH layers of the surface if the internal space is large enough to permit this. Such binding is much weaker.

The quantity of water on the external surfaces and in the open internal spaces bound by adsorptive forces depends on the size and shape of these surfaces and accordingly change the binding forces in the above mentioned ways. Water is bound with different energies, and thermo-analytical curves usually reveal its steady loss over a wide temperature range as poorly defined features. The quantity of water, bound in this way, often cannot be expressed by stoichiometric means, although the quantity is fairly well determined by the type of the structure.

Water bound by adsorption does not have a structure determining function. The loss of adsorbed water from the internal spaces due to the heating is not an equilibrium reaction and the dehydration curves from quasi-isothermal heating techniques are always essentially non-isothermal (Fig. 1).

The influence of the size of internal spaces over the force of binding may be followed in the case of water distribution curves of perlites with different water contents as a function of temperature (Fig. 2). Here it may be seen that the water of samples with higher water content escapes at a lower temperature, i.e. the smaller internal spaces bind the water stronger.

The loss of water from opals of different origins and from different places are rather dissimilar (Fig. 3) and may be divided into four groups. According to the previous figures, perlites have a more regular internal space system than opals. A similar form of the water distribution curves of

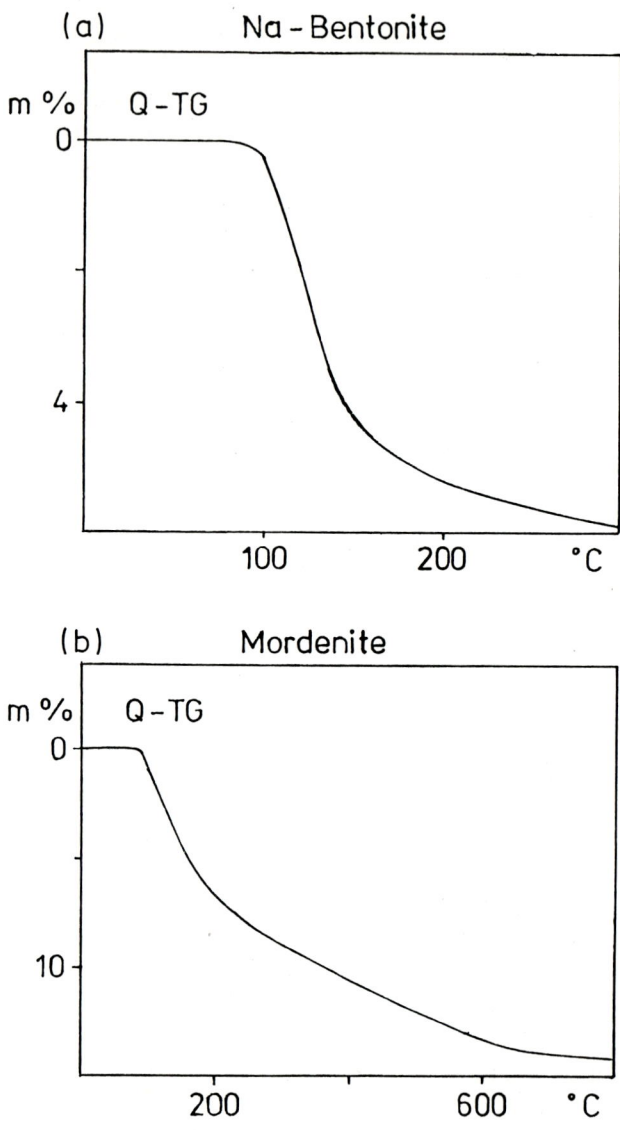

Fig. 1: Q-TG curves of water bound on surface of internal spaces, (a) interlayer water of montmorillonite; (b) zeolitic water in mordenite.

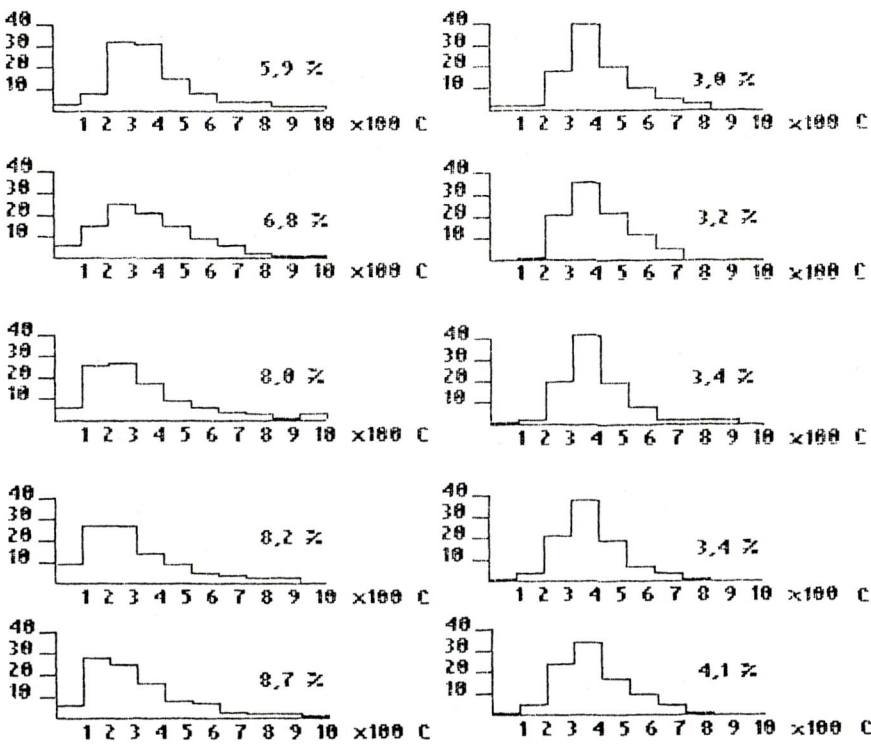

Fig. 2: Water distribution curves of perlites.

perlites takes place at a higher temperature than for opals, because of the presence of AlO_4 tetrahedra, which has a greater polarity and thereby a stronger binding effect.

Water Present in Completely Confined Internal Spaces

Inclusion water involves the most primitive form of water in minerals. The included water at its emergence, totally filled the cavity, and then, upon

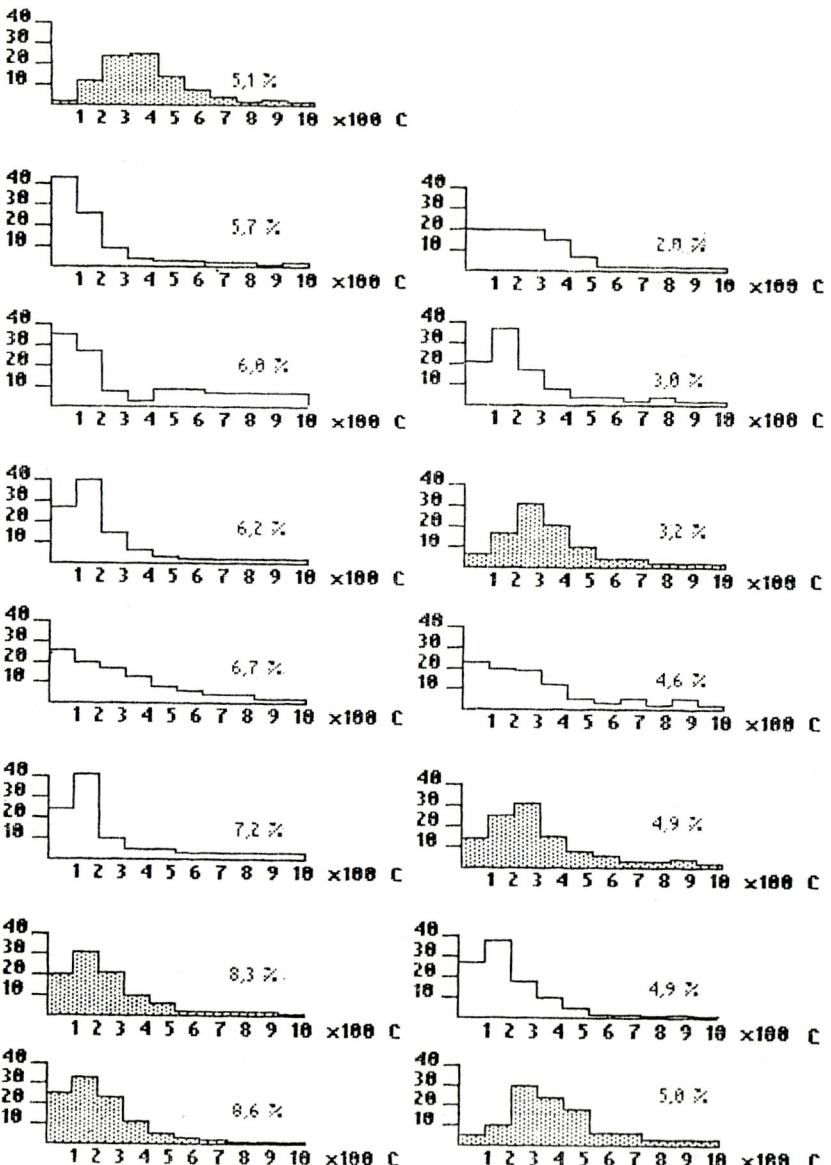

Fig. 3: Water distribution curves of opals.

cooling, its volume was decreased, allowing it to move freely, like a level. In the course of heating at the temperature of the emergence of the inclusion, the liquid expands again to fill the whole space, then, on further heating, its volume increases further and the crystal bursts (decrepitation) (see baryte in Fig. 4).

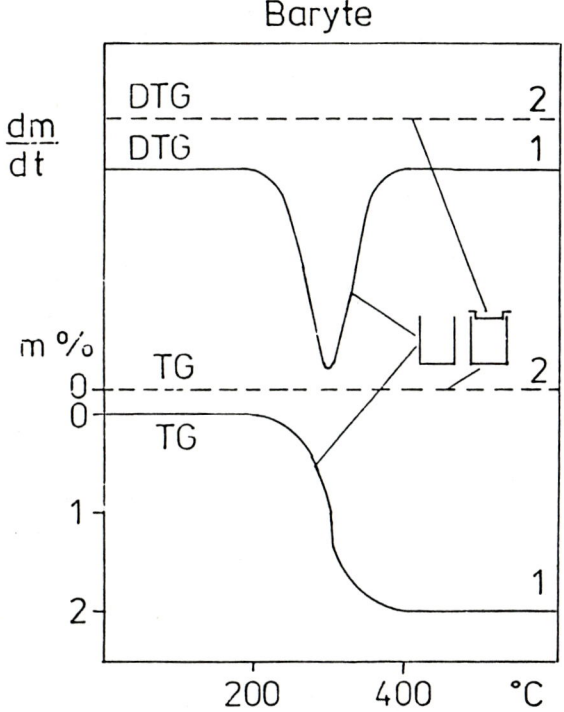

Fig. 4: Decomposition of unpulverized baryte. Curve 1 determined with uncovered crucible, Curve 2 with covered crucible.

Water Bound in Solid Solution

Here, the ions do not occupy the entire space, thereby permitting the presence of 2-3 water molecules. Escape from their fixed position is possible only if the process of self-diffusion within the crystal achieves a considerable rate. The rapid departure of such water can also be caused by a structural modification of the mineral; (see aragonite in Fig. 5). During the rearrangement of the lattice elements, water bound by occlusion is liberated.

In the case of zeolites and perlites some water can also be retained at the decomposition of the structure due to heating, as in a silicate melt. In this case the water escapes at temperatures higher than 900°C.

Crystal Water

Formation of water bound by coordination forces, can be explained by the fact that cations within an aqueous solution coordinate water molecules, which are incorporated into the lattice as aquo complexes in the course of mineral precipitation. They often play a rôle in size control, e.g. sulphate minerals can form with a smaller cation, if the size difference will be compensated by water. Crystal water is an integral and stoichiometric part of the structure. It does not have a structure determining function by itself, but its escape can result in the rearrangement of the structure. On heating, this water generally escapes at low temperatures in several, more or less overlapping stages. The dehydration of crystal water in most cases is an isothermal process striving for equilibrium, i.e. equilibrium of the reaction depends only on the gaseous product of the decomposition, i.e. in this case, on the partial pressure of water (Fig. 6).

Thermal Dissociation

The components escaping in the course of thermal dissociation are an integral part of the structure which they influence. They are not present in molecular form in the structure, which explains why the mechanism of the thermal dissociation is a process of two phases:

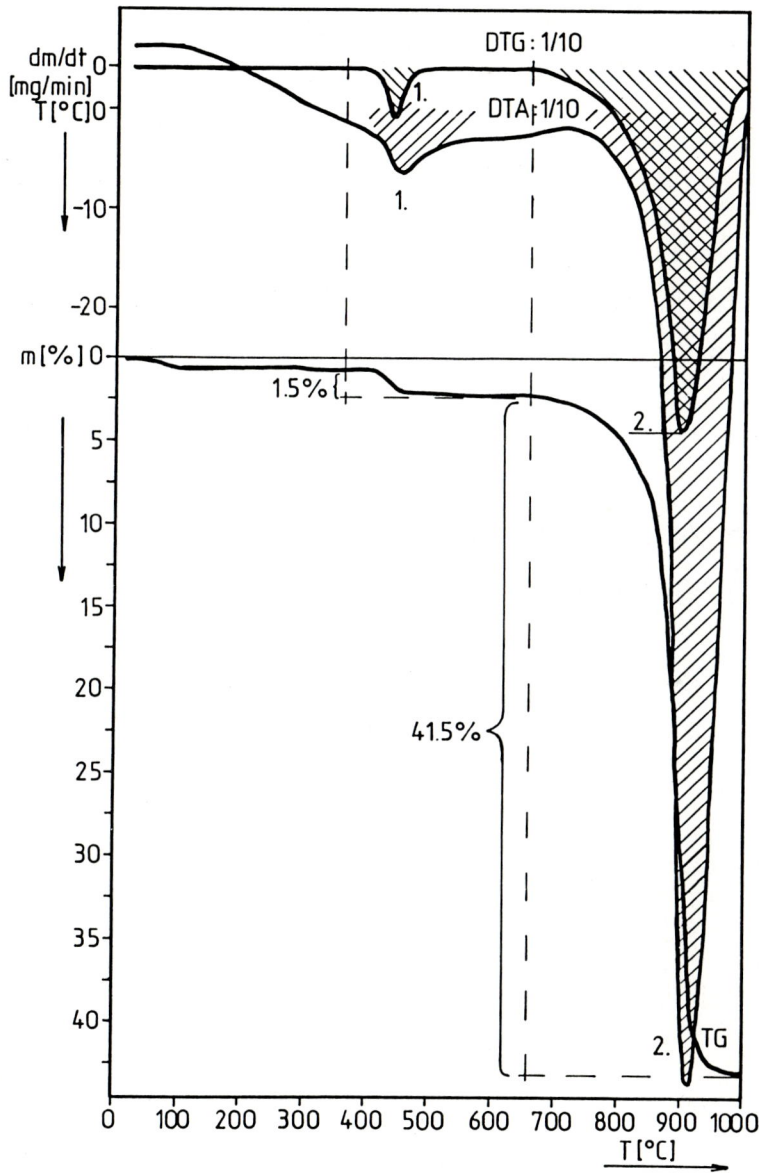

Fig. 5: TG, DTG and DTA curves of aragonite.

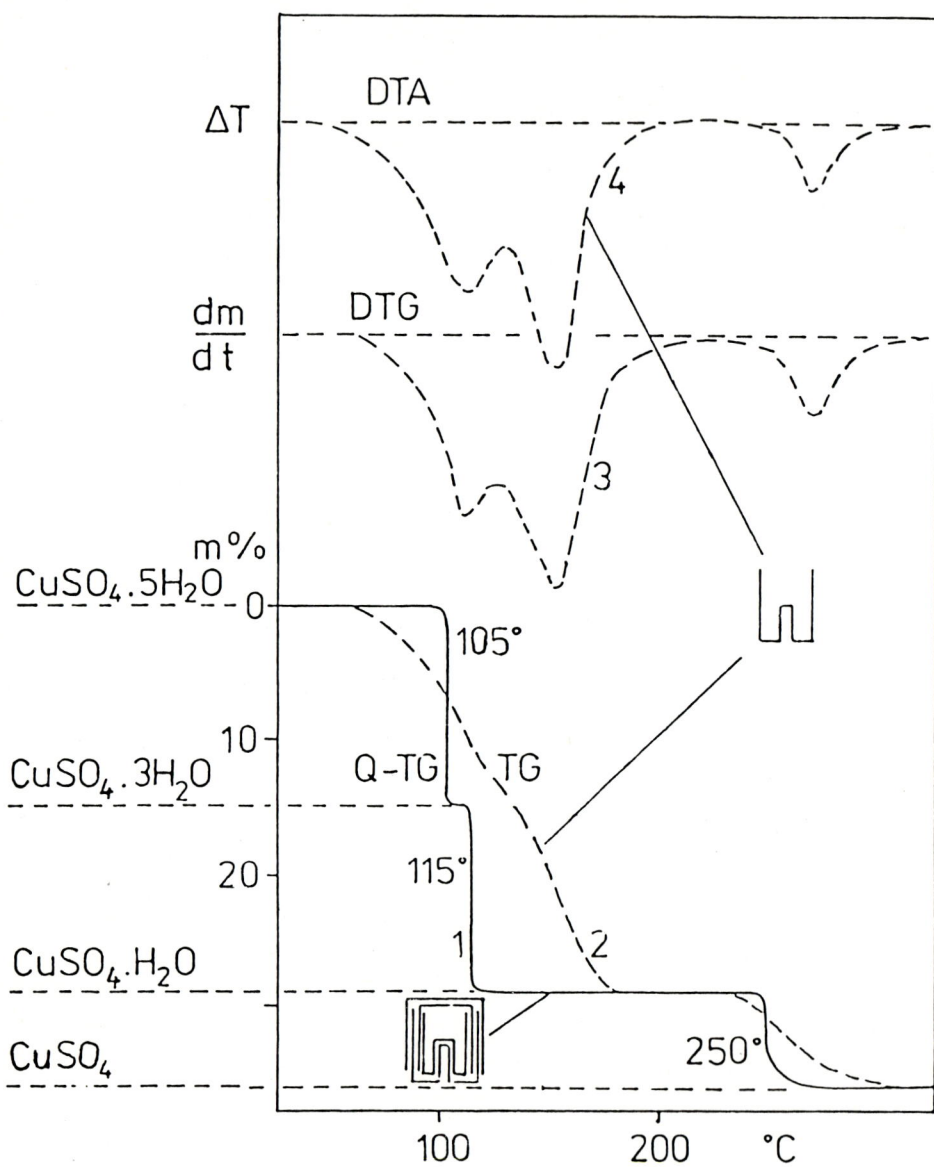

Fig. 6: Dynamic TG, DTG, DTA and Q-TG curves of chalcanthite.

1. Formation of the escaping components - in case of hydroxides: proton accommodation;

2. Loss of molecular water from the crystal lattice.

In case of thermal dissociation reactions, because of the much stronger ionic or covalent bonds in the structure the process is influenced by lattice structure features and by the compound instead of the thermodynamic phase-equilibrium. Among these factors bonding strength should be mentioned first. Table 2 clearly shows that a decrease in electronegativity of the cation, in connection with the same anion results in an increase in bonding strength and accordingly in an increase in the dissociation temperature.

The decomposition temperature of a given bonding is also determined by its position within the structure of the mineral. Table 2 shows that the temperature of the release of OH-groups is dependent on the mineral structure. On the above basis it can be stated that the same bonding decomposes at a higher temperature if it is within a complex structure.

Essentially, thermal dissociation is an isothermal process, striving for equilibrium (see portlandite, Q-TG curve in Fig. 7). In more complicated structures, because of the hindering processes the order of the reaction is increasingly further from 0 (non-isothermal). This phenomenon can be followed well in the Q-TG curves of boehmite, kaolinite and pyrophyllite (Fig. 7).

Besides the mentioned factors, i.e. the temperature of the reaction, or the type of water removed under quasi-equilibrium circumstances, other thermal data are also suitable for the characterisation of the different water types. For the description of the shape of thermoanalytical curves the terms breadth, sharpness or asymmetry are generally used (Table 3). Beside these empirical parameters the constants of the kinetic equation were calculated

Table 2: Dehydroxylation temperature for different minerals

	B	Fe^{3+}	Al	Mn^{3+}	Mg	Ca
Electronegativity, after Pauling 1964	2.0	1.8	1.5	1.5	1.2	1.0
Simple *Hydroxides* (H_2O = 18 - 43%)	Sassolite 120-210°C		Gibbsite 270-380°C		Brucite 410-540°C	Portlandite 480-610°C
Oxyhydroxides (H_2O = 10 - 15%)		Goethite 300-410°C	Boehmite 450-580°C	Manganite 300-410°C		
Sulphates with OH-group (H_2O = 11 - 13%)		Jarosite 350-460°C	Alunite 450-580°C			
Silicates Phyllosilicates <u>1:1 layer type</u> (H_2O = 11 - 13%)		Kaolinite group 490-630°C Kaolinite / 610-760°C Dickite			Al-Mg Mg-Serpentines 550-690°C Chrysotile / 620-770°C Antigorite	
<u>2:1 layer type</u> (H_2O = 4 - 5%) With interlayer cation and water 1. Smectites		Nontronite 400-510°C	Beidellite 490-610°C	Montmo-rillonite 620-780°C	Saponite 730-900°C	
2. Hydromicas		Glauconite 500-630°C	Illite 520-650°C		Ledikite 770-930°C	
Free of interlayer cation and water			Pyrophyllite 640-800°C		Talc 820-1000°C	
Free of interlayer water (Micas)			Muscovite 780-950°C		Phlogopite 1100-1300 °C	

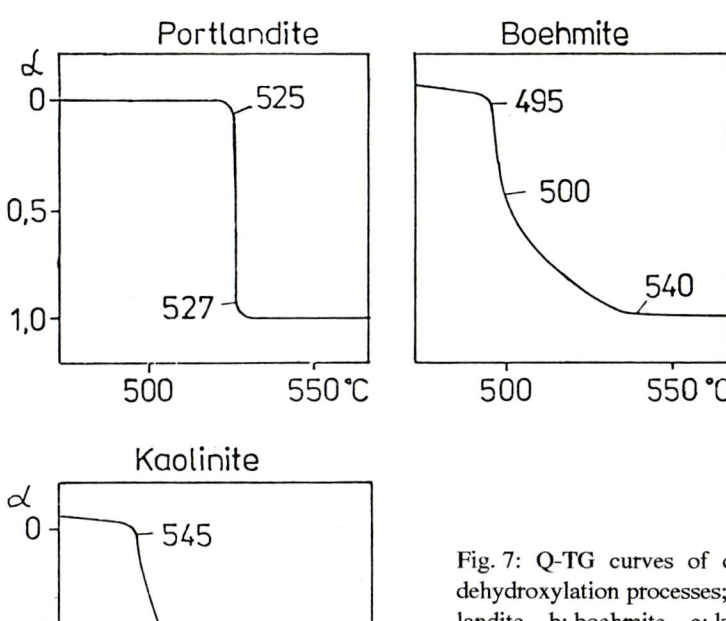
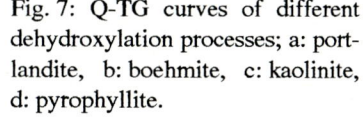
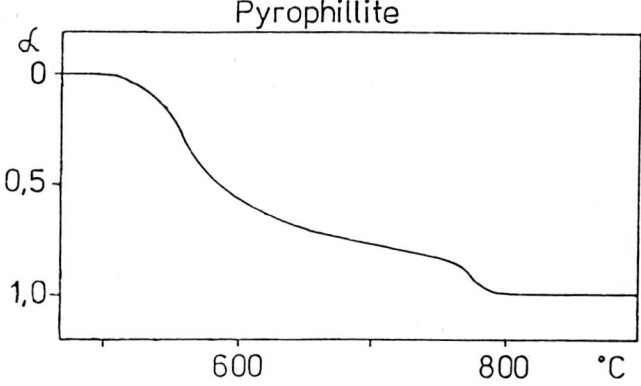

Fig. 7: Q-TG curves of different dehydroxylation processes; a: portlandite, b: boehmite, c: kaolinite, d: pyrophyllite.

by means of the program of derivatograph-c from the TG and DTG curves as a function of time on the following basis:

n = estimated order of reaction (after KISSINGER)

$$E = \frac{n \cdot (\ln(1-\alpha_1) - \ln(1-\alpha_2))}{(1/T_1 - 1/T_2)}$$

A = pre-exponential factor (after DOYLE, with a conversion of 0.2)

Table 4 summarizes the measured and calculated data of different water types. It is possible to see, that the data reflect well the above mentioned energy relations among the different types of bonding. These parameters are usefull not only for the characterisation of different water types, but for the identification of the reactions, which are found in similar temperature ranges i.e. the adsorptive, zeolitic or crystal water.

Table 3

Breadth of peak:	1. $T = T(0.8; 0.2)$	
	2. $\omega = \dfrac{1}{(da/dt)_{max}}$	(after POKOL et al., 1985)
Sharpness of peak:	$\dfrac{\Delta T(0.8; 0.2)}{\omega}$	

Asymmetry: 1. % of the decomposed part at the maximum rate of decomposition (after POKOL et al., 1985).

    2. T1/T2 ratio between the area of peak before and after of maximum rate of decomposition (calculated by derivatograph-c).

    3. $R_{max} = \dfrac{\Delta T(peak\ max; 0.2)}{\Delta T(0.8; 0.2)}$ (after POKOL et al., 1985)

Table 4

Water types	Number of samples	T_{max}	Breadth of peak $T(0.8;0.2)$		Sharpness of peak	%	Asymmetry T_1/T_2	R_{max}	Kinetic parameters n	E	A
Crystal water											
(epsomite)	1	142 (185)	61.5	0.17	361.7	49	1.011	0.37	1.28	37	3.7317
last molecule	3	331	35	0.51	72.6	50	1.12	0.45	1.23	166.3	13.77
Interlayer water											
montmorillonite	20	122	70	1.006	85	47	0.875	0.41	1.36	30.12	3.057
mixture of mont. and illite	8	119	72.5	1.09	97.76	43	0.735	0.29	1.5	32.5	3.4
halloysite	1	111	59	1.41	41.8	42	0.74	0.29	1.43	38.2	4.3166
Zeolitic water											
clinoptilolite	1	139.5	212	1.39	152.5	23	0.311	0.04	2.28	22.2	1.4764
Water of amorphous											
opal	1	116.5	221	6.67	33.13	32	0.481	0.12	1.79	14.7	0.4918
Simple hydroxides											
gibbsite	3	301	33	0.77	88.9	53	1.32	0.74	1.15	159.4	14.0
jarosite	1	443.5	43.6	0.27	161.5	65	2.014	1.32	0.93	210.6	15.1
Oxyhydroxides											
böhmite	4	536	46.6	0.53	109.5	54	1.334	0.58	1.11	235.6	14.62
goethite	4	306	53.7	1.005	80.26	50.5	1.165	0.5	1.17	86.7	7.07
Phyllosilicates											
kaolinite	16	565	86	0.885	111.23	50.2	1.1	0.49	1.24	118.6	6.33
halloysite	1	566	68	0.39	174.4	68	2.223	0.82	0.84	131.8	7.5
serpentine	1	680	93	0.63	147.6	72	2.722	0.9	0.76	99.4	4.504
montmorillonite	4	546	106.9	3.095	67.67	50	1.2495	0.5	1.275	81.4	4.191
mixture of mont. and illite	5	531	104.9	2.456	74.3	64	1.833	0.76	1.07	69.7	3.243

References

BREGER I. A., CHANDLER J. C. & ZUBOVICH P. (1970) - An infrared study of water in heulandite and clinoptilolite.- Amer. Mineral., 55, 825-840

FÖLDVARI M., PAULIK F. & PAULIK J. (1988) - Possibility of thermal analysis of different types of bonding of water in minerals.- Jour. Therm. Anal., 33, 121-132

HEIDE K. (1974) - Untersuchung der Hochvakuumentgasung bei dynamischer Temperaturänderung bis 1200°C von natürlichen Gläsern unterschiedlicher Genese.- Chemie der Erde, 34, 195-214

JOHNSON G. K., FLOTOW W. E., O'HARE P. A. G. & WISE S. (1983) - Thermodynamic studies of zeolites: natrolite, mesolite, and scolecite.- Amer. Mineral., 68, 1134-1145

NEMECZ E. (1962) - Thermal behavior of the adsorbed and interlaminar water content of montmorillonite.- Acta Geol. Hung., Budapest, 365-388

NEMECZ E. (1981) - Clay Minerals.- Budapest.

PERLAKI E. & SZÖŐR GY. (1973) - The perlites of the Tokaj Mountains.- Acta Geologica Academiae Scientiarium Hungaricae, Budapest, 85-106

POKOL GY., GAL S. & PUNGOR E. (1985) - The application of empirical quantities describing the shape of thermoanalytical curves.- Proc. 8th ICTA, Thermochimica Acta, 89-92

ULLRICH B., ADOLPHI P., SCHOMBURG J. & ZWAHR H. (1988) - Kombinierte thermoanalytische Untersuchungen an Zeolithen. Teil II.: Heulandit-Stilbit-Gruppe.- Chemie der Erde, 48, 141-154

THE DETERMINATION OF HYDRATED SULPHATES IN THE WEATHERED CRYSTALLINE ROCKS BY MEANS OF THERMAL ANALYSIS

Czesław August

Institute of Geological Sciences
University of Wrocław
Poland

Abstract

Under heating sulphates either (1) melting at a certain temperature and then crystallization under cooling occurs, or (2) dissociation giving high temperature reaction products (mainly oxides) occurs.

The latter is typical for hydrated sulphates. Usually, they contain considerable amounts of molecular (H_2O) or ionic (OH^-, H_3O^+) water, which can occupy different sites within or outside the crystalline structure. The water is bound by different types of bonds or by similar bonds but of different force and this is observed in thermograms of sulphates as multistage dehydration and dehydroxylation effects.

The water is removed from sulphates at a temperature and in a manner which are characteristic for each particular mineral. This is a diagnostic indicator which makes easier both qualitative and quantitative identification.

Similarly, high temperature effects of sulphate decomposition are specific for different varieties of the hydrated sulphate group and they are also helpful for their identification. The present author is inclined to question the view of some research workers that thermal effects on thermograms of some sulphates do not come from the original minerals but rather reflect transformations of transitional phases which are formed during the thermal analysis. During heating, however, particular sulphates produce characteristic transitional phases giving information on the mineral being analyzed.

This paper gives the thermal analysis results for several hydrated sulphates with special attention given to dehydration and dehydroxylation.

Geological Background

Minerals of the sulphate group first of all originate under weathering conditions. They are often found to cover weathered rock surfaces, particularly in mines and quarries (e.g. Wiesciszowice, Zarów, Strzegom, Borów). Genetically, they are connected with sulfides which are susceptible to oxidation and hydrolysis.

This paper presents the results of thermal studies on hydrated sulphates which occur on surfaces of two types of crystalline rocks in the Sudetes (SW Poland):

- quartz-sericite-chlorite schists impregnated with pyrite and subordinate chalcopyrite, galena and sphalerite (Wiesciszowice quarry);
- granitic rocks with hydrothermally altered zones and kaolinized granites (Strzegom-Sobótka massif).

The pyrite content in both rock types is the factor required for the appearance of secondary hydrated sulphates. The sulphates are found in the form of polymineral skins and cauliflower - like efflorescences up to several m^2 in size on strongly weathered rock surfaces in abandoned quarries. They have variable colours: from snow-white through various shades of yellow (greenish-yellow, lemon, orange, honey) to brown.

The following nine minerals of the hydrated sulphate group (simple and complex) have been identified: gypsum, alunogen, copiapite, epsomite, fibroferrite, halotrichite, hexahydrite, meta-aluminite and slavikite.

Analytical Technique

The sulphate minerals were analysed using derivatograph type 1500 Q and having DTA, DTG, TG and Q-TG graphs recorded. The analytical conditions: The temperature range 20-1000°C, heating rate 5°C/min (occasionally 10°C/min), air atmosphere, ceramic crucible and labyrinth (the Q-TG curve recorded), weight of samples about 200 mg, reference material Al_2O_3. A binocular microscope was used in sample preparation and purity was controlled by X-ray diffraction. Before being ground the samples were stored at a room temperature of about 20-25°C, and after grinding they were kept in a desiccator.

The Water Problem in Hydrated Sulphates

Structural studies on hydrated sulphates (e.g. KUBISZ, 1964 and 1967; FANFANI et al., 1973) indicated that three types of water are present: H_2O, OH^- and H_3O^+. The latter was ascertained in this mineral group by KUBISZ (op.cit.) and was also found in copiapite from Strzegom (AUGUST, 1986).

In Fe- and Al-sulphates water occupies the following sites (according to FANFANI et al., 1971 and 1973):

- in octahedra $Fe(OH)(H_2O)_2O_3$ constituting together with SO_3 chains,

- in isolated octahedra Al^{3+}, $Fe^{3+}(H_2O)_6$

- as molecular water with no direct bonds with cations but bound within the structure by hydrogen bonds.

The H_3O^+ group may occur in the sulphate structure in the form of octahedra $H_3O^+ \cdot 6\,H_2O$ (KUBISZ, 1967).

The strength of particular bonds varies depending on their lengths (FANFANI et al., 1973).

X (cation position) -	H_2O	1.93 Å
	H_2O	1.97 Å
	H_2O	1.90 Å
Fe (1)	H_2O	2.02 Å
	H_2O	2.05 Å
	OH	1.96 Å
Fe (2)	H_2O	1.94 Å
	H_2O	2.04 Å
	OH	1.96 Å
H_2O	H_2O	2.58 Å
	H_2O	2.71 Å
	H_2O	2.95 Å

These differences in types and bonding lengths are mainly responsible for multistage dehydration and dehydroxylation reactions observed in hydrated sulphates.

Sulphates Characterized by Thermal Analysis

Alunogen $Al_2[SO_4]_3 \cdot 18\,H_2O$

   Locality: Żarów
   Form: light beige fibrous aggregates

DTA (Fig. 1) shows three dehydration stages with peak temperatures at 130°C, 140°C and 320°C. They correspond to the three weight loss steps on the Q-TG curve which displays also an additional loss at 115°C. The dehydration follows the scheme:

I. $Al_2[SO_4]_3 \cdot 18\ H_2O \quad \xrightarrow[-5\ H_2O]{115°C} \quad Al_2[SO_4]_3 \cdot 13\ H_2O$

II. $Al_2[SO_4]_3 \cdot 13\ H_2O \quad \xrightarrow[-5\ H_2O]{130°C} \quad Al_2[SO_4]_3 \cdot 8\ H_2O$

III. $Al_2[SO_4]_3 \cdot 8\ H_2O \quad \xrightarrow[-5\ H_2O]{140°C} \quad Al_2[SO_4]_3 \cdot 3\ H_2O$

IV. $Al_2[SO_4]_3 \cdot 3\ H_2O \quad \xrightarrow[-3\ H_2O]{320°C} \quad Al_2[SO_4]_3$

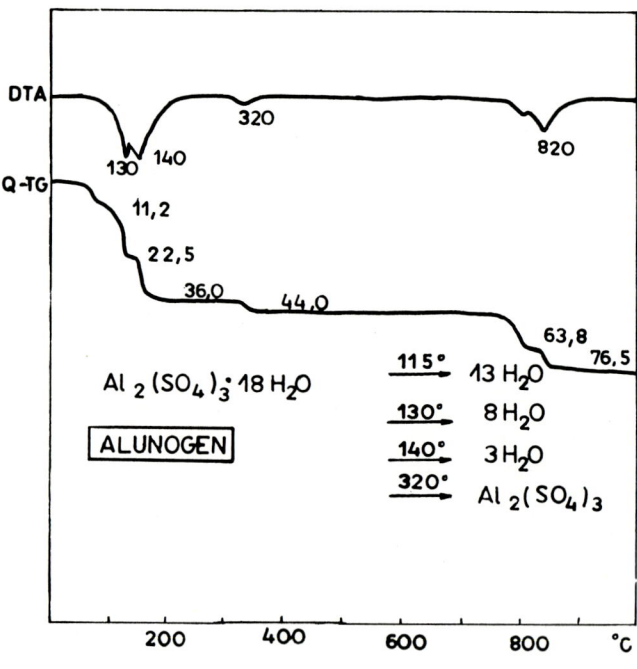

Fig. 1

The decomposition of $Al_2[SO_4]_3$ occurs in two stages at 800°C and 820°C with accompanied weight loss of 19.8 weight% and 17.7 weight% respectively. This thermal analysis curve for the alunogen from Zarów is different from the data presented by CESBRON (1964) and SADRZADEH (1973), according to whom the dehydration of the mineral involves two stages (132°C and 323°C) and the $Al_2[SO_4]_3$ decomposition is a single process (at 827°C).

Copiapite

$$H_3O^+_{0.1} Mg_{0.16} Fe^{2+}_{0.3} Al_{0.4} (Fe^{3+}_{0.88} Al_{0.12})_4 [(SO_4)_6(OH)_2] \cdot 20\ H_2O$$

Locality: Borów and Wiesciszowice
Forms: lemon coloured botryoidal aggregates

Several characteristic endothermic peaks in the DTA curves reflect dehydration and dehydroxylation (Fig. 2). The loss of molecular water takes place in four steps at 130°C, 150°C, 190°C and 240°C. The dehydroxylation is a two-stage process, involving the dissociation of compounds of the $Fe(OH)_3$ type at 340°C and of the γ-AlOOH type at 540°C.

Basing on Q-TG data the processes are as follows:

I.
$$H_3O^+_{0.1} Mg_{0.16} Fe^{2+}_{0.3} Al_{0.4} (Fe^{3+}_{0.88} Al_{0.12})_4 [(SO_4)_6(OH)_2] \cdot 20\ H_2O$$
$$\xrightarrow[-3\ H_2O]{130°C}$$

II.
$$H_3O^+_{0.1} Mg_{0.16} Fe^{2+}_{0.3} Al_{0.4} (Fe^{3+}_{0.88} Al_{0.12})_4 [(SO_4)_6(OH)_2] \cdot 17\ H_2O$$
$$\xrightarrow[-7\ H_2O]{150°C}$$

$$H_3O^+_{0.1} Mg_{0.16} Fe^{2+}_{0.3} Al_{0.4} (Fe^{3+}_{0.88} Al_{0.12})_4 [(SO_4)_6(OH)_2] \cdot 10\ H_2O$$

III. $H_3O^+_{0.1} Mg_{0.16} Fe^{2+}_{0.3} Al_{0.4} (Fe^{3+}_{0.88} Al_{0.12})_4 [(SO_4)_6(OH)_2] \cdot 10\ H_2O$

$$\xrightarrow[-6\ H_2O]{190°C}$$

IV. $H_3O^+_{0.1} Mg_{0.16} Fe^{2+}_{0.3} Al_{0.4} (Fe^{3+}_{0.88} Al_{0.12})_4 [(SO_4)_6(OH)_2] \cdot 4\ H_2O$

$$\xrightarrow[-2\ H_2O]{240°C}$$

V. $H_3O^+_{0.1} Mg_{0.16} Fe^{2+}_{0.3} Al_{0.4} (Fe^{3+}_{0.88} Al_{0.12})_4 [(SO_4)_6(OH)_2] \cdot 2\ H_2O$

$$\xrightarrow[-2\ H_2O,\ -1\ (OH)]{340°C}$$

VI. $H_3O^+_{0.1} Mg_{0.16} Fe^{2+}_{0.3} Al_{0.4} (Fe^{3+}_{0.88} Al_{0.12})_4 [(SO_4)_6(OH)]$

$$\xrightarrow[-1\ (OH)]{540°C}$$

$H_3O^+_{0.1} Mg_{0.16} Fe^{2+}_{0.3} Al_{0.4} (Fe^{3+}_{0.88} Al_{0.12})_4 [(SO_4)_6]$

The dissociation of the sulphate compounds in copiapite produces a strong endothermic peak with a double maximum at 740°C and 810°C, and a weaker but clear peak at 960°C. The weight loss corresponding to these dissociations is 30 wt-% and 2.4 wt-%, respectively.

Fibroferrite $Fe^{3+}[SO_4/OH] \cdot 5\ H_2O$

Locality: Wiesciszowice
Form: light grey fibrous aggregates

DTA displays two endothermic effects related to dehydration at 170°C and 230°C with the loss of 3 and 2 molecules of H_2O, respectively (Fig. 3). The dehydroxylation occurs around 530°C. The water losses are as follows:

I. $Fe^{3+}[SO_4/OH] \cdot 5 H_2O \xrightarrow[-3 H_2O]{170°C} Fe^{3+}[SO_4/OH] \cdot 2 H_2O$

II. $Fe^{3+}[SO_4/OH] \cdot 2 H_2O \xrightarrow[-2 H_2O]{230°C} Fe^{3+}[SO_4/OH]$

III. $Fe^{3+}[SO_4/OH] \xrightarrow[-1 (OH)]{530°C} [Fe^{3+}SO_4]$

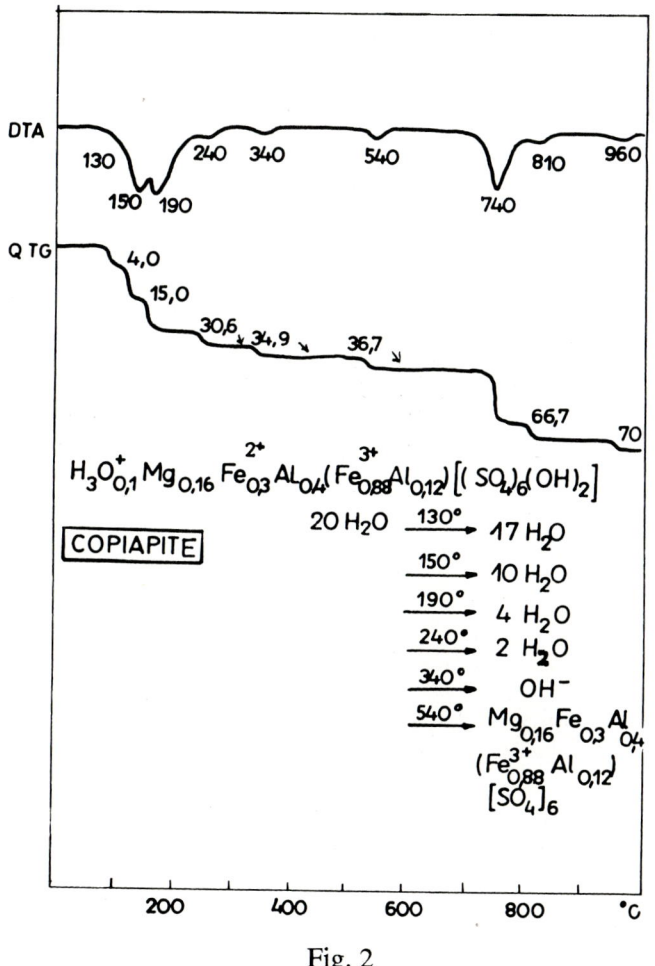

Fig. 2

The decomposition of FeSO$_4$ takes place at 750°C. The thermal analysis curves of the sample from Wiesciszowice (Fig. 3) resembles that for fibroferrite from Chile described by CESBRON (1964).

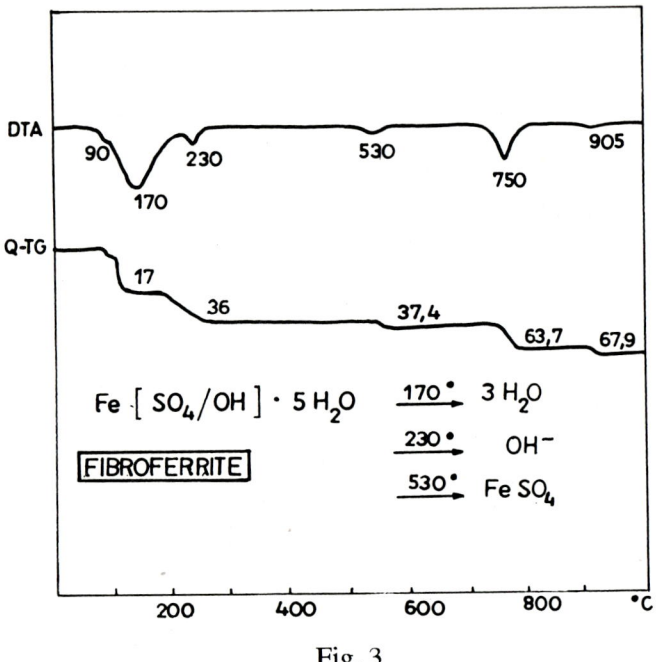

Fig. 3

Slavikite   MgFe$_3^{3+}$[(SO$_4$)$_4$(OH)$_3$] · 18 H$_2$O

Locality: Wiesciszowice
Forms: light green skins on weathered rock surfaces.
       Under the microscope - round microplates.

Thermal analysis curves show many endothermic peaks (Fig. 4). DTA indicates three steps of dehydration, at 110°C, 150°C and 250°C. The dehydroxylation involves two stages, at 450°C and 530°C. TG allows to determine both processes as follows:

I.
$$MgFe_3^{3+}[(SO_4)_4(OH)_3] \cdot 18\ H_2O$$
$$\xrightarrow[-3\ H_2O]{110°C}$$

II.
$$MgFe_3^{3+}[(SO_4)_4(OH)_3] \cdot 15\ H_2O$$
$$\xrightarrow[-11\ H_2O]{150°C}$$

III.
$$MgFe_3^{3+}[(SO_4)_4(OH)_3] \cdot 4\ H_2O$$
$$\xrightarrow[-4\ H_2O]{250°C}$$

IV.
$$MgFe_3^{3+}[(SO_4)_4(OH)_3]$$
$$\xrightarrow[-1\ (OH)]{450°C}$$

V.
$$MgFe_3^{3+}[(SO_4)_4(OH)_2]$$
$$\xrightarrow[-2\ (OH)]{530°C}$$

$$[MgFe_3^{3+}(SO_4)_4]$$

The decomposition of $[MgFe_3^{3+}(SO_4)_4]$ occurs at 790°C and 910°C. The DTA curves of the slavikite from Wiesciszowice are similar to those of the slavikite from Medzev, Czechoslovakia (MAKOVICKY & STRESKO, 1967).

Conclusion

The thermal analyses conducted showed that the investigated minerals belong to the group of hydrated sulphates which contain Fe, Al and Mg in the octahedra.

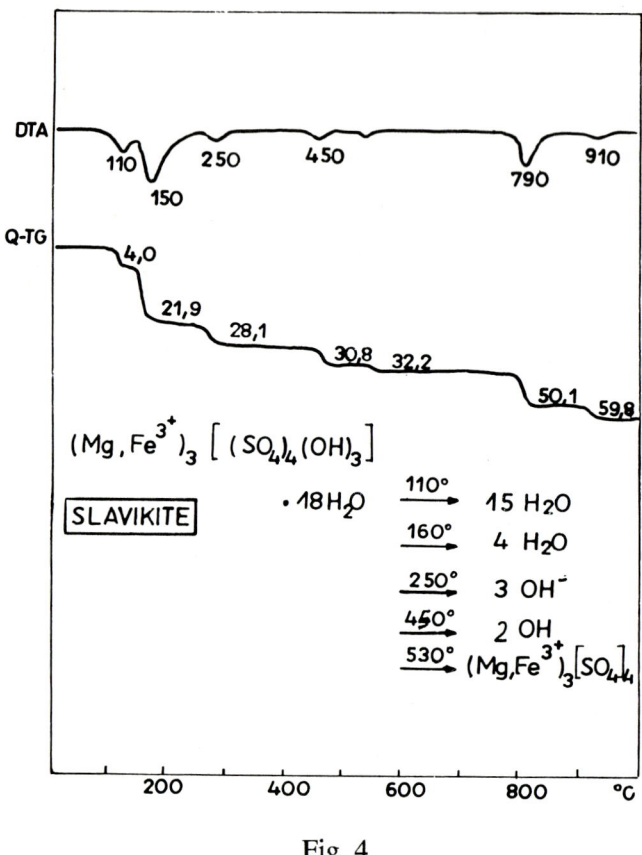

Fig. 4

On the basis of the results of structural studies of hydrated sulphates (KUBISZ, 1964; FANFANI et al., 1970, 1971 and 1973) the assumption can be made that the multistage dehydration is caused by differences in the length (strength) of bonds and the position of water in the structure.

The earliest reactions (temperature of 90-130°C) are due to the loss of the molecular water not linked directly to cations ("free" water), but bound by hydrogen bonding with the structure.

Later, at a somewhat higher temperature (at about 150°C) occurs the loss of the H_2O molecules present in the so-called isolated octahedra that are often empty (without cations on X-position) or contain H_3O^+ ions in the position of the central cation (e.g. copiapite).

The molecular water co-ordinating cations Fe or Al in the octahedra of the $Fe(OH)(H_2O)_2O_3$ type is more strongly bound. Its loss occurs around 230°C, and in the case of alunogen at the temperature of 320°C.

The dehydroxylation occurs depending on the number of cations. In the structures containing one cation (e.g. Fe) this process is a single-stage (e.g. fibroferrite) and in the structures containing two various cations (e.g. Fe and Al or Fe and Mg) dehydroxylation is a two-stage process (e.g. copiapite and slavikite).

Moreover, the investigations conducted showed that on the basis of the Q-TG curve the number of the molecules of water present in the structure of hydrated sulphates can be estimated.

References

AUGUST C. (1986) - Copiapite from Borków near Strzegom (Lower Silesia).- Miner. Polon., vol. 15, 1-2

CESBRON F. (1964) - Contribution á la minéralogie des sulfates de fer hydratés.- Bull. Soc. Fr. Mineral. Cristallogr., 87

FANFANI L. et al. (1970) - The crystal structure of roemerite.- Am. Miner., 55, 1-2

FANFANI L. et al. (1971) - The crystal structure of butlerite.- Am. Miner., 56, 5-6

FANFANI L. et al. (1973) - The copiapite problem: The crystal structure of a ferrian copiapite.- Am. Miner., 58, 3-4

KUBISZ J. (1964) - The study of the hypergenetic sulphate minerals of Poland.- Pr. Geol. PAN, 26 (in Polish).

KUBISZ J. (1967) - The meaning of positive hydrogen-oxygen ions in minerals.- Pr. Geol. PAN, 11, (in Polish).

MAKOVICKY E. & STRESKO V. (1967) - Slavikite from Medzev near Kosice, Czechoslovakia.- Tchermaks Miner. Petr. Mitt., XII, Heft 1

Physico-Chemical Mineralogy

INTERNAL THERMAL REACTIONS OF MINERALS

L. Stoch

Academy of Mining and Metallurgy
Kraków, Poland

Abstract

A group of thermal decomposition and internal reorganisation reactions, which takes place within the structural framework of minerals are considered. The rules governing these processes and their usefulness in the interpretation of thermal analysis results are discussed.

Introduction

Thermal analysis has until today been applied as a qualitative and quantitative method for the compositional determination of mineral and rock samples. It is often considered as a method supplementing X-ray analysis (XRD): Differential thermalanalysis (DTA) and thermogravimetry (TG) help to detect and to determine quantitatively the components whose degree of crystallinity is so low that they remain undetected by the diffraction method, e.g. some ferrohydroxides or some smectites.

Lately considerable progress has been made with investigations of the high temperature reaction minerals. Advanced techniques for crystal structure investigations permit the study of the actual mechanism of thermal reactions. Also high resolution electron microscopy, XRD, neutron mass resonance (NMR), infra-red (IR) and other spectroscopic methods, provide data which considerably extend our knowledge of the thermal reactions of minerals on an atomic scale.

Methods have also been developed for the investigation of kinetics and the measurement of enthalpy of reactions, by means of DTA, TG and differential scanning calorimetry (DSC). The availability of these measurements as well as their applications have increased.

The kinetic and thermodynamic parameters of thermal reactions of solids are structurally sensitive values, as shown by the decomposition of Workson kandites (STOCH & WACŁAWSKA, 1981; HORVATH, 1989). These relationships may be easily detectable by the application of the Constant Rate Thermal Analysis method, e.g. PAULIK et al. (1987), Q-derivatography or change of gaseous decomposition product pressure (READING et al., 1984).

All of this creates an opportunity for the utilization of data mentioned above to detect even small differences in chemical composition, structural defects, degree of purity, crystal perfection, grain size and other specific features of minerals. It brings therefore a new direction of application and practical meaning to the thermal investigation of minerals. For these reasons a better understanding of the influence of structural factors on the thermal reactions of solids and the development of structural thermochemistry is an important goal for the thermal investigation of minerals. It helps also in the proper interpretation of thermal analysis plots. A few examples for these interrelationships are given below.

Internal Thermal Reactions

Solid bodies differ from gases and liquids in that they have a relatively rigid internal framework lattice. The geometry of this crystal lattice, the nature and direction of the chemical bonds, the anisotropy of the crystal structures and its properties are the main factors in governing the thermal reactions of solids and determining their mechanisms. The displacement of atoms and ions taking part in thermal reactions proceeds usually by way of diffusional transport which is slow and directional in crystalline solids. The directions depend on the anisotropy of the internal structure. The defects of the structure are responsible for the fact that the medium in which this transport is taking place has often a discontinuous nature.

When the temperature of solids is close to the reaction temperature the atoms and ions in the structure become mobile. This has been demonstrated by MACIEJEWSKI & OSWALD (1985) for calcite where near the decomposition temperature its crystal edges and corners become rounded, and crystals are linked together similarly as in the sintering process. On the other hand, dehydroxylation of the layer silicate kaolinite is preceded by increased mobility and displacement of only a part of the OH groups, while the structural framework remains essentially unchanged (STOCH, 1984).

It is well known, that the mobility of the atoms and the degree of freedom of their displacements vary. But even when the freedom is considerable, as in case of calcite, the influence of the crystal structure is still visible. This is the essential difference between the thermal reactions of solids which are localized and the reactions in gases, which are not.

It is due to these factors that often no direct relation can be found between the reaction rate of solids and their concentration. These reactions are also influenced by the specific surface area, the morphology of the crystal grains, macro- and micro-defects and others. Long lists can be found in literature for these factors.

So the course of reactions of solid bodies at constant pressure and temperature is determined beside concentration by the intrinsic factors connected with the structure, i.e. particularly with the degree of its perfection and the kinds of its defects (f_{intri}). However, in the case of some reactions (decomposition) these are influenced by morphology and grain size (f_{morph}).

Taking into consideration the influence of the above mentioned intrinsic and morphological factors, as well as the influence of the gaseous products of decomposition with the pressure P_{gp} which determine together with the concentration x the reaction rate V, its formula can be put down as follows:

$$V = dx / dt = f(x, f_{intri}, f_{morph}, P_{gp}) \qquad [1]$$
when:
P, T = const.

From among the specific features of thermal reactions of solids the leading one is the reaction relating parent substance (precursor) to solid product. When taking this factor as a criterion the reactions can be divided into: 1. topochemical reactions and 2. intraframework or internal reactions (Table 1). The disadvantage of this relation is the structural and textural dependence of the product and precursor which is occasionally defined as topotaxy (STOCH, 1987a).

Examples of these processes are:

1. displacement of the elements of the structural framework over short distances, such as polymorphic (β-quartz → α-quartz) and martensitic transformations (baddeleyite);

2. breaking part of the crystalline solid bonds to pass into an X-ray amorphous state (transformation of quartz into the amorphous phase - metacristobalite);

3. breaking part of the bonds with splitting off and subsequent liberation of the molecules of gaseous products (thermal decomposition of hydroborates, dehydroxylation of kaolinite, dehydration of hydrated phosphates etc.);

4. diffusional displacement of atoms and ions over distances greater than the interatomic distances, by a) formation of new structures (polymorphic transformations of reconstructive and order-disorder types), b) the exchange of components between the domains, leading to their segregation and/or between solid and surrounding phases and synthesis of new compounds: Also the recrystallization of crystalline

minerals, crystallization of amorphous solids, multicomponent glasses and gels, exsolution of solid solutions, sponoidal effect in glasses.

Table 1: Classification of heterogeneous processes according to the topological relation parent substance - solid product.

	Topochemical	Internal
Solid-gas systems		
Sublimation and condensation	+	-
Thermal dissociation and recombining of gas products	+	+
Oxidation and reduction	+	+
Exchange of chemical components	+	+
Solid-liquid systems		
Melting	-	+
Crystallization, precipitation	+	-
Simple dissolution	+	-
Incongruent dissolution	+	-
Ion exchange	-	+
Intercalation	-	+
Recrystallization (transformation)	-	+
Hydration	+	+
Solid-solid system		
Phase transitions		
displacive, rotational, orientational	-	+
reconstructive	-	+
substitutional	-	+
Amorphisation of crystalline solids	+	+
Crystallization of amorphous solids	+	+
Reactions of new compounds synthesis by		
exchange of components	+	-
segregation of components	-	+

5. incorporation of ions and molecules, forming new layers in the structure (ion exchange in silicates, formation of intercalation compounds of graphite and aluminium).

6. change of valency of ions (oxidation and reduction reactions).

The course of these processes is considerably influenced by the structure of the parent substance (precursor). They take place in the bulk of the precursor grain due to the diffusion of chemical components, while the structure of the product remains in a topotatic relation with respect to the primary structure.

Specifically these processes are discussed further in relation to the reactions of thermal decomposition and rebuilding of the internal structures of minerals.

Internal processes include those phase transformations as well as chemical reactions which proceed simultaneously within the whole bulk of the crystal grain, occasionally even without any visible changes in their outer form (pseudomorphism). The solid products of these processes are formed inside the structural framework of the parent substance (precursor) and usually remain in a topotactic relation with respect to the primary structure (STOCH, 1987a; STOCH, 1989).

Internal Thermal Decomposition

The internal thermal dissociation is characteristic for substances the structural framework of which remains preserved after the process is fully or only partly completed. The substances undergoing internal thermal decomposition are usually those whose structure contains elements composed of polymerized coordination polyhedra with strong bonds (layers, chains) such as silicates, borates, phosphates and organic polymers. The process takes place in the bulk volume of the substance and the decomposition centres are distributed uniformly within the whole grain.

Here we should also consider cases where reaction fronts move from the edge to the centre of the grain, but the essential part of the structural framework remains unchanged during the process (dehydration of some phosphates, layer silicates etc.).

This reaction is made up of two partial processes:

1. breaking away of atoms or ions and molecules (H, OH, H_2O, CO_2 etc.) and the formation of free molecules of the gaseous decomposition products,

2. removal of the gas molecules from the framework of solid product.

When decomposition is of an internal character the temperature of both processes may be considerably shifted with respect to each other (STOCH, 1989; STOCH, 1987b).

Topochemical decomposition proceeds usually according to the contracting disc model. The closed box, which bursts under the pressure of gaseous products seems to be the model of internal decomposition process.

The removal of the gas molecules occurs then when its intraframework pressure exceeds the value necessary to disrupt the weaker chemical bonds of the structural framework. The volume and shape of the solid often does not change during this process, which occasionally occurs in a rapid way. In these cases it is connected with a step-like increase in volume. The intraframework decomposition process is also a little sensitive to changes in pressure of the gaseous decomposition products.

The decomposition of colemanite, a borate with a chain structure, is an example for such a multistage thermal decomposition (Fig. 1). It proceeds as follows:

$$Ca_2(B_6O_8)(OH)_6 \cdot 2\,H_2O \xrightarrow{340°C} Ca_2(B_6O_{11})(OH)_6 \cdot 2\,H_2O\ (3\,H_2O_{if})$$

$$\xrightarrow{368°C} Ca_2(B_6O_{11})\,(5\,H_2O_{if}) \xrightarrow{386°C} Ca_2(B_6O_{11})_{amorphous} + H_2O_{vapour}$$

The decomposition begins with splitting-off OH groups and the formation of H_2O molecules. Next, H_2O bound with strong hydrogen bonds to borate chains is splitt off. The water molecules liberated in both stages remain enclosed in the voids of the framework (H_2O_{if}). When the intraframework pressure of H_2O exceeds the strength of the bonds, the framework breaks and the water vapour escapes in an explosive manner (WACŁAWSKA et al., 1988).

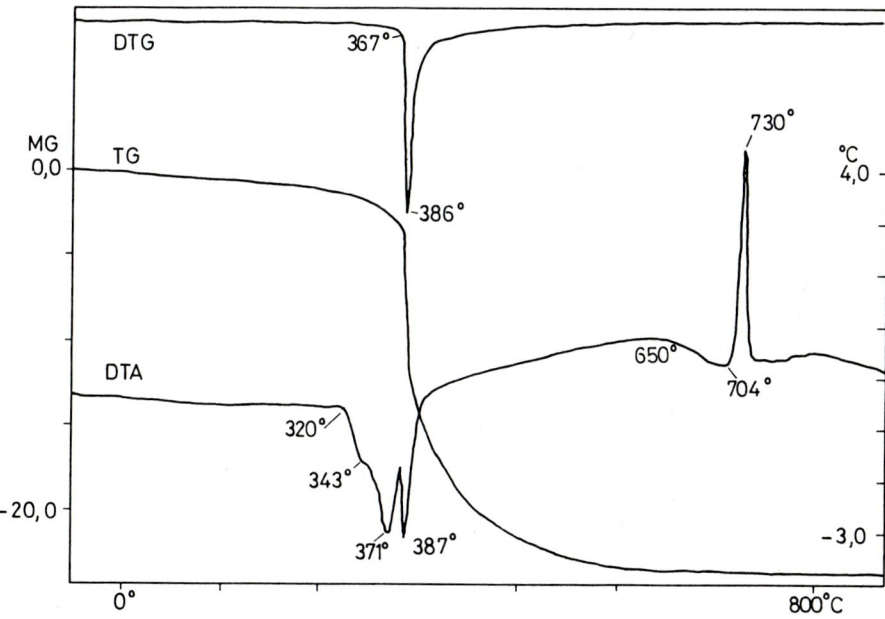

Fig. 1: DTA, TG and DTG curves of colemanite.

The number of substances which undergo thermal decomposition of an intracrystalline character appears to be considerable.

The layer borate pandermite $Ca_2(B_5O_8)(OH)_3 \cdot 2\,H_2O$ becomes decomposed in several stages loosing H_2O at temperatures of 300°C and 360°C. The first stages of dehydroxylation do not result in any structural change. At 390°C recrystallization (exothermic) takes place and afterwards the rest of OH groups as H_2O are removed at 465°C (endothermic). These processes are accompanied by the delamination of crystals and an increase in volume of the heated sample.

The decomposition of the system $PbO - P_2O_5 - H_2O$, investigated by WORZALA & JOST (1989), follows a similar mechanism. It is termed by the authors a topotactic reaction due to the close relation of the structure of the products from the succeeding stages of decomposition.

The dehydroxylation of kaolinite is accompanied by a gradual contraction of sample volume as follows from TG and DTG curves. Sometimes however, the dehydroxylation of kaolin minerals is accompanied by a rapid increase in sample volume. It appears at the beginning of the second stage of dickite dehydroxylation. The use of this phenomenon to distinguish between dickite and kaolinite has been proposed (SCHOMBURG & STÖRR, 1978; SCHOMBURG, 1984). Also an explosive increase of volume accompanies the dehydroxylation of kaolinite with elevated dehydroxylation temperatures (Keokuk kaolinite, USA).

In both the cases there occur varieties with perfect crystals and a very well ordered structure free of any discontinuities. Consequently the removal of water molecules from the interior of crystals is particularly difficult. However, when the intracrystalline water vapour pressure exceeds the critical value, violent separation of layers and water liberation occurs, which is shown by rapid increase in sample volume.

The swelling of micas which accompanies its dehydroxylation is based on the same principle. It is observed when their plates are sufficiently large.

Some chlorites (prochlorite) also increase in volume during dehydroxylation (SMYKATZ-KLOSS, 1982).

Swelling or even explosion of samples is a characteristic feature indicating the internal mechanism of the mineral decomposition.

Since the intracrystalline pressure is very high, the changes of water vapour partial pressure in the reaction environment influence only slightly the decomposition rate and temperature. This seems to be the next feature of the internal decomposition processes.

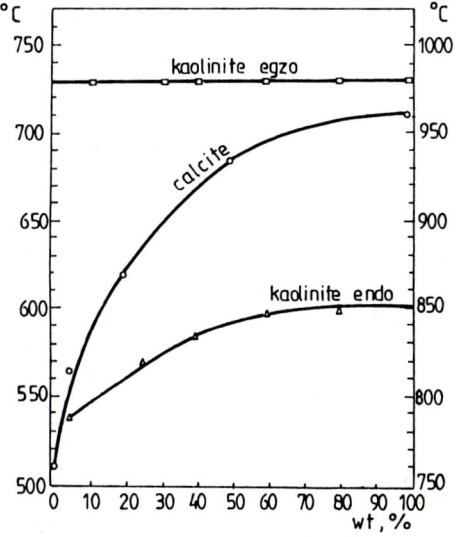

Fig. 2: Shifting DTA peak temperature of calcite and kaolinite with increase of their content in the mixtures with Al_2O_3 (heating rate 10°C per minute).

Contrary to internal decomposition the surface chemical thermal dissociation is strongly dependent on the partial pressure of the gaseous products of decomposition. It is visible in a spectacular manner as a shifting DTA peak temperature with increasing amounts of the decomposing substance. Calcite decomposition is an example (Fig. 2). This phenomenon may be used in DTA practice to distinguish internal and topochemical decomposition. Curves of peak temperature vs. log of sample amount introduced by SMYKATZ-KLOSS (1974) is a helpful tool. High curve slopes indicate that the decomposition of a given mineral is sensitive to variations in the partial pressure of the specific gas evolved.

The Re-Constitution of Crystal Structures

The heating of minerals often involves the synthesis of new crystal phases with compositions different from those of the parent substance. The new compound is sometimes formed within the framework of the parent substance (precursor). It is proceeded by displacement and sometimes even by segregation of its components.

The data about the mechanism of these processes so far collected indicates that they proceed according to some general principles which are as follows (STOCH, 1989):

1. The first product of the synthesis or crystallization of new phases produced inside the framework of the parent substance is close in chemical composition and structure to those of the microregions (domains) of the original crystal.

2. The system has a tendency to reach an equilibrium state corresponding to the chemical composition of the parent substance by the gradual re-arrangement of the structure of newly formed phases and/or the segregation of their chemical components, while the crystallization of the next generation of compounds is usually of a simpler chemical composition.

3. The formation of a new compound within the framework of the precursor occurs through the displacement of cations and the mutual adjustment of their positions together with the framework elements corresponding to the anion sub-lattice of the new compound. Simple compounds made up of chemical elements which are less strongly bound in the structure of the precursor (modifiers of glass), and are more mobile, tend to be the first to crystallize. The formation of

compounds is followed by the formation of compounds with more complex anions.

A good example of the above is the formation and reconstruction of quartz-like solid solutions in glasses of the system $MgO - Al_2O_3 - SiO_2$. At higher temperatures its chemical composition is changed and it passes into a cordierite type equilibrium crystal phase (SCHREYER & SCHAIRER, 1961). The processes of crystallization and reconstruction of the crystal phases in the system $Li_2O - CaO - MgO - Al_2O_3 - SiO_2$ are also good examples.

These are completely analogous with the phenomena of framework rearrangement and the formation of new phases in the amorphous substances formed as a result of the reactions of thermal decomposition of many crystalline substances. This can be demonstrated using the example of certain layer silicates and borates.

In the octahedral sheet of the structure of montmorillonite, the distribution of the Mg^{2+} and Al^{3+} cations has an ordered character. At high temperature they locally form $MgAl_2Si_2O_8$ units which create a solid solution having the structure of high-temperature quartz. During further heating this structure is more ordered and Mg-spinel and then cordierite together with mullite and cristobalite are formed (STOCH et al., 1985).

The investigation of the hydrated layer and chain borates has also confirmed the validity of these regularities.

As a result of thermal decomposition the structures of colemanite and pandermite become amorphous. In the DTA curve of both amorphous borates there occurs a weak endothermic peak (650°C). It corresponds to the rebuilding of persistent elements of primary framework and probably the formation of submicroregions of new crystalline phases. The exothermic peak of the growth of their crystals appears at 740-745°C. During the heating of colemanite the compound $2\,CaO \cdot B_2O_3$ is formed, while pandermite gives $CaO \cdot B_2O_3$ (745°C peak). Subsequently the rest of the amorphous matrix crystallizes and CaO · 2 B_2O_3 is formed (800°C peak).

When the artificial glass of anhydrous pandermite composition is heated, the resultant DTA shows an endothermic peak for glass network rebuilding at 650°C, an endothermic peak of nucleation (750°C) and then small quantities CaO · 2 B_2O_3 followed by Ca · B_2O_3 as a main crystalline phase are formed (800°C exothermic peak). This demonstrates the influence of the amorphous precursor structure on the succession and temperature of the formation of crystals during their recrystallization.

Thermal amorphisation which results in the re-arrangement of the internal structure of solids is characteristic for minerals with polymerized anions such as silicates.

In the temperature range between 1100-1200°C pure quartz becomes unstable. It transforms into the X-ray amorphous phase metacristobalite. This amorphous substance is formed considerably below the melting temperature of quartz, after prolonged heating for the time necessary to break off parts of the oxygen bridges between the SiO_4 tetrahedra. After a longer period of heating cristobalite crystallizes (Fig. 3).

The rigid strong network of quartz consisting of six and eight-membered rings of SiO_4 is not able to transform at once into the structure of cristobalite which is build of six-membered rings. A intermediate glassy phase appears first and next cristobalite is formed by simple crystallization of the glass. The α-quartz → α-cristobalite transition is classified as a polymorphic transformation of reconstructive type. It is a transformation which exhibits a very specific mechanism with an intermediate state of structural disorder, which remains stable for a relatively long period.

The stability of the intermediate state is greatly influenced by small quantities of isomorphous substitution in the precursor quartz grains. The alkalies depolymerize the silicon - oxygen network and accelerate its formation. Aluminium inhibits the crystallization of cristobalite, increasing the range of existence of the intermediate state. Cristobalite does not allow isomorphic substitutions in its structure, while the redistribution of chemical components and segregation of Al^{3+} must precede the cristobalite formation, thus hampering the process (STOCH et al., 1985).

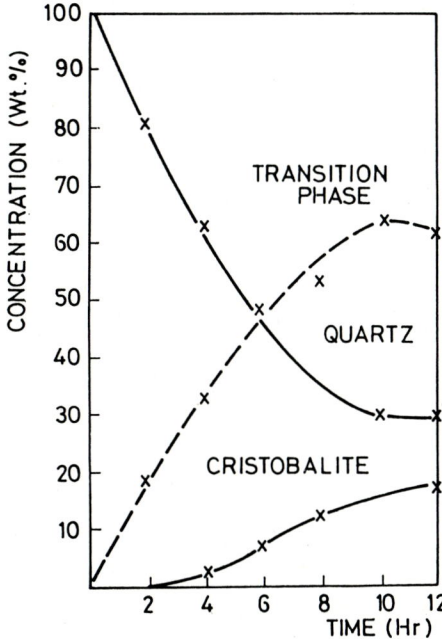

Fig. 3: Internal rebuilding process of the quartz structure into transitional amorphous substance and cristobalite crystallization; temperature 1500°C (STOCH et al., 1985).

The progress of crystallization of amorphous solids corresponds to the model of the process: nucleation-growth. This model assumes that the nuclei of crystallization grow uniformly in two or three dimensions at rates still describable by an f(x) function. Taking into account the structure of the precursor and the newly formed compound relation, the rate of the process may be expressed as follows:

$$V - k \cdot f(x, f_{int}) \qquad P, T - constant$$

The significance of other factors influencing the decomposition process is much smaller compared with thermal dissociation.

Conclusion

Investigations of thermal reactions of minerals supplemented with data on the structural transformations accompanying these reactions make it possible to indicate some factors which govern these reactions. These help in understanding the phenomena recorded by the thermal analysis curves, and can also facilitate the prediction of the course of technological processes utilizing mineral raw materials.

References

HORVÁTH I. (1989) - Reactivity of Solids, 7, 173

MACIEJEWSKI M. & OSWALD H. R. (1985) - Thermochim. Acta, 85, 39

PAULIK J., PAULIK F. & ARNOLD M. (1987) - J. Thermal Anal., 32, 301

READING M., DOLLIMORE D., ROUQUEROL J. & ROUQUEROL F. (1984) - J. Thermal Anal., 29, 775

SCHOMBURG J. (1984) - 9th Conference on Clay Mineralogy and Petrology, Zvolen 1982 (KONTA J. (ed.)), 150, Univerzita Karlova, Praha.

SCHOMBURG J. & STÖRR M. (1978) - Chemie der Erde, 37, 107

SCHREYER W. & SCHAIRER J. F. (1961) - Z. Krist., 116, 60

SMYKATZ-KLOSS W. (1974) - Differential Thermal Analysis - Application and Results in Mineralogy.- Springer Verl., Berlin-Heidelberg-New York, 188 p.

SMYKATZ-KLOSS W. (1982) - J. Thermal Anal., 23, 15

STOCH L. (1984) - J. Thermal Anal., 29, 919

STOCH L. (1987a) - J. Thermochim. Acta, 110, 359

STOCH L. (1987b) - J. Thermal Anal., 32, 1651

STOCH L. (1989) - Thermochim. Acta, 148, 149

STOCH L., LACZKA M. & WACŁAWSKA I. (1985) - Mineralogia Polonica, 16, 43

STOCH L. & WACŁAWSKA I. (1981) - J. Thermal Anal., 20, 291

WACŁAWSKA I., STOCH L., PAULIK J. & PAULIK F. (1988) - Thermochim. Acta, 126, 307

WORZALA H. & JOST K. H. (1989) - Advanced Solid State Chemistry Proc., 2nd International Symposium on Solid State Chemistry, Pardubice 1989.- (FRUMARY M., CERNY V. & TICHY L. (eds.)), 52, Elsevier.

KINETICAL STUDY OF MINERAL REACTIONS BY MEANS OF CONTROLLED TRANSFORMATION RATE THERMAL ANALYSIS (CRTA)

J. Rouquerol[1], S. Bordère[1] & F. Rouquerol[2]

[1] Centre de Thermodynamique et de Microcalorimétrie du CNRS
26 rue du 141ème RIA, 13003 Marseille, France

[2] Université de Provence, Case 2,
Place Victor Hugo, 13331 Marseille Cedex 3, France

Abstract

After presenting the aims and principle of Controlled transformation Rate Thermal Analysis (CRTA) which is, in some respect, the "image" or "opposite" of conventional Thermal Analysis, this paper reviews a few applications to kinetics of thermal decompositions, namely (I) the enhanced separation of successive steps, (II) the use of the rate-jump method for the "assumptionless" determination of the apparent activation energies and (III) the easier discrimination between the possible mechanisms. Examples given deal with sepiolite, dolomite, kerogens and hexahydrated uranyle nitrate.

Introduction

From the very beginning of Thermal Analysis, geosciences have been one of its privileged fields of application, since Thermal Analysis proved to be quite useful (I) to *characterize and analyse* the rocks and minerals (for instance clays), (II) to *study their thermal processing* (for instance, to get a reactive oxide from a natural carbonate, hydroxide or sulphate...) and (III) to model and *understand their genesis* along the geologic ages, under the combined action of temperature and pressure. It follows that geosciences very well know both the possibilities and limitations of present Thermal Analysis and offer an excellent field to probe any methodological improvement.

We shall here describe one of them, namely the general approach of Controlled transformation Rate Thermal Analysis (CRTA) and illustrate its interest in view of understanding the mechanism of thermolysis of various minerals.

The CRTA Approach

Any kinetical study carried out by Thermal Analysis aims to first establish (and then explain by an appropriate mechanism) the formal relationship allowing, at any time of the thermolysis, to express the rate of reaction $d\alpha/dt$ as a function of the degree of reaction α and temperature T (and also, for studies going a step further, as a function of partial pressures, grain size, impurities etc. ...):

$$\frac{d\alpha}{dt} = f(\alpha) \cdot k(T) \cdot g(p) \qquad [1]$$

where the functions f, k and g must be determined.

To reach that goal, an oversimplification is usually introduced, i.e. the assumption that, at a given instant, the temperature of the sample may be either measured (from the experiment) or predicted (from the mechanism and corresponding theoretical model). The same holds for the composition of the surrounding atmosphere and for the degree of reaction. In many cases (if not in most) the reality is far from that, simply because of temperature gradients and pressure gradients occurring within the sample itself and mainly due to the reaction under study. For instance, any endothermal dehydration or decarbonation gives rise to a self-cooling of the sample which may produce a temperature difference (between the central and peripheral parts of the sample) ranging between a few Kelvins and more than 50 Kelvins, depending on the heating rate and sample mass (LIPTAY, 1973). The overpressures due to gas evolution are also extremely dependent on the experimental conditions and may even produce a spurting out of the sample. Without going further, one easily understands it is hopeless to try to derive a mechanism by assuming that, at any time, the whole system is characterized by one temperature and one atmosphere composition. Three ways are available to cope with this difficulty:

a) *to carry out the experiment on so small a sample* (for instance, ca. 5 mg (CRIADO et al., 1980)) *that although the temperature gradients remain similar to those observed with large samples, the temperature differences are small.* This way is of course extremely demanding for the sensitivity of the equipment and for the homogeneity of the sample (which is not always possible to achieve with natural samples). Moreover, the small amount of material does not lend itself to easy characterizations (specially of their crystalline and porous structure) during the course of the thermal decomposition.

b) *to effectively measure* during the experiment and calculate during the modelling *the temperature and pressure gradients* within the samples. This approach suffers from being both experimentally and conceptually heavy and quite difficult to carry out.

c) *to keep the rate of the reaction at such a slow rate that the temperature and pressure gradients may be neglected.* This may be achieved in a "blind" way, by simply selecting an extremely slow heating rate (say, a few Kelvins per hour) which may be of course quite time-consuming and not really justified in the temperature ranges where nothing happens in the sample. Hence the interest of controlling directly the rate of the reaction (instead of the heating rate), which is the basis of "Controlled transformation Rate Thermal Analysis", which we shall designate, in the following, by its abbreviation "CRTA".

The general principle of CRTA is easily understood from the principle of conventional Thermal Analysis, which is recalled in Fig. 1a: the controller heats the sample in such a way that it follows a predetermined temperature programme (checked from the data delivered by the "thermometer" and "chronometer"). The resulting variation, against time, of the parameter "X" (the one measured by the equipment of thermal analysis under consideration and which may be a mass, an enthalpy, a length, a flow of evolved gas etc. ...) is recorded and makes the thermal analysis trace. To achieve now a control of the rate of reaction we need exactly the same basic equipment, simply connected in a completely different way, as represented in Fig. 1b. Here, it is the degree of the reaction - as measured by the parameter "X" - which must follow a predetermined programme (most often, but not necessarily, corresponding to a *constant* rate of reaction) by the appropriate action of the controller on the heating of the sample. The a-priori unknown thermal analysis trace is here simply the recording of temperature vs time. As we see from Figures 1a and 1b, CRTA is, by its principle, the "opposite" or the "mirror image" of conventional Thermal Analysis and there is no continuous nor progressive way to go from one to the other. Being "the image" of Thermal Analysis, CRTA extends over the same variety of equipments and fields of applications, simply improving, as we shall see hereafter, the quality of the information enclosed in the experimental data.

The Enhanced Separation of the Steps

The first improvement brought by CRTA immediately appears when comparing the Thermal Analysis traces, like in Fig. 2, which deals with the dehydration of a natural sepiolite from Vallecas (Spain). The dashed curve corresponds to a conventional TG curve (RAUTUREAU et al., 1977) whereas the plain one was obtained by Controlled transformation Rate Evolved Gas Detection, which is probably one of the simplest and most rewarding ways to carry out CRTA experiments (ROUQUEROL, 1989). The separation of the

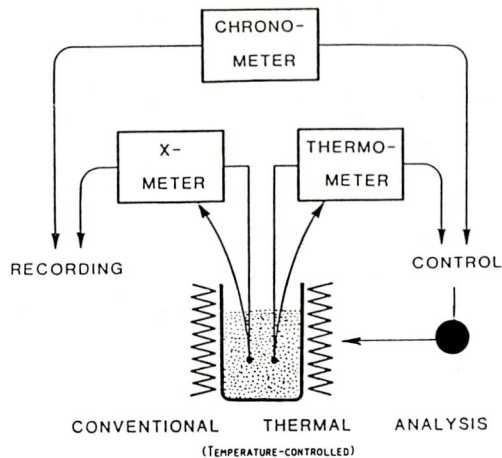

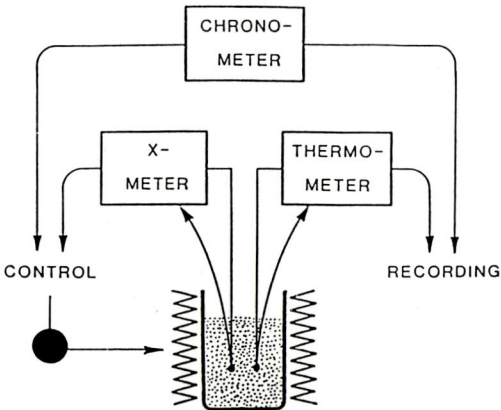

Fig. 1: General principle of conventional Thermal Analysis (Fig. 1a) and Controlled transformation Rate Thermal Analysis (Fig. 1b).

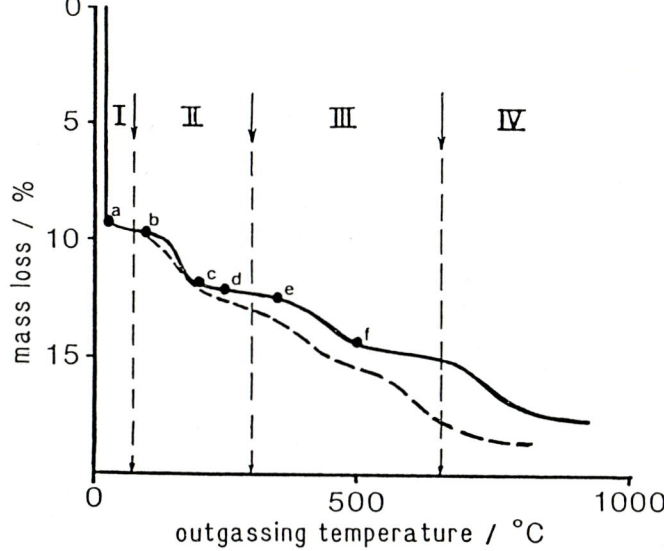

Fig. 2: Dehydration of *sepiolite* followed by conventional Thermogravimetry (dashed curve) and by Controlled transformation Rate Evolved Gas Detection (plain curve).

four steps is clearer on the CRTA curve, where we successively distinguish (GRILLET et al., 1988) the departure of zeolitic water (I), then that of water molecules bound to the Mg atoms situated on the edges of the structural sheets (in two successive halves, II and III) before a final dehydroxylation of the mineral (IV). *The clear distinction of the steps is actually a prerequisite for any safe kinetical study* since the basic concept of "degree of reaction" may be clearly defined for an individual and well isolated step of a reaction, and certainly not for overlapping steps when one has no means to determine, at any time, the part of each step to the overall, macroscopic, change recorded for the parameter "X".

Still more striking is the comparison of the curves obtained for the dehydration and denitration of hexahydrated uranyle nitrate $UO_2(NO_3)_2 \cdot 6H_2O$ (BORDÈRE et al., 1990), as shown in Fig. 3. The dashed curve was obtained on a 500 mg sample by conventional TG, under atmospheric pressure, with a flow of pure nitrogen (4 l/h), at a heating rate of 1 K/min. The plain curve draws benefit partly from the smaller mass

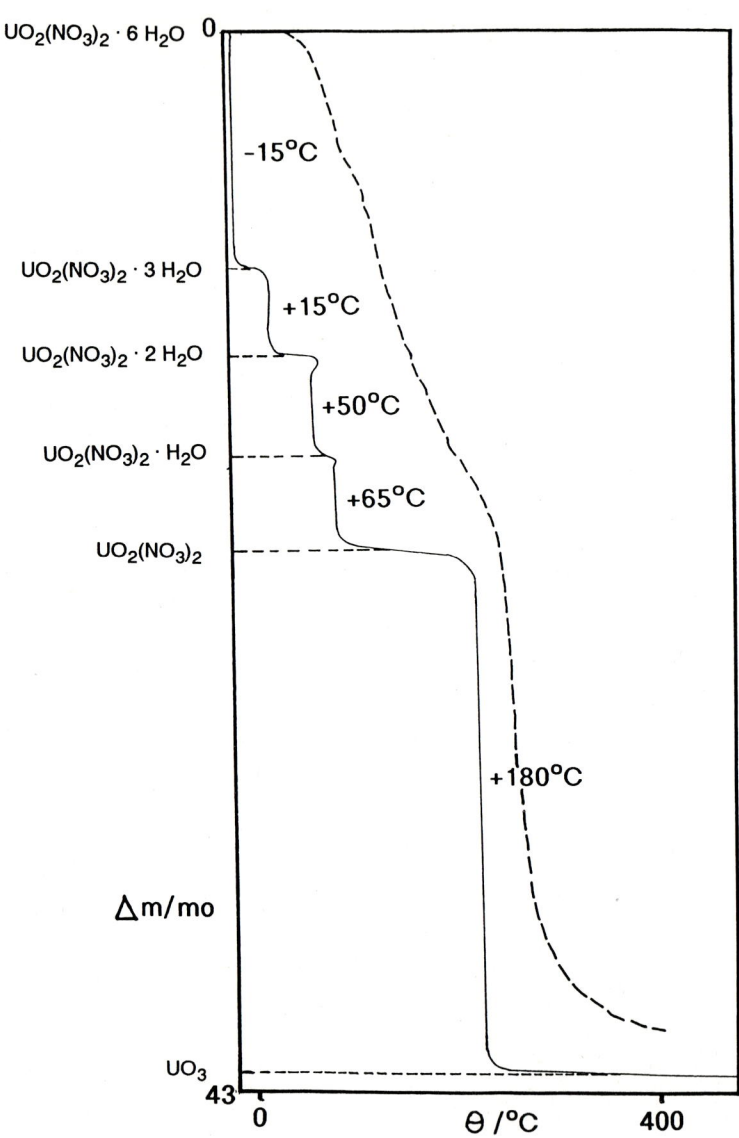

Fig. 3: Dehydration and denitration of *hexahydrated uranyle nitrate* followed by conventional Thermogravimetry (dashed curve) and by Controlled transformation Rate Evolved Gas Detection (plain curve).

(60 mg) but also from the use of Controlled transformation Rate Evolved Gas Detection (Rate of gas evolution: 2 mg·h^{-1}·g^{-1}) and from the control of the pressure (self-generated atmosphere) at a value enhancing the separation (here: 5·10^{-2} mbar).

The simple set up of Controlled transformation Rate Evolved Gas Detection (CR-EGD) above is represented in Fig. 4. The parameter "X" is here the gas flow evolving from the thermal decomposition of the sample (located in the glass bulb) and is measured from the pressure drop through the diaphragm D. The pressure signal is used to control the heating of the sample, so that both the residual pressure over the sample (10^{-5} mbar to 100 mbar, depending on the set up and pressure gauge) and the rate of the reaction are simultaneously controlled. Let us notice that *the glass bulb lends itself to an easy and safe transfer of the sample, under vacuum*, from the CRTA set up to any equipment which may be used to characterize the sample; for example a gas adsorption equipment, to determine the surface area and pore size distribution of the product (ROUQUEROL et al., 1985) or a NMR equipment to study the mobility of the remaining water molecules (ROUQUEROL et al., 1966).

The above experiment may be converted into Controlled transformation Rate Evolved Gas *Analysis* (CR-EGA) by simply replacing the pressure gauge by a partial pressure gauge like a quadrupole gas analyser, as is represented in Fig. 5. The gas phase may be continuously monitored in excellent conditions since the rate of production of the gas (which is controlled) may be kept at a value allowing to send *all the gas flow through the source of the quadrupole* without any risk of overpressure and burning of the filament, although the discrimination usually brought by the leak valve used in standard set ups is here totally eliminated. Moreover, the set up may be operated in two ways, by controlling either the *total* residual pressure over the sample or the *partial* pressure corresponding to any peak selected by the experimentalist (because of its expected part in the studied mechanism) in the mass spectrum. The latter procedure was used to record the curves of Fig. 6, corresponding to the thermolysis of 3 samples of kerogen under a controlled partial pressure of water. The 3 samples were selected from the same evolution path (type III, following the usual classification (TISSOT et al.,1974)) but at increasing depths of burial and therefore at increasing degrees of natural degradation in the order A, B, C.

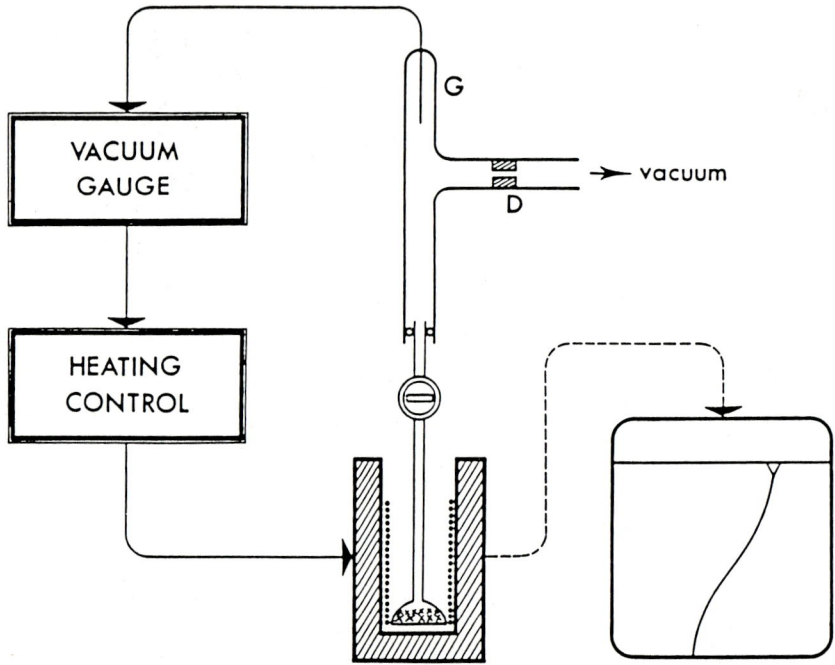

Fig. 4: Principle of a simple set up of Controlled transformation Rate Evolved Gas *Detection* (CR-EGD) under constant residual pressure and self-generated atmosphere.

The plain curves are the temperature recordings obtained by CR-EGA for a constant rate of evolution of H_2O from the samples. These curves may be called "partial TG curves" since they give us *the mass loss of water* (directly proportional to the time elapsed) *as a function of temperature*. The dotted curves represent the corresponding flow of CO_2 evolving from the samples. It is interesting to see that, for the youngest sample (A) this production of CO_2 remains, during a good while, at half the production of H_2O, which suggests that both gases are produced by the same family of reactions (ROUQUEROL, 1985).

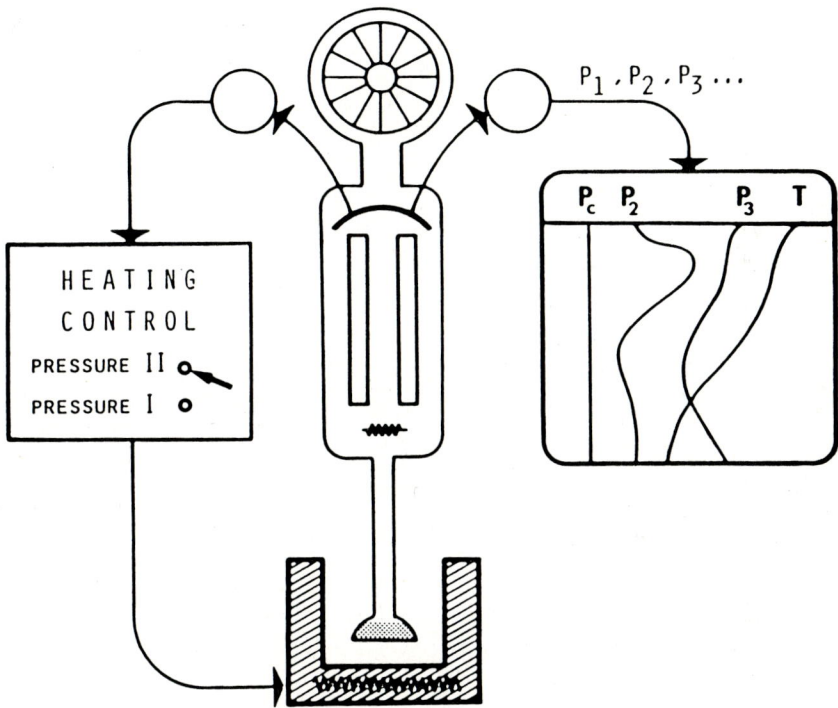

Fig. 5: Principle of a set up of *Controlled transformation Rate Evolved Gas Analysis* (CR-EGA).

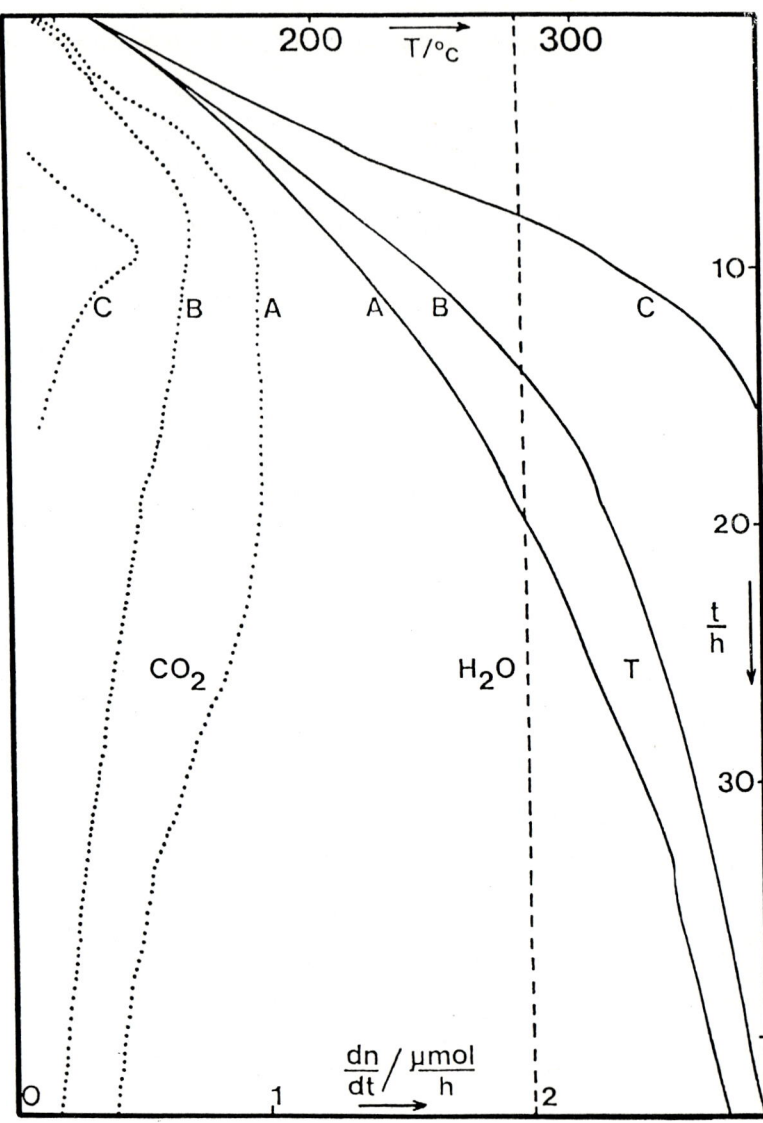

Fig. 6: Experimental curve for a CR-EGA experiment carried out on three different samples of kerogen.

The Rate-Jump Method for the "Assumptionless" Determination of the Apparent Activation Energy

The word "assumptionless" means here that, as we shall see, the method does not need any assumption about the so-called "function of α" represented as $f(\alpha)$ in equation [1]. Nevertheless, we assume that the Arrhenius definition of the activation energy still holds, so that the "function of temperature" $k(T)$ in equation [1] may be written as:

$$k(T) = A \cdot \exp(-E/RT) \qquad [2]$$

When using conventional Thermal Analysis, the above equation may be applied in various ways, in order to determine the activation energy E, once incorporated into equation [1]:

$$\frac{d\alpha}{dt} = A \cdot f(\alpha) \cdot \exp(-E/RT) \cdot g(p) \qquad [3]$$

In any case, the experiments must of course lead to a knowledge of a relationship between α, t and T. Basically, the available routes are the following:

a) *Comparison of 2 isothermal experiments*: strictly speaking, this is out of the scope of Thermal Analysis proper (which cannot be isothermal, by definition) but remains a kind of reference experiment for kinetical studies. Here, equation [3] is applied to a couple of "iso-conversion" points corresponding to the same *apparent* degree of reaction in both experiments. We say "apparent" because although this degree (measured for instance by the mass loss divided by the total mass loss expected) may be formally identical, it may not correspond to exactly the same reaction, since for instance the solid product of the reaction may be different from one temperature to the other. It may for instance differ in its crystalline or porous structure, in its grain size, in its strains and residual surface energy, so that the higher the degree of reaction, the larger the actual difference of mechanism at the two temperatures under consideration. The underlying (and usually unchecked) assumption is therefore that the dependence of the solid product on its temperature of formation may be forgotten.

b) *Analysis of a genuine thermal analysis experiment*: here the comparison of two successive states of the sample involves a change both in temperature and in degree of reaction. A function $f(\alpha)$ must therefore be introduced and tried before a value of E is derived, which of course weakens the significance of the latter.

c) *Application of the temperature-jump method*: this method, proposed by FLYNN (1962) still compares isothermal experiments but carried out alternatively, during short times, on the same sample, so that the assumption mentioned at the end of paragraph a) is not any more needed. The limitation of the method lyes in the selection of the two temperatures which are usually well suited for studying only part of the transformation: before, the rates are too high, whereas afterwards they are too slow. They must be, therefore, in some respect, manually and progressively adapted by the experimentalist.

CRTA provides a simple way to extend and automate Flynn's approach: instead of modulating the temperature we may now *modulate the rate of reaction*, which is periodically brought to jump from value 1 to value 2 and, a while later, to drop back to 1 (ROUQUEROL et al., 1975). The corresponding recording of T vs time, i.e. vs α (since the rate of transformation is constant) is represented in Fig. 7. Along the five successive steps of the thermolysis of hexahydrated uranyle nitrate, *over 100 such successive rate jumps could be performed on a sample of 60 mg*. This means that the change in the degree of reaction α during one jump may be neglected, so that the function $f(\alpha)$ may be eliminated when applying equation [3] to successive states of the sample, corresponding to reaction rates 1 and 2 and, as a consequence, to temperatures T_1 and T_2. *This method is fully automatic* and does not need any previous knowledge about the temperature range or number of steps involved. We applied it recently, quite successfully to the thermolysis of dolomite $CaMg(CO_3)_2$ (ORTEGA et al., 1990) and, as said above, to that of $UO_2(NO_3)_2 \cdot 6\,H_2O$.

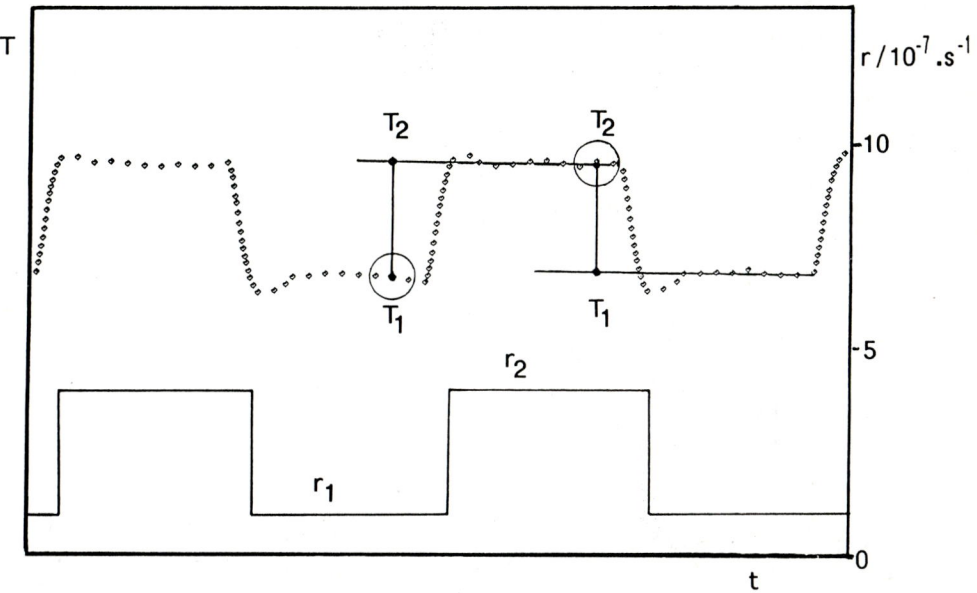

Fig. 7: Experimental recording of temperature vs time (dotted curve) during a rate-jump experiment. Controlled rate of reaction represented by plain curve.

CRTA for an Easier Discrimination Between the Possible Mechanisms

We are listing in Table I, with slight modifications, the main mechanisms found in the thermal decomposition of solids, as they are given by SHARP et al. (1966). Any of these mechanisms may be expressed in terms of a relationship, either differential (directly involving the rate of transformation $d\alpha/dt$) or integral (involving, instead, the degree of reaction α) between α and T. Each such relationship may lead - with the input of the appropriate energy of activation E and pre-exponential factor A - to a "theoretical curve", i.e. the one which must fit the experimental curve to prove that the mechanism assumed is the real one.

The main mechanisms of heterogeneous kinetics (not including catalysis)

$$\frac{d\alpha}{dt} = f(\alpha) \cdot A \cdot \exp(- E / RT) \qquad \alpha = \text{degree of reaction for \underline{a} given step}$$

Usual name	Symbol	$f(\alpha)$
Mechanisms controlled by growth of nuclei (Avrami-Erofe'ev)		
Two-dimensional growth	A_2	$2(1-\alpha)[-\ln(1-\alpha)]^{1/2}$
Three-dimensional growth	A_3	$3(1-\alpha)[-\ln(1-\alpha)]^{2/3}$
Mechanism controlled by random nucleation and/or growth of nuclei in bulk of reactant		
First "order"	F_1	$-\ln(1-\alpha)$
Mechanisms controlled by extent of reacting interface		
Contracting surface	R_2	$(1-\alpha)^{1/2}$
Contracting volume	R_3	$(1-\alpha)^{2/3}$
Diffusion-controlled mechanisms		
One-dimensional diffusion	D_1	$1/2 \, \alpha^{-1}$
Two-dimensional diffusion	D_2	$[-\ln(1-\alpha)]^{-1}$
Three-dimensional diffusion (Jander)	D_3	$3/2 \, [1 - (1-\alpha)^{1/3}]^{-1} (1-\alpha)^{2/3}$
Three-dimensional diffusion (Ginstling-Brounshtein)	D_4	$3/2 \, [(1-\alpha)^{-1/3} - 1]^{-1}$

Table I

Nevertheless, most unfortunately, all such theoretical curves have the same general S-shape, with an initial part corresponding to an increasing rate of reaction, a main central part and then a final part with a decay of the rate of reaction. This of course is of no help at all for the selection of the mechanism and it is often the case - most often, even - that several mechanisms provide similar theoretical curves which give a satisfactory fit with the experimental curve under study.

A major interest of CRTA is that, when considering the same list of mechanisms of Table I and when deriving the corresponding theoretical curves one discovers a completely new picture. Indeed, the restriction which we now have to consider that the rate of transformation is kept constant leads to *three different families of isokinetical curves* (CRIADO et al., 1990), quite easy to distinguish, which are represented in Fig. 8.

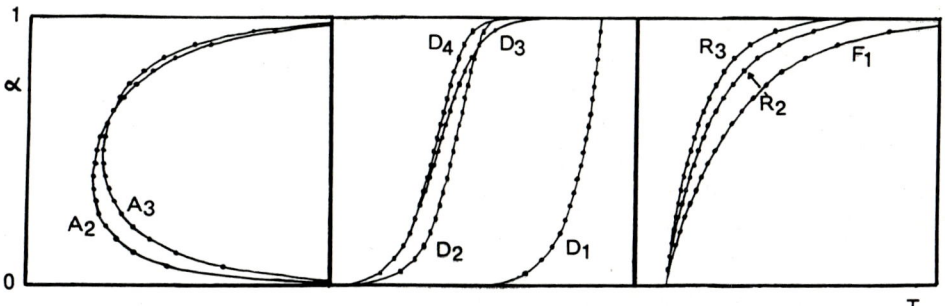

Fig. 8: The three main shapes of theoretical α vs T curves obtained for CRTA.

- A typical *U-shape* corresponds to the mechanisms controlled by *nucleation and growth of nuclei* (A_2 and A_3 in Table I).

- The *S-shape* is now limited to the mechanisms *controlled by a diffusion* (D_2 to D_3).

- Finally, the *"decelerating decay"* shape corresponds to the mechanisms controlled by a boundary (the "reacting interface") (R_2 and R_3).

This means that, from the first glance to the experimental curve α vs T (provided it was obtained by CRTA) one may determine the type of the mechanism involved. For instance, in the case of the thermal decomposition of hexahydrated uranyle nitrate, CRTA clearly shows that the first four steps (which give rise to U-shaped α vs T curves) are controlled by a mechanism of nucleation and growth of nuclei, whereas the denitration step (which provides an S-shaped curve) is controlled by the diffusion of the gas evolved through an array of pores which could be characterized by gas adsorption experiments (BOURDINEAUD-BORDÈRE, 1989). To go one step further, i.e. to discriminate between the various mechanisms forming one given family, one must of course try to fit a theoretical curve with the experimental one. Nevertheless, here again, the job is made easier because of the availability of the energies of activation obtained by the "assumptionless" rate-jump method.

Conclusion

This paper, which was focused on the kinetical possibilities of CRTA, shows how the increase in control brought by CRTA makes it much easier and safer to derive an energy of activation and to determine the most probable mechanism.

Another important interest of CRTA is that it may be considered as a general method of thermal treatment, which may be used for instance to process ores, clays or minerals in carefully defined conditions.

Finally, the principle of CRTA applies to any thermal treatment, whatever the surrounding atmosphere and pressure, the mass of material, the overall duration of the treatment and the parameter "X" selected to control the rate of transformation (ROUQUEROL, 1989): a great deal of novel applications and developments are still to be devised.

References

BORDÈRE S., FOURCADE R., ROUQUEROL F., FLOREANCIG A. & ROUQUEROL J. (1990) - J. Chim. Phys., 87, 1233

BOURDINEAUD-BORDÈRE S. (1989) - Thèse Université de Provence, Marseille.

CRIADO J. M., ORTEGA A. & GOTOR F. (1990) - Thermochim. Acta, 157, 171

CRIADO J. M., ROUQUEROL F. & ROUQUEROL J. (1980) - Thermochim. Acta, 38, 117

FLYNN J. H. (1962) - Thermal Anal., 2, 1111

GRILLET Y., CASES J. M., FRANCOIS M., ROUQUEROL J. & POIRIER J. E. (1988) - Clays and Clay Minerals, 36, 233

LIPTAY G. (1973) - Atlas of Thermoanalytical Curves.- Akademiai Kiado, Budapest, Vol.2

ORTEGA A., AKHOUAYRI S., ROUQUEROL F. & ROUQUEROL J. (1990) - Thermochim. Acta, 163, 25

RAUTUREAU M. & MIFSUD A. (1977) - Clay Minerals, 12, 309

ROUQUEROL F., REGNIER S. & ROUQUEROL J. (1975) - in Thermal Analysis. Vol. 1.- (BUZAGH E. (ed.)), Akademiai Kiado, Budapest, p. 313

ROUQUEROL F., ROUQUEROL J. & IMELIK B. (1985) - in Principles and Applications of Pore Structural Characterization.- (HAYNES J. M. & ROSSI-DORIA P. (editors)), J. W. Arrowsmith, Bristol (1985), p. 213

ROUQUEROL J. (1985) - Pure and Appl. Chem., 57, 69

ROUQUEROL J. (1989) - Thermochim. Acta, 144, 209

ROUQUEROL J., FRAISSARD J., ELSTON J. & IMELIK B. (1966) - J. Chim. Phys., 4, 607

SHARP J. H., BRINDLEY G. W. & NARAHARI ACHAR B. M. (1966) - J. Amer. Ceram. Soc., 49, 379

TISSOT B., DURAND B., ESPITALIE J. & COMBAZ A. (1974) - Am. Assoc. Pet. Geol. Bull., 54, 499

THERMOANALYTICAL INVESTIGATIONS OF BINARY OXIDE SYSTEMS

W. Eysel

Mineralogisch-Petrographisches Institut, University of Heidelberg
Heidelberg, Germany

Dedicated to Hans Seifert on the occasion of his 60th birthday

Abstract

The possibilities and limits of DSC and DTA for the determination of binary phase diagrams are discussed. The major factors, responsible for the experimentally measured signal shapes and their interpretation are structural relationships of the phases (nucleation) and chemical reaction rates (diffusion). Examples for a variety of signals are given, using several oxide systems.

Introduction

For earth scientists, particularly mineralogists, the understanding and application of multicomponent systems is essential. The earth crust may be considered as a system with a practically infinite number of components, more than 99% of which are oxides (including >96% silicates). Even small subsystems like special rocks or individual minerals are usually very complex and difficult to characterize in detail.

Since mineralogists are educated and experienced in handling such problems, they also play a significant role in the investigation of industrial oxide systems (refractories, cement, gypsum, glass, ceramics etc.), including high tech materials (ferrites, fast ionic conductors, high temperature superconductors and coatings of ceramics).

Even though the influence of pressure on oxide systems becomes more and more important for the solution of problems of petrogenesis, the majority of phase diagram investigations was and is carried out at atmospheric pressure, i.e. at $p = constant$. Only such systems are considered here. For them the classic parameters are temperature and composition. The determination of such phase diagrams includes the investigation of phase transitions (polymorphism and melting), chemical reactions and solid solutions with the aim to construct equilibrium diagrams (phase boundaries and compatibility relations).

The major difficulties in investigating oxide systems are based on their high melting temperatures, sluggish reaction rates and resulting metastable states. How to recognize and to overcome these difficulties, particularly with DTA and DSC investigations, is subsequently discussed using some binary systems as examples.

Methods of Investigation

There are two basically different types of experiments by which high temperature states can be investigated:

a) Static methods. A sample is annealed at the desired temperature until a stable state is reached (usually considered as the equilibrium state). Then properties of the sample are measured at this temperature. Of particular interest is the phase analysis, which may be obtained for example by high temperature X-ray diffraction, hot stage microscopy or high temperature spectroscopy. Usually the system is investigated this way using several samples in one or more isothermal sections.

In suitable cases it is possible to quench the attained high temperature state to room temperature without changes to eventually different low temperature states. In these cases the samples can be investigated much easier at room temperature. The sluggishness of many oxide transitions and reactions fortunately enables this possibility very often.

b) Dynamic methods. These investigations are carried out under programmed and controlled temperature changes and thus represent nothing else but the various methods of thermal analysis. Compared to the static methods their advantage is, that they cover a preselected temperature range in a short time. Their disadvantage is, that sluggish effects are overheated or undercooled, i.e. kinetic influences may affect the results significantly and slow events may even not be detected at all. For a given system this is, in addition, even more complicated by the fact, that the reaction speeds depend not only on the nature of the phases involved but also on temperature. Thermal analysis methods and results, therefore, have to be used very critically.

For the study of many systems a combination of static and dynamic methods has not only turned out very valuable but even essential.

Construction of DTA Signals

Assuming equilibrium conditions, DTA runs in multicomponent systems yield three types of signals:

a) Isothermal transitions (melting and first order polymorphic transformations of pure phases) and eutectic and peritectic reactions.

b) Transitions and reactions which extend over a finite temperature interval and are called non-isothermal or interval transitions in this paper. In these one or more solution phases (melts, solid solutions) take part. The distribution of the substituting atoms between the coexisting phases depends on temperature.

c) Some second order transitions may be considered to start at 0 K and end at a well defined upper temperature. (λ point, order-disorder transitions). For experimental investigations these have no low temperature limit. They are not considered here, even though their upper limits may be excellent temperature calibration points on heating and cooling.

Whether transitions of types a) and b) proceed isothermally or over a temperature interval can be derived with Gibb's Phase Rule.

For the non-isothermal reactions it is fortunately possible to approximate the shape of *equilibrium* DTA signals by a simple calculation if the phase diagram is known. Vice versa such knowledge helps to interprete experimental signals in order to get informations on an unknown phase diagram.

The calculation and construction of interval signals was described and applied by various authors (e.g. ETTER & WITTENBERG, 1963; GÄUMANN, 1966; ETTER et al., 1969; EYSEL, 1971; KRÄMER, 1979). A summarizing review with additional literature was given by GUTT & MAJUMDAR (1972). With the help of the lever rule the increase in the amount of the high temperature phase is calculated and plotted against the temperature. The resulting curve represents the expected DTA signal. To compare the experimental and calculated signal shapes, it is reasonable to normalize both with respect to their peak heights and their temperature scales.

Since equilibrium conditions are the basis of this calculation, the endotherm on heating corresponds to an exactly mirror symmetric exotherm on cooling.

It should be mentioned, that the calculation of the signal shape only on the basis of the transformed amount neglects the changes in specific heats and in the heats of solution of the phases involved. Nevertheless the method has proved to be a very useful approximation.

Other properties of the expected DTA curves can be obtained from isothermal eutectic (or peritectic) peaks by dividing their signal area A through the sample masses m and plotting A/m against the composition (c.f. GUTT & MAJUMDAR, 1972, and the example in Fig. 4). This allows to determine the exact composition of the eutectic point and of the limits of solutions at the eutectic temperature.

All these constructions predict DTA and DSC signals for the equilibrium case. The true measured curves may agree with them rather well or deviate significantly, depending on the influence of kinetic properties.

In the following chapters three examples of oxide systems are given, which demonstrate how reliable DTA and DSC results can be in some cases and how misleading in others. The examples are ordered with decreasing applicability of dynamic methods.

The System Na_2SO_4-K_2SO_4

This system was investigated with various methods (EYSEL, 1971, 1973) and is shown in Fig. 1. Only those experimental results are shown, that were obtained by DSC. Na_2SO_4 exhibits four modifications, numbered V (thenardite), III, II and I. The small stability range of II is omitted in the figure. Other phases in the system are L (low K_2SO_4 = arcanite), H (high K_2SO_4) and G (glaserite or aphthitalite), the latter being a limited solid solution $(Na,K)_2SO_4$ in the medium part of the system. I and H are isostructural and form an unlimited solid solution series. G, L, II and III are structurally related and form limited solid solutions. The low form V is structurally different and exhibits no measurable solution of K for Na.

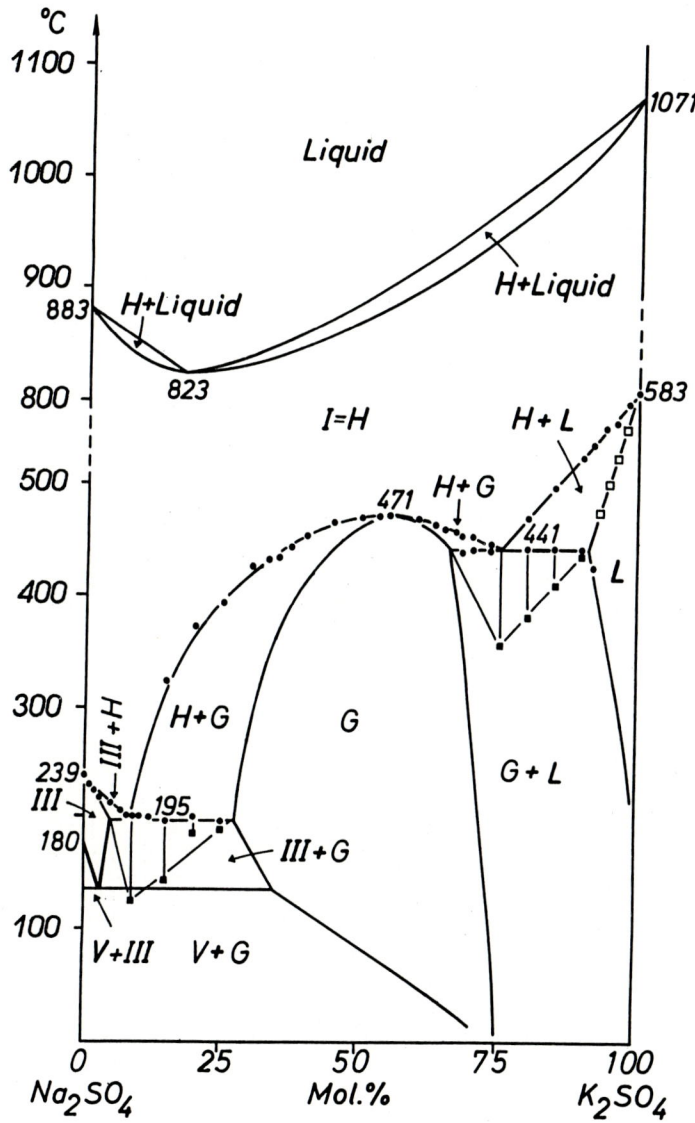

Fig. 1: Phase diagram of the system Na_2SO_4-K_2SO_4. DSC results indicated by measured points. Results of static methods not included.

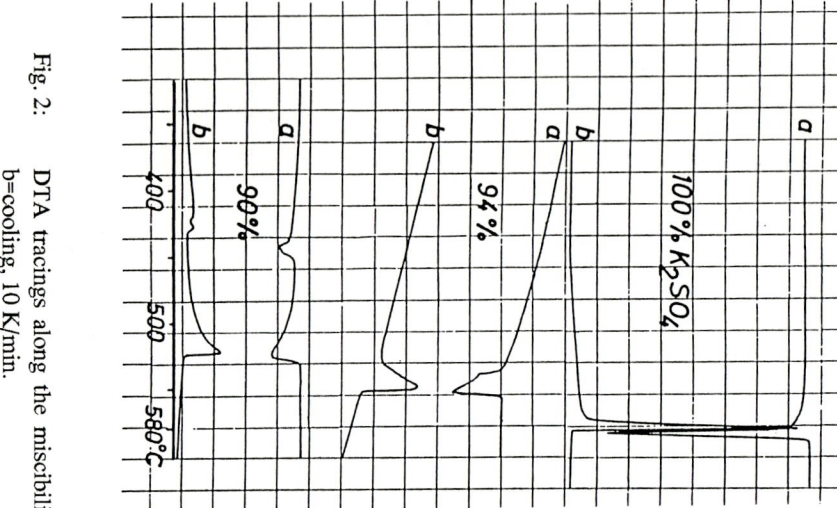
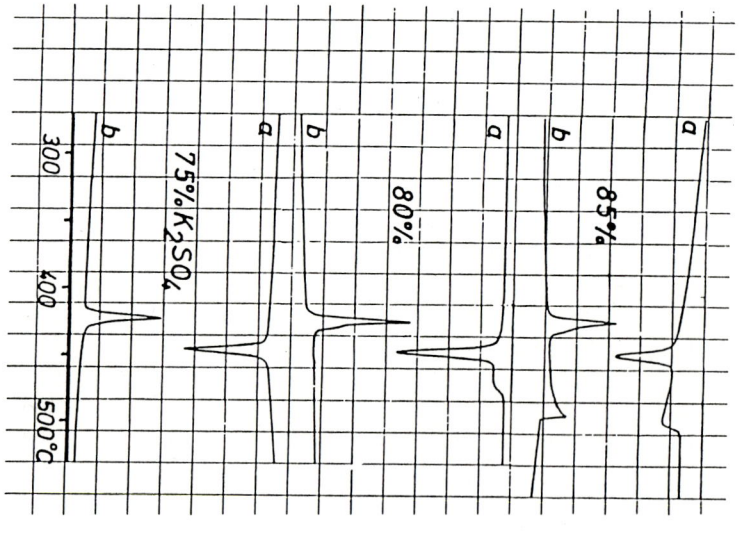

Fig. 2: DTA tracings along the miscibility gap H+L in the system Na$_2$SO$_4$-K$_2$SO$_4$. a=heating, b=cooling, 10 K/min.

Particularly at high temperatures the structural relationships allow fast nucleations, exsolutions and re-reactions of most phases involved. Thus the DSC signals along the miscibility gap H+L (>75% K_2SO_4) are reflecting the equilibrium states rather closely, even though all events are based on solid state reactions. The close similarity of the heating and cooling signals in Fig. 2 demonstrate that isothermal and interval signals can be clearly distinguished. Moreover the calculated heating curves (crosses in Fig. 3) are in good agreement with the experimentally obtained tracing. The open squares in Fig. 3 were calculated by a method, described in the previous paper (EYSEL, 1971).

Fig. 4 shows the determination of the limit of solid solution (91%) for phase L at the eutectic temperature at 441°C, as determined with the help of the normalized peak sizes A/m of the eutectic signals.

Thus the part of the system Na_2SO_4-K_2SO_4 represented in Fig. 3 is an excellent example for the case, that the dynamic DSC signals follow the equilibrium reaction closely and allow a rather reliable construction of the phase boundaries.

The sodium rich part at lower temperatures is dominated by the very complicated polymorphism of Na_2SO_4 ss with several transitions of different speed in a very narrow temperature range. Na_2SO_4 IIIss samples, e.g., on heating transform directly to Na_2SO_4 Iss as shown by triangles (Fig. 5). On cooling (black dots) two transitions I→II→III are observed and the temperature range of II increases with the content in K.

Obviously the transition III↔II is more sluggish than II↔I. On heating, therefore III→II is shifted into II→I and both take place together. On cooling they are clearly separated. Similar observations are known also for other solid solution series with several neighbouring transitions.

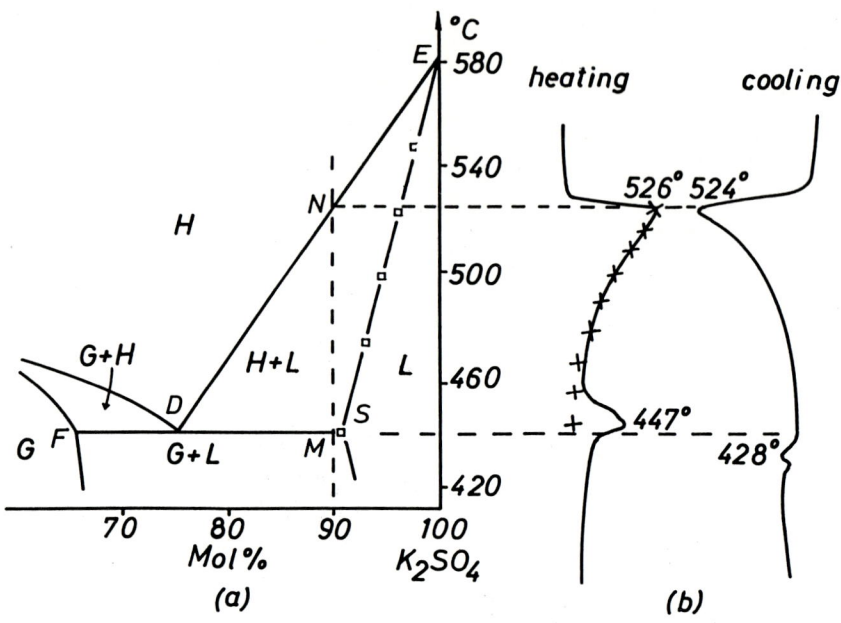

Fig. 3: (a) Miscibility gap H+L in the system Na_2SO_4-K_2SO_4 and (b) DTA signals of the sample with 90% K_2SO_4.

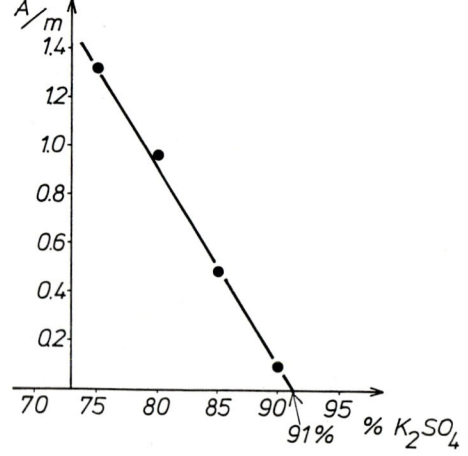

Fig. 4: Limit of solid solution of Na_2SO_4 in low K_2SO_4 at the eutectic temperature in Fig. 2.

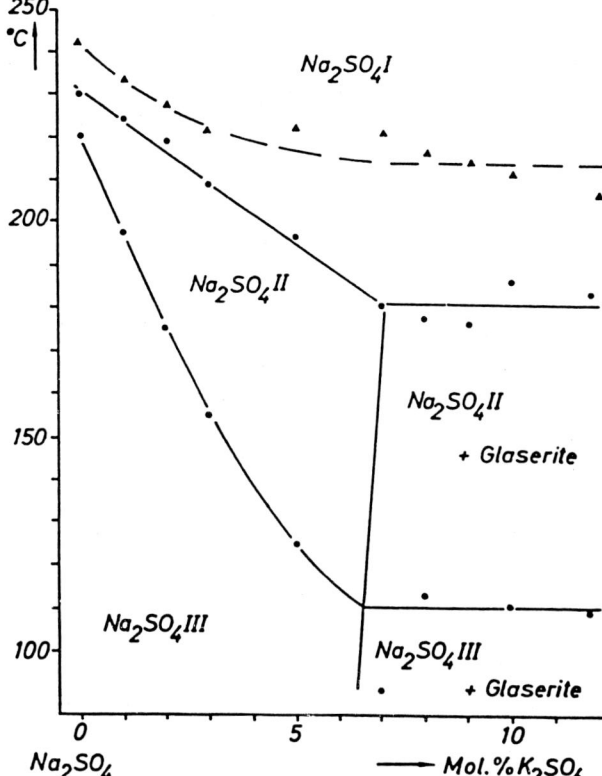

Fig. 5: Effect of substitution of Na by K in Na_2SO_4 on the transition temperatures III→I (dashed line) and I↔II→III (full lines).

Due to the structural differences and the low temperatures all transitions and reactions including Na_2SO_4 V are extremely sluggish. Their DSC and DTA signals are far off the equilibrium state and can only be interpreted with the help of static investigations.

The System Ca_2SiO_4-Ca_2GeO_4

This system (EYSEL & HAHN, 1970) contains four solid phases α, α', β and γ, which form solid solutions. DTA tracings and their evaluation are shown in Figs. 6 and 7. The final equilibrium diagram (Fig. 8) was constructed, using these DTA curves and the results of various static methods.

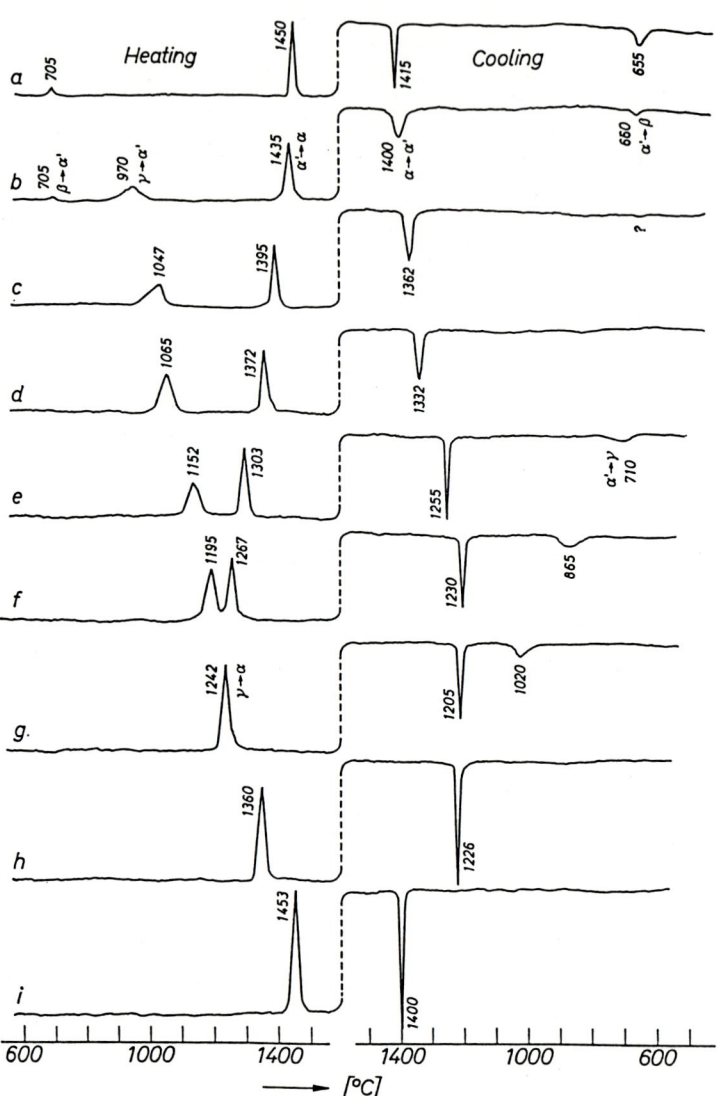

Fig. 6: DTA tracings of heating and cooling runs in the system Ca_2SiO_4-Ca_2GeO_4 (10 K/min, Pt/PtRh10).
a) Ca_2SiO_4,
b) 5% Ca_2GeO_4,
c) 10% Ca_2GeO_4,
d) 15% Ca_2GeO_4,
e) 25% Ca_2GeO_4,
f) 30% Ca_2GeO_4,
g) 35% Ca_2GeO_4,
h) 70% Ca_2GeO_4,
i) Ca_2GeO_4.

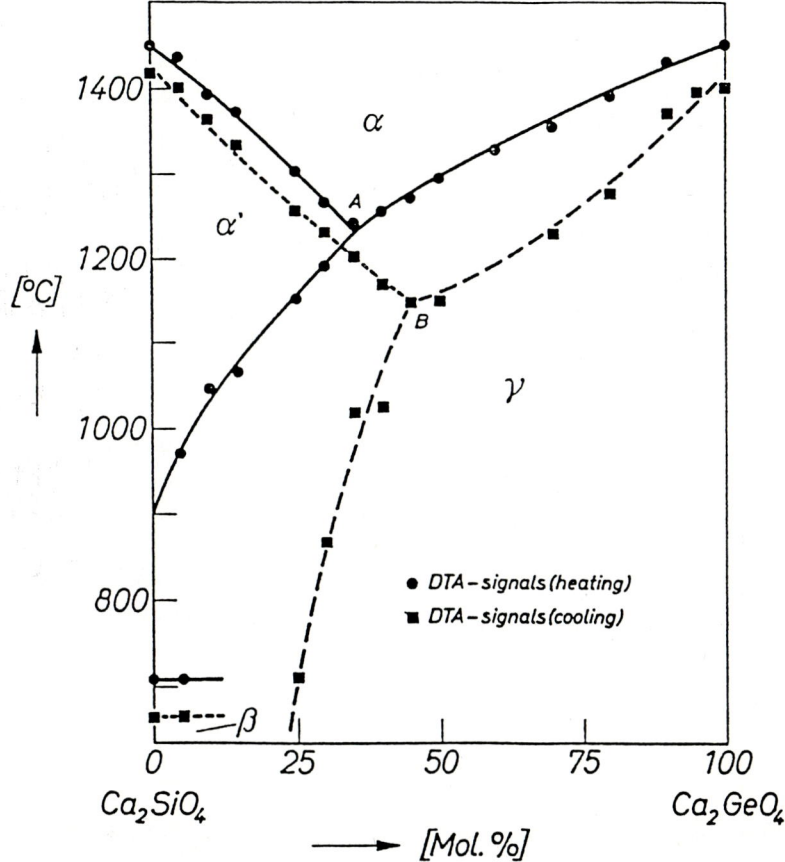

Fig. 7: Transition curves in the system Ca_2SiO_4-Ca_2GeO_4 as obtained from the DTA runs in Fig. 6. Solid lines: heating. Dashed lines: cooling.

It is evident, that in this system the DTA signals are based on overheating and undercooling effects and exhibit no typical shapes of interval signals as to be expected from the equilibrium diagram. Instead the solid solutions are overheated (or undercooled) into the field of the next stable phase and then, on nucleation, transform "in one step"; thus the peak shapes simulate isothermal effects.

The widths of the hystereses (Fig. 7) demonstrate very well the different relationships of the crystal structures: α, α' and β are closely related, γ is different. Fig. 7 also shows clearly the influence of temperature on the reaction speeds:

The hysteresis of the transition $\alpha' \leftrightarrow \gamma$ increases dramatically with falling transition temperatures (i.e. decreasing Ge content).

β-Ca_2SiO_4 is one of the four major constituents of Portland cement clinker. From the results in the system Ca_2SiO_4-Ca_2GeO_4 it was concluded, that β is a metastable substitute. It is formed by distortion of the α' crystal structure, since the reconstructive and sluggish equilibrium transition $\alpha' \rightarrow \gamma$ is suppressed at low transition temperature of Ge-free Ca_2SiO_4.

Obviously in this system the DTA signals do not give clear indications about details of the phase boundaries. The peak shapes are very similar to those of the pure compounds and their more or less evident broadening is easily interpreted as a kinetic effect.

The System Pb_3SiO_5-Pb_3GeO_5

Only the Ge-rich part of this system (BREUER & EYSEL, 1980) is of true binary nature and shall be considered here. In this system (Fig. 9) the high temperature solution phase is a melt. Usually melts support material transport and accelerate reactions. Here the melt is highly viscous with the opposite effect: Lead silicate melts are known as excellent glass formers. This applies also to germanates, even though their glass forming properties are smaller than those of the silicates. The strong undercooling of the melts is evident from the retrograde inclination of the cooling signal in Fig. 10.

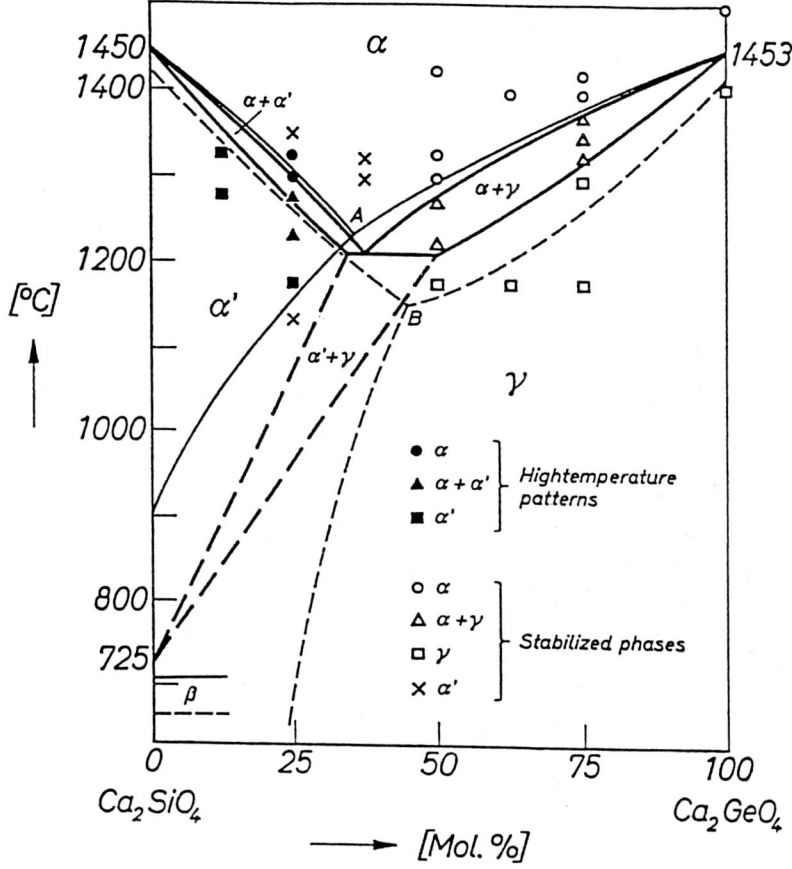

Fig. 8: Construction of the phase diagram Ca_2SiO_4-Ca_2GeO_4, including X-ray diffraction results. Thin lines: DTA results from Fig. 7. Heavy lines: Finally accepted phase diagram.

The evaluation of the DTA results (small black dots and black squares in Fig. 9) allows only a very rough estimation of the boundaries of the two phase region Pb_3GeO_5 ss+L.

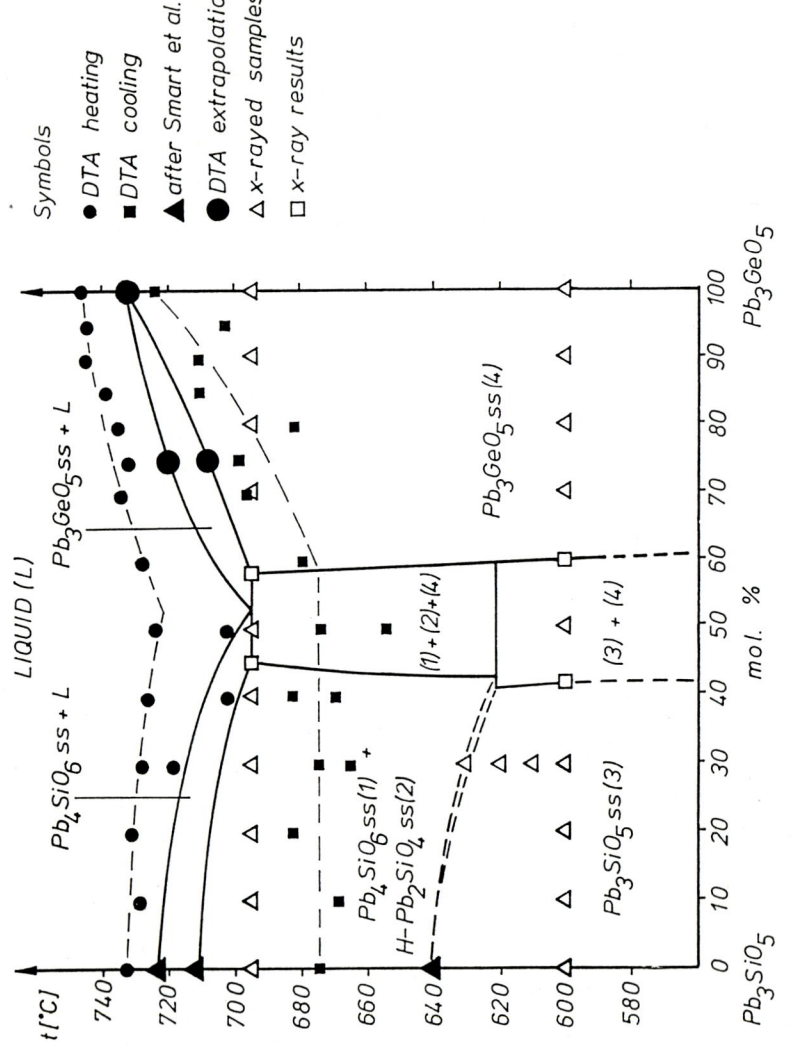

Fig. 9: The system Pb_3SiO_5-Pb_3GeO_5.

A much better determination of the melting point of Pb_3GeO_5 and of the width of the two phase lens Pb_3GeO_5 ss+L was obtained by varying the heating rate β and extrapolating it to zero (Fig. 11 and large black circles in Fig. 9). Such extrapolations have proved to be the only reliable method for temperature calibration of DSC and DTA (EYSEL & BREUER, 1984; HÖHNE et al., 1990). As shown here, the method can be also very helpful for the determination of phase diagrams.

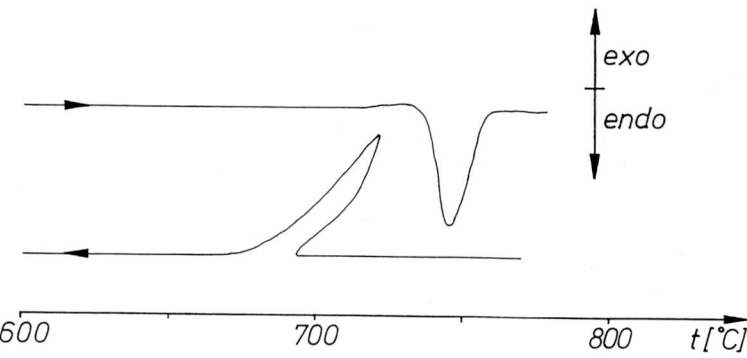

Fig. 10: DTA curve of Pb_3GeO_5 (20 K/min).

Summarizing Discussion

The applicability of dynamic methods for the investigation of phase diagrams is strongly depending on the kinetic behaviour of the various physical events and chemical reactions considered. Major influencing factors are:

1) Nucleation rates of new phases during polymorphic transitions, crystallization, melting, exsolution etc. Nucleation depends strongly on the structural relationship of the phase involved.

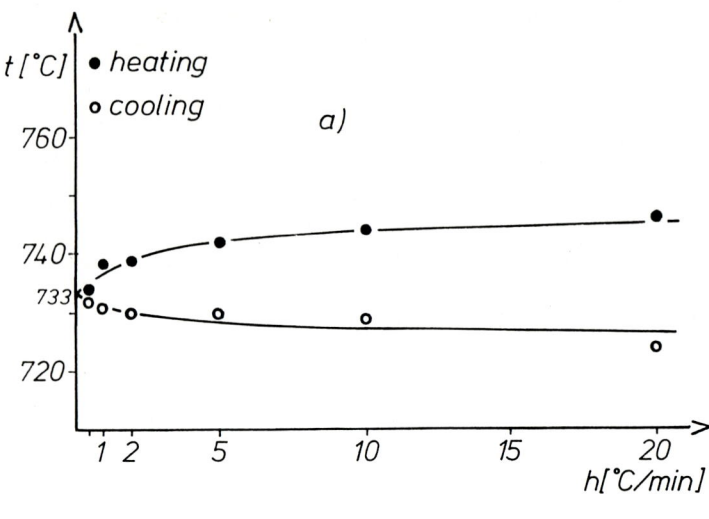

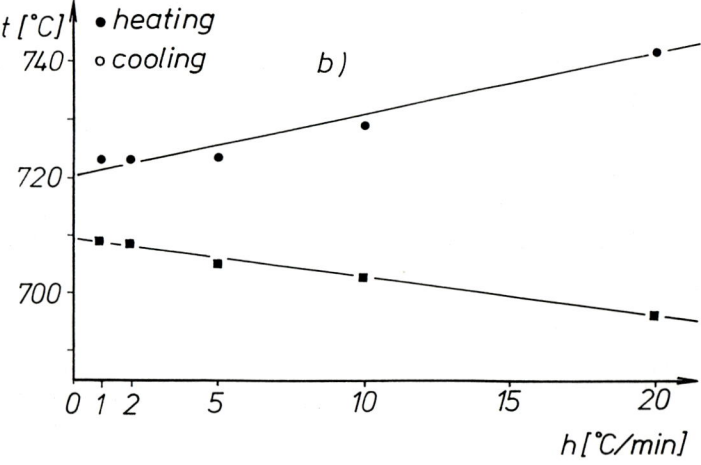

Fig. 11: Approximation of equilibrium temperatures by extrapolating the heating (cooling) rate β towards zero. a) Pb_3GeO_5, b) $Pb_3(Si_{0.25}Ge_{0.75})O_5$.

2) Transformation or reaction speed (diffusion rate), once a nucleus has been formed.

3) Relative temperature. In general it can be said, that physical or chemical events take place and proceed the faster, the closer they are to the melting temperature.

All these factors can result in strong deviations of the experimental results from the equilibrium state to be determined. Only very critical evaluation of the measured data can help to keep errors of interpretation as small as possible. Helpful procedures are:

- Compare signal shapes and temperatures of heating and cooling signals. The more mirror symmetric they are, the more they represent the equilibrium state.

- The results of sluggish events can be improved by repeated runs with different heating rates β and extrapolation of β towards zero.

- Try to estimate how sluggish the measured event may be. If several crystalline phases are involved, consider their structural relations. Take also into account how far below the melting point the event occurs.

- Apply static methods wherever possible.

Thousands of phase diagrams have been published so far, among them a huge number of oxide systems (c.f. Phase Diagrams for Ceramists, Vols. I-VI), For the investigation of many of them DTA and DSC have been applied as major tools. Unfortunately very often the results were interpreted without taking the limits of these dynamic methods into account. Before applying such diagrams, it is recommended to go back to the original literature and check their reliability.

References

BREUER, K.-H. & EYSEL, W. (1980) - Thermal Analysis.- Proc. 6th ICTA, Vol. 2, pp. 263-268. Birkhäuser Verlag, Basel.

ETTER, D. E., TUCKER, P. A. & WITTENBERG, L. J. (1969) - Thermal Analysis.- Proc. 2nd ICTA, Vol. 2, pp. 829-850. Academic Press, New York.

ETTER, D. E. & WITTENBERG, L. J. (1963) - USAEC Document MLM-1172, Avail. CESTJ-1-J-14.

EYSEL, W. (1971) - Thermal Analysis.- Proc. 3rd ICTA, Vol. 2, pp. 179-192. Birkhäuser Verlag, Basel.

EYSEL, W. (1973) - Am. Mineral., 58, 736-747

EYSEL, W. & BREUER, K.-H. (1984) - Analytical Calorimetry.- Vol. 5, pp. 67-80. Plenum Press.

EYSEL, W. & HAHN, TH. (1970) - Z. Kristallogr., 131, 322-341

GÄUMANN, A. (1966) - Chimia, 20, 82-88

GUTT, W. & MAJUMDAR, A. J. (1972) - in "Differential Thermal Analysis".- (MACKENZIE R. C. (ed)), Vol. 2, pp. 97-117. Academic Press, New York.

HÖHNE, G. W. H., CAMMENGA, H. K., EYSEL, W. & HEMMINGER, W. (1990) - Thermochimica Acta, 160, 1-12

KRÄMER, V. (1979) - J. Thermal Analysis, 16, 295-306

Thermoanalytical Investigations of Binary Oxide Systems

KINETIC ANALYSIS OF THE CRYSTALLIZATION OF SILICATE MELTS BY MEANS OF DSC, DTA AND THERMAL OPTICAL METHODS

K. Heide

Otto-Schott-Institut, Universität Jena
Fraunhoferstr. 6, Jena, Germany

Abstract

The paper is concerned with the possibilities and limitations of the kinetic analysis of silicate melt crystallization by means of thermal methods (DSC, DTA, thermal optical methods). The theoretical background is given, the experimental possibilities are demonstrated. Experimentally, the investigation of the crystallization of a melt includes the determination of the nucleation process, of the nucleation rate and its dependence on temperature and other influencing factors, and of the mechanism of crystal growth.

Introduction

Geology has always involved time as a central concept in describing geoscience phenomena. Therefore to understand the origin and evolution of the geological problems the kinetic factor is important. But the application of kinetics treatment to geological problems is "still embryonic" (LASAGA, 1981). The reasons for this are quite different. There are problems e.g. in respect to:

- the time scale (evolutionary processes are possible in a range between ca. 10^{-8} seconds to more than 10^{17} seconds),

- the volume (evolutionary processes occur in scale between atomic and astronomic dimensions),

but also in respect to:

- measurements and
- theoretical interpretation of results.

The kinetic analysis of the crystallization of a simple silicate melt gives a good example for the numerous difficulties involved. In this paper I want to discuss problems which arise by using thermoanalytical methods together with the interpretation of results in the field of silicate formation from a melt.

A comprehensive review about the kinetics and crystallization of igneous rocks was shortly given by KIRKPATRICK (1981). Other papers in the Advances in Physical Geochemistry were summarized in "Kinetics and Equilibrium in Mineral Reaction" edited by SAXENA (1983) and in respect to Geochemical Transport and Kinetics in the papers of the Carnegie Institution of Washington 1973.

All these papers are focused mainly on the interpretation of results, e.g. in respect to the value of activation energy (E_a). This means that the potential barriers, e.g. breaking bonds or distorting molecular structures, which result from the exponential dependence on the temperature of reaction rate "constant" following e.g. the "classic" equation proposed by Arrhenius in 1889:

$$k = A \cdot \exp(-E_a / RT)$$

In this way it was concluded that a low activation energy (E_a less than 20 kJ/mol) offers an important clue for diffusion-controlled reaction mechanisms. Therefore a high activation energy value gives a hint for the solution-transport mechanism. As was shown already 1968 in the case of the metamorphic transformation in salt deposits, the experimentally determined value of the activation energy is strongly determined by the procedure for the evaluation of the experiments (HEIDE, 1968).

In the case of the transformation of kieserite, $MgSO_4 \cdot H_2O$ there was determined, by means of thermoanalytical methods, an activation energy for an induction period of 575 kJ/mol and for the main period only a value of 160 kJ/mol. The first value is in the range expected from breaking bonds in crystals (e.g. the average value for the bond energy of Si-O is given with 465 kJ/mol (LINKE, 1977)), but the second is in the range of the bulk diffusion activation energy between ca. 80 kJ/mol and 350 kJ/mol (for silicates see LASAGA, 1981).

There are a lot of examples from such experiments and interpretations. A theoretical concept for describing the kinetics in terms of energy and concentration was given by the chemists for elementary reactions and this theory was successful especially in the characterization of the homogeneous gas-phase reaction (LEIDLER, 1950).

But in the case of heterogeneous reactions, which are of high interest in the geosciences, the theoretical basis is at present under discussion. A critical review of the methods in relation to the theoretical treatment is necessary and, in my opinion, understated today.

Theoretical Basis

In general the crystallisation of a melt occurs in two different ways. At first a crystallization results by cooling of a melt and at second by heating of a frozen melt, a glass (Fig. 1).

In terms of the classification of reaction order by Ehrenfest the crystallization of a melt by cooling is a first-order transformation from a homogeneous system to a heterogeneous system, that means the formation of a isochemical crystalline solid in the melt and vice versa by heating during the formation of a homogeneous melt from the crystal. The rate of the process is determined by the cooling rate and the realized undercooling (ΔT), the viscosity (η), fusion heat (ΔH_f) and melting temperature (T_m).

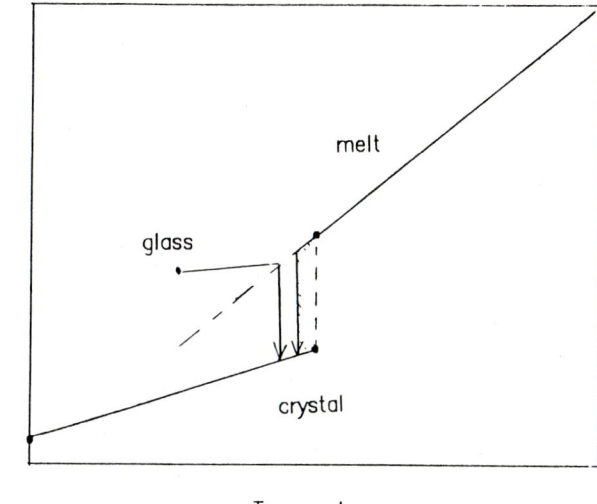

Fig. 1: Enthalpy change during the transformation of a liquid and of a glass in a crystal by cooling and heating respectively.

UHLMANN (1971) and coworkers have found the following relation:

$$\frac{dV_{melt}}{dt} = \frac{c}{\eta} \left(1 - \exp\left[\frac{-\Delta H_m \, \Delta T}{RT \cdot T_m} \right] \right)$$

As TAMMANN (1933) has shown previously the rate of formation of nuclei is characterized by a maximum which results from the crossing effects of the influences of undercooling and increasing of viscosity (Fig. 2).

In any case the rate of nucleation is the factor limiting crystallization. According to the theory of nucleation (see VOLMER, 1939; or TURNBULL, 1965) two borderline cases must be taken into consideration:

- homogeneous steady state nucleation and

- heterogeneous nucleation.

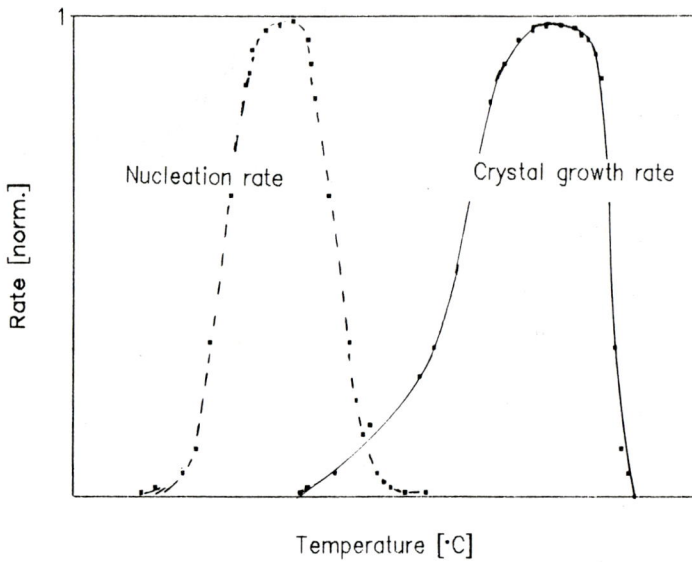

Fig. 2: Nucleation rate and rate of crystal growth in a metastable melt by heating a glass.

The formation of a homogeneous nucleus is based on the classical theory (VOLMER & WEBER, 1926), on the concept of heterophase fluctuation (fluctuations in density, composition and structure).

Essential to the formation of a nucleus is the increasing fluctuation of the cluster dimension to a critical radius (Fig. 3). In the case of a spherical cluster the critical radius is the state with the maximum free energy ΔG^*, which results from the chemical free energy change per unit volume transformed ΔG_v and the surface energy per unit area σ.

$$\Delta G^* = 4/3\, \pi r^3\, \Delta G_v + 4\, \pi r^2\, \sigma$$

$$\Delta G_v \approx \Delta H_f\, \Delta T\, /\, T_s$$

ΔH_f = heat of fusion, ΔT = undercooling

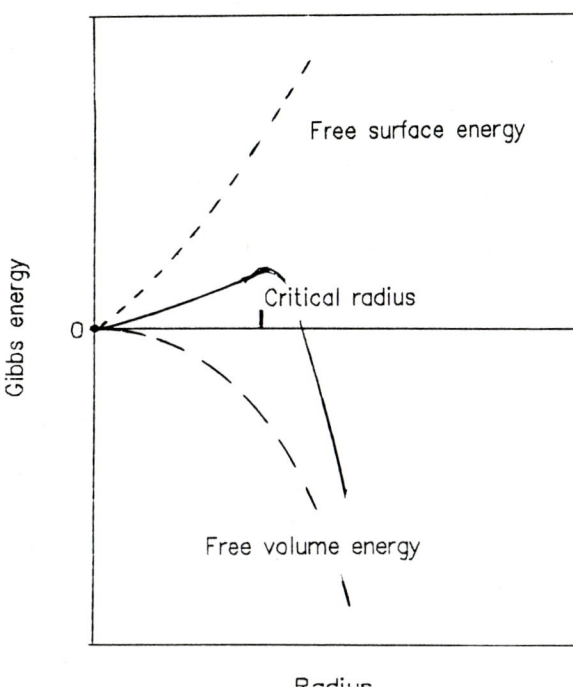

Fig. 3: The Gibbs enthalpy relation by forming of a critical nucleus in a metastable melt.

In the case of the homogeneous nucleation in the undercooled metastable melt a formation of nuclei results spontaneously in the bulk.

This classical treatment is a macroscopic concept, which separates unequivocally the effects of the volume and surface. However, in microscopic scale we must take into consideration that the surface energy is a function of the cluster size (MÜLLER E. et al., 1988) and the definition of volume is arbitrary. Furthermore, the free enthalpy in an inhomogeneous system is a function of the gradient of concentration, as was demonstrated by CAHN & CHARLES (1965).

In general the clusters require a period to reach critical size. Nucleation at a given undercooling is rate controlled by an activation energy which depends on differences between melt, crystal structure and the number of bonds to be broken, i.e. this is an explanation why the anorthite from the melt is easily formed. In this case there exists a great similarity between the structure of the melt and the crystal and a homogeneous nucleation results (MÜLLER et al., 1991). On the other hand the nucleation of albite is almost impossible in a melt, because it would require the breaking of two T-O bonds to transform 6-rings of the melt into 4-rings of the crystal. Pre-existing defects and their boundaries with the surrounding system may drastically decrease the activation energy and a heterogeneous nucleation results. The heterogeneous nucleation occurs sooner and with smaller undercooling and in many cases from external or internal surfaces (surface nucleation).

At a critical undercooling rate the nucleation may not occur and the melt can be quenched to a glass.

In glass forming systems can exist, in an analogous way to the stable liquid-solid transformation, a metastable liquid-liquid phase separation. As we can see in Fig. 4 there exists a range of composition where the formation of a heterogeneity occurs without activation, the so-called sponoidal decomposition results.

In all these cases the nucleation rate is the product of the number of critical nuclei and the rate at which atoms form the critical clusters. This term is determined by the activation energy (the energy barrier that must be overcome by an atom to become attached to a cluster). The frequency of the attachment of an atom to a cluster may be approximated by a term involving a diffusion coefficient.

After the formation of a critical nucleus it generally continues to increase in size to become a crystal.

The crystal growth rates are controlled by (see Fig. 5):

- phase boundary kinetics at the crystal-liquid interface,

- diffusion of material in the liquid towards and away from the interface,

- the amount and the removal of the latent heat of crystallization from the interface into the melt (a factor of probably little significance, because the thermal diffusivities are several orders of magnitude larger than the largest diffusion coefficient, i.e. 10^{-2} cm^2/s to 10^{-3} cm^2/s for heat flow, and at most 10^{-6} cm^2/s for diffusion).

- volume difference between liquid and crystal.

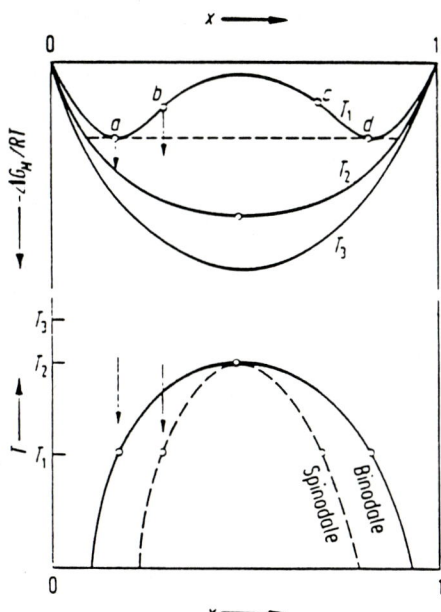

Fig. 4: Phase diagram with a subliquid immiscibility. In the region between the binodales ("instable region") a separation occurs without any nucleation energy.

In the case of single crystals a theoretical treatment was developed in terms of crystal surface steps, screw dislocations and also by computer simulations (e.g. the Monte Carlo simulations of GILMER, 1977). However, for real systems a theoretical treatment of crystal-growth phenomena is difficult and strongly influenced by external conditions such as the purity, homogeneity and perfection of structure.

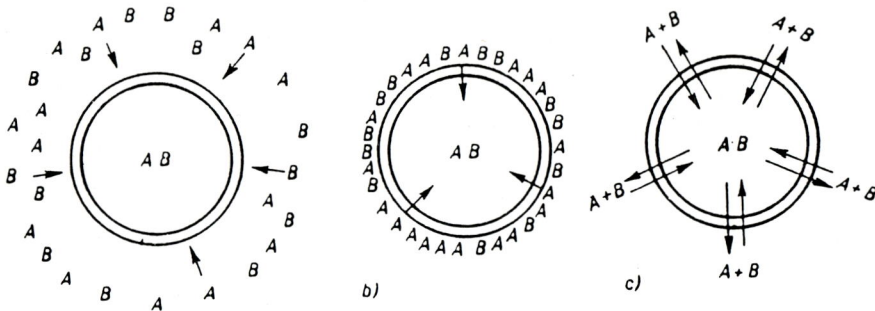

Fig. 5: The main processes controlling the rate of crystallization: a) the bulk diffusion; b) the phase boundary diffusion; c) the chemical reaction.

Experimental Possibilities for the Investigation of Crystallization by Use of Thermal Methods

The experimental tasks for the investigation of the crystallization of a melt are as follow:

- determination of the nucleation process (homogeneous or heterogeneous, stable or metastable, stationary or mobile),

- determination of the nucleation rate,

- determination of the dependence of the nucleation rate upon the temperature, concentration of the heterogeneities, total and partial pressure and the redox conditions,

- determination of the crystal-growth mechanism.

These functions control the volume of crystals which are formed in unit time from a melt of a given volume and at constant temperature.

$$V(t) = \omega \left[\int_{t_w}^{t} W(t)\, dt \right]^{n} (dN/dt)/dt$$

n = term of dimension of nuclei
W = rate of growth
ω = shape factor (4π/3 for sphere)

How can thermal methods be used to solve these questions? In respect to the geological processes we must take into consideration that the processes are not under isothermal conditions. This means the non-isothermal techniques such as DTA and DSC could be suitable for the study of transformation processes under non-isothermal conditions. The handycap here is that time scale in nature is quite different to those possible for experiments which can be completed in the laboratory. In respect to the crystallization of silicates there exist a lot of papers which establish great disagreement between the observed and calculated transformation rates. The experimental method mainly used was to make a melt of a desired composition, quench it to a glass, reheat pieces of the glass at the desired nucleation temperature for a known time, then heat the same sample for a longer time to grow the nuclei to become visible crystals. In DTA or DSC experiments we observed in any case the process of glass transformation and crystal-growth by a negative and positive deflection of the DTA / DSC curve (Fig. 6).

By such experiments it is quite difficult to separate the process of nucleation and crystal growth. As we can see in a lot of thermoanalytical curves there exists a small endothermal deflection in the curve immediately before the exothermal crystallization peak. This effect was also discussed in the literature in terms of endothermic nucleation energy. Systematic investigations are necessary for the explanation of this effect. The exothermic effect of crystallization is also used for the quantitative description of crystallization. But as we can show by simultaneous microscopic observations it is very complicated to separate this quantitatively from the thermal effect of the change in crystal volume. At present optical observation of the number and dimensions of crystals is the most common and successful way for the evaluation of quantitative data.

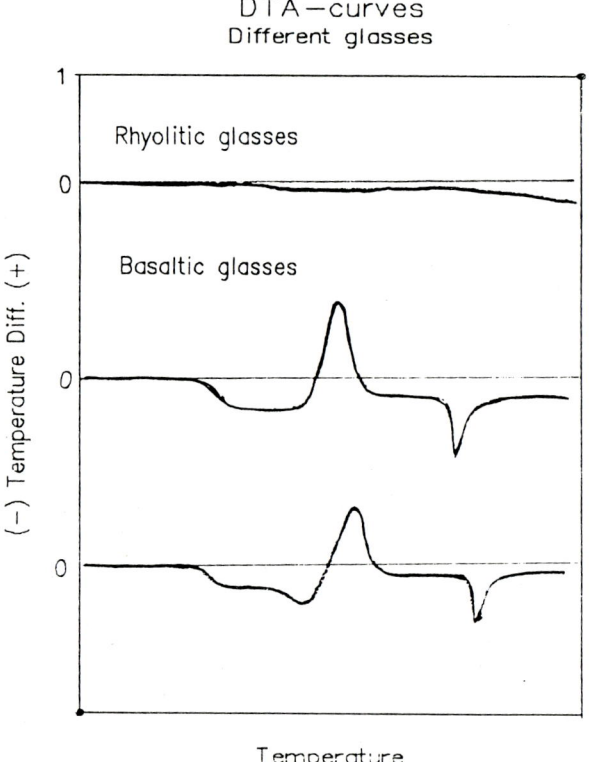

Fig. 6: Characteristic DTA curves of natural glasses: a) without thermal effect (rhyolitic glasses); b) and c) with an endothermic deflection at the glass transition and the melting range, and a positive deflection at the crystallization range. The nature of the endothermic deflection immediately before the crystallization (formation of nuclei?) is open.

With hot-stage microscopy it is possible to combine light-microscopy with non-isothermal conditions. It was demonstrated (HEIDE, 1982) that a way exists to automatise optical measurements, which allows the determination of data for the calculation of kinetic parameters which correspond to the conventional theoretical treatment.

Conclusion

The use of optical effects in thermal analysis gives new possibilities for the characterization of phase transformation phenomena. The quantities of optical properties could be quite different in depending on the temperature in comparison to calorific or mechanical effects. There is a special opportunity for geosciences to open this field to chemists and physicists for the investigation of solid state reactions.

References

ARRHENIUS S. (1889) - Z. f. phys. Chemie, 1, 110

CAHN J. W. & CHARLES R. J. (1965) - The initial stages of phase separation in glasses.- Phys. Chem. Glasses, 6, 181-191

GILMER G. H. (1977) - Computer simulation of crystal growth.- J. Cryst. Growth, 42, 3-10

HEIDE K. (1968) - Zum Mechanismus der Umbildungsvorgänge in Salzgesteinen.- Chem. der Erde, 27, 353

HEIDE K. (1982) - Dynamische thermische Analysenverfahren.- Verlag für Grundstoffindustrie, Leipzig.

KIRKPATRICK R. J. (1981) - Kinetics of crystallization of igneous rocks.- in Kinetics of Geochemical Processes. (LASAGA A. C. & KIRKPATRICK R. J. (eds.)), Rev. Min., Vol. 8

LASAGA A. C. (1981) - Rate laws of chemical reactions.- in Kinetics of Geochemical Processes. (LASAGA A. C. & KIRKPATRICK R. J. (eds.)), Rev. Min., Vol. 8, 1-67

LEIDLER K. J. (1950) - Chemical Kinetics.- McGraw-Hill Book Comp., New York.

LINKE D. (1977) - Proc. XIth Int. Congr. Glass Prag, Vol. I, 149

MÜLLER E., VOGELBERGER W. & FRITSCHE H. G. (1988) - The dependence of surface energy of regular clusters and small crystallites on particle size.- Cryst. Res. Techn., 23, 1153-1159

MÜLLER E., HEIDE K. & ZANOTTO E. (1991) - Strukturchemische Aspekte der Keimbildung in Silikatgläsern.- Z. Krist.

SAXENA S. K. (1983) - Kinetics and Equilibrium in Mineral Reactions.- Advances in Physical Geochemistry, Vol. 3, Springer Verlag, Heidelberg.

TAMMANN G. (1933) - Der Glaszustand.- L. Voß, Leipzig.

TURNBULL D. (1965) - Thermodynamics and kinetics of formation of the glass state and initial devitrification.- in Physics of non-crystalline solids, North-Holland, Amsterdam, 41-56

UHLMANN D. R. (1971) - Crystal growth in glass-forming systems - a review.- in Advances in nucleation and crystallization in glasses, Am. Ceram. Soc., Columbus (Ohio), 91-115

VOLMER M. (1939) - Kinetik der Phasenbildung.- Th. Steinkopf Verlag, Dresden.

Technical Mineralogy

APPLICATION OF THERMAL ANALYSIS IN MINERAL TECHNOLOGY

A. M. Abdel Rehim

Geology Department, Alexandria University
Alexandria, Egypt

Abstract

The present work illustrates the importance of the application of thermal analysis in various fields of mineral technology, including the industrial chemical processing of minerals, manufacture of refractories, ceramics, glass, cement, mineral synthesis and metallurgical processes. Selected examples of applications in this field are given.

Introduction

Differential thermal analysis (DTA) and thermogravimetry (TG) have become the most important methods of thermal analysis by their frequent applications in science, industry, research and quality control.

Thermal analysis is applied to various mineral technological processes, including industrial chemical processing of minerals, manufacture of refractories, ceramics, glass, cement, mineral pigments, fluxes, metallurgical processes, chemical industry, mineral synthesis and thermo-analytical testing of minerals and the end products of processing.

DTA is one of the basic techniques for qualitative and quantitative analysis of minerals. It has primarily been used for the identification of minerals i.e. clays, bauxites, carbonate minerals and the determination of mineral composition of raw materials. The technique is highly useful for the classification of mineral raw materials of different industries. In thermoanalytical investigations, the physical and chemical transformation of minerals appear more or less separated from one another, at different temperatures. On these bases, DTA makes possible the detection and determination of mineral constituents more accurately than the other thermal methods (PAULIK F. et al., 1966).

DTA is also used in the analysis of cements, for investigation of raw materials, production processes, the formation of calcium and aluminium silicates and the influence of additives on their formation (PAULIK F. et al., 1966; LACH, 1976; BLAZEK, 1973; MIDGLEY, 1976).

Extensive DTA studies were carried out on the thermal decomposition of coal, coke, lignites, peats and peat constituents and their burning process (PAULIK F. et al., 1966; BLAZEK, 1973).

In metallurgy DTA has been used for the qualitative evaluation of heat changes in metallurgical reactions and their mechanisms. Numerous investigations have been made on ores, slags and metals under reducing and oxidizing conditions (TSIDLER, 1958; IVANOVA, 1961; LOSKOTOV, 1965; MIRSON & ZELIKMAN, 1965; BLAZEK, 1973).

The above brief introduction shows some aspects of the application of thermal analysis in mineral technology. Time does not permit an exhaustive discussion and this review shall only cover selected areas and examples, to illustrate a range of applications of DTA in mineral technological processes.

Industrial Chemical Processing of Minerals

Thermal analysis is a valuable tool in the study of chemical processing of minerals. This study includes the thermal investigation of mineral raw materials, thermal reactions taking place during production processes and their mechanisms, determination of the optimum temperature and thermo-analytical testing of the end products. Some selected examples of the application of DTA in the chemical processing of minerals are demonstrated.

1. *Oxidizing Roasting of Some Sulphide Ore Minerals*

The technological processing of sulphide minerals after their benefication, begins with the oxidized roasting of the ore to convert the metal sulphides into the oxides. Each sulphide mineral has a particular temperature of ignition, at which its oxidation takes place intensively due to the evolution of heat of the reaction of oxidation. This temperature shows the beginning of roasting and depends on the physical properties of sulphide minerals, heat capacity, grain size, surface area, mineral structure and its oxidation products. In industry, roasting of sulphide minerals is carried out in cylindrical furnaces with mechanical stirring, in suspended-state furnaces and fluidized-bed furnaces. The roasting degree is characterized by the sulphur content remaining in the roasted ore (TSIDLER, 1958; LOSKOTOV, 1965; MIRSON & ZELIKMAN, 1965).

Oxidation of divalent metal sulphide minerals takes place according to oxide theory, as the following:

$$2\,MS + 3\,O_2 \rightarrow 2\,MO + 2\,SO_2$$

$$2\,SO_2 + O_2 \rightarrow 2\,SO_3$$

$$MO + SO_3 \rightarrow MSO_4$$

According to sulphate theory, oxidation of sulphide minerals takes place as the following:

$$MS + 2\,O_2 \rightarrow MSO_4$$

$$MSO_4 \rightarrow MO + SO_3$$

$$2\,SO_3 \rightarrow 2\,SO_2 + O_2$$

$$MO + SO_3 \rightarrow MSO_4$$

a) Roasting of Galena in the Lead Industry (in air)

The DTA curves (Fig. 1, SMYKATZ-KLOSS, 1974) show that galena oxidizes during roasting at high temperatures, 700-900°C, by an exothermic reaction. The mechanism of the reaction of oxidation of galena is a complicated one. According to some DTA data galena begins oxidation at 337°C as represented by the shallow exothermic peak and oxidation interrupted at 412°C and then continues further (BLAZEK, 1973).

Other data show that oxidation of lead sulphide proceeds via lead sulphate and basic lead sulphate before lead oxide (JAYAWEERA & SLEEMAN, 1976). Lead sulphide is converted to lead sulphate at a low temperature (200°C) and to basic sulphates at higher temperatures (400-600°C to $PbSO_4 \cdot PbO$, at 800-900°C to $PbSO_4 \cdot 2\,PbO$, at 900-950°C to $PbSO_4 \cdot 4\,PbO$) before being completely desulphurized to lead oxide. Pure lead sulphate is stable at temperatures up to 900°C, but above 400°C will react with lead oxide.

During roasting, the resulted lead sulphate begins to dissociate in a current of air at 637°C and intensive dissociation takes place at 705°C with the formation of basic lead sulphate. The latter decomposes at higher temperature to lead oxide (LOSKOTOV, 1965). The oxidized roasting of galena takes place at 700-900°C according to the following simple reactions:

$$2\,PbS + 3.5\,O_2 \rightarrow PbO + PbSO_4 + SO_2$$

$$3\,PbS + 5\,O_2 \rightarrow 2\,PbO + PbSO_4 + 2\,SO_2$$

The resultant lead sulphate reacts with galena at 550°C by:

$$3\,PbSO_4 + PbS \rightarrow 4\,PbO + 4\,SO_2$$

At different ratios of lead sulphate, sulphide and oxide, lead metal is produced as a result of their reactions. The resulting lead again oxidizes with air to PbO.

The thermal behaviour of cerussite (Fig. 2), which may be found associated with galena, shows that its dissociation takes place in two distinct steps in air as represented by the double endothermic peaks at 330-400°C and 390-440°C according to the following (IVANOVA, 1961; BLAZEK, 1973; SMYKATZ-KLOSS, 1974):

$$PbCO_3 \xrightarrow{330\text{-}400°C} PbO \cdot PbCO_3 + CO_2$$

$$PbO \cdot PbCO_3 \xrightarrow{390\text{-}440°C} PbO + CO_2$$

DTA data (Fig. 1) shows that the pyrite associated with galena oxidizes during roasting at 406-443°C by an exothermic reaction and may be represented as follows (KOPP & KERR, 1957; MACKENZIE, 1962; BLAZEK, 1973; SMYKATZ-KLOSS, 1974):

$$4\,FeS_2 + 11\,O_2 \rightarrow 2\,Fe_2O_3 + 8\,SO_2$$

$$3\,FeS + 5\,O_2 \rightarrow Fe_3O_4 + 3\,SO_2$$

At lower temperature, pyrite may oxidize by the following reactions:

$$FeS_2 + 3\,O_2 \rightarrow FeSO_4 + SO_2$$

$$FeS + 2\,O_2 \rightarrow FeSO_4$$

The resultant iron sulphate dissociates completely under roasting conditions into ferric oxide (LOSKOTOV, 1965).

Application of Thermal Analysis in Mineral Technology

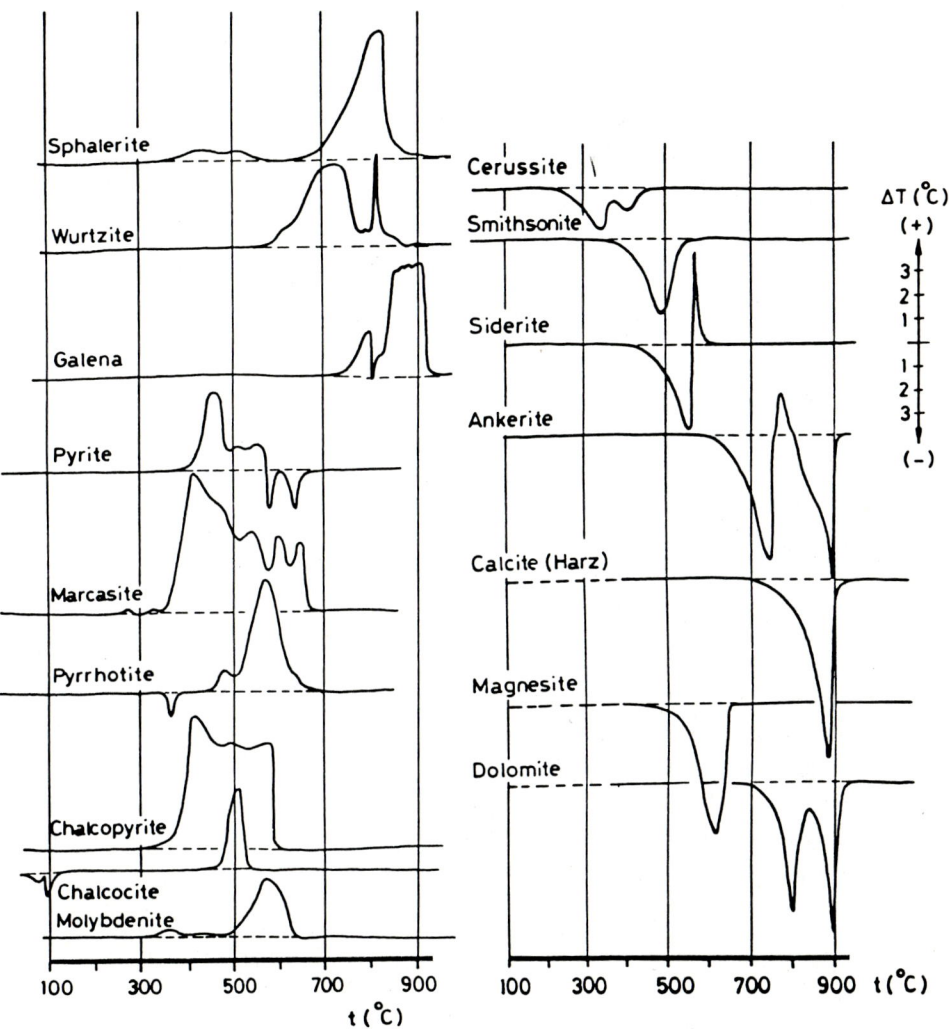

Fig. 1: DTA curves of some sulphide minerals (SMYKATZ-KLOSS, 1974).

Fig. 2: DTA curves of some carbonate minerals (SMYKATZ-KLOSS, 1974).

b) Roasting of Sphalerite in Zinc Industry (in air)

Oxidized roasting of sphalerite is carried out at 800-900°C, which is consistent with the DTA data (Kopp & Kerr, 1957; Ivanova, 1961; Blazek, 1973; Smykatz-Kloss, 1974). Fig. 1 shows that oxidation of the surface layer of sphalerite takes place at 480-500°C by an exothermic reaction. The main oxidation of sphalerite takes place at 800-900°C as represented by the large exothermic peak at 820°C. The following reactions are considered applicable:

$$ZnS + 2\,O_2 \rightarrow ZnSO_4$$

$$3\,ZnSO_4 + ZnS \rightarrow 4\,ZnO + 4\,SO_2$$

$$3\,ZnSO_4 \xrightarrow{720°C} 3\,ZnO \cdot 2\,SO_3 + SO_2 + \tfrac{1}{2}\,O_2$$

$$3\,ZnO \cdot 2\,SO_3 \xrightarrow{845°C} 3\,ZnO + 2\,SO_2 + O_2$$

The products of roasting contain zinc oxide and sulphate in different proportions, depending on roasting conditions and temperature. The associated smithsonite ($ZnCO_3$) with sphalerite decomposes during roasting at 500-525°C by an endothermic reaction (Blazek, 1973; Smykatz-Kloss, 1974), as shown in Fig. 2 and represented by the following reaction:

$$ZnCO_3 \xrightarrow{500\text{-}525°C} ZnO + CO_2$$

Limestone is added to the charge as a flux for roasting temperature regulation and to prevent sintering of the roasted ore. Calcite dissociates during roasting of the ore at 860-900°C by an endothermic reaction (Ivanova, 1961; Waters, 1967; Blazek, 1973; Smykatz-Kloss, 1974), that is why it absorbs the evolved heat from the reaction of sulphide oxidation.

c) Roasting of Chalcopyrite in the Copper Industry (in air)

DTA data (Fig. 1) shows that oxidation of chalcopyrite during roasting begins at 350-400°C and takes place intensively at 450°C as represented by the large and wide exothermic peak (TSIDLER, 1958; IVANOVA, 1961; BLAZEK, 1973; SMYKATZ-KLOSS, 1974). The basic exothermic effect is broad and the ascending and descending parts reflect the irregularities of the oxidation of large particles.

On heating chalcopyrite in absence of air, it dissociates at 550°C by the following reaction (TSIDLER, 1958):

$$4\ CuFeS_2 \xrightarrow{550°C} 2\ Cu_2S + 4\ FeS + S_2$$

The copper concentrate also contains other copper sulphides such as chalcocite (Cu_2S) and covellite (CuS), associated with chalcopyrite. The DTA curve of chalcocite (Fig. 1) shows a small endothermic peak at 100°C, representing phase transformation. Oxidation of chalcocite takes place at 520°C as represented by the sharp exothermic peak.

Oxidation of copper sulphide minerals takes place at the temperature of roasting, by the following reactions (TSIDLER, 1958):

$$6\ CuFeS_2 + 17.5\ O_2 \rightarrow 3\ Cu_2O + 2\ Fe_2O_4 + 12\ SO_2$$

$$Cu_2S + 1.5\ O_2 \rightarrow Cu_2O + SO_2$$

$$Cu_2S + 2\ O_2 \rightarrow 2\ CuO + SO_2$$

$$2\ CuS + 2.5\ O_2 \rightarrow Cu_2O + 2\ SO_2$$

Copper and iron sulphate are also formed as intermediate products. Oxidized roasting of copper concentrate is carried out at a maximum temperature of about 900°C, at which copper and iron sulphates dissociate with the formation of the corresponding oxides.

d) Roasting of Molybdenite in the Molybdenum Industry (in air)

Molybdenite is considered as the essential ore mineral for production of ferromolybdenum and pure chemical molybdenum compounds, as molybdenum trioxide, ammonium paramolybdate, sodium and calcium molybdates. Processing of molybdenite begins by its oxidized roasting of molybdenite to produce molybdenum trioxide (MIRSON & ZELIKMAN, 1965).

The DTA data, Fig. 1, shows that intensive oxidized roasting of molybdenite in an air atmosphere takes place at 570°C by an exothermic reaction with the formation of molybdenum trioxide by:

$$MoS_2 + 3.5\ O_2 \xrightarrow{570°C} MoO_3 + 2\ SO_2$$

At the lower temperature of 500°C, a thick layer of the oxide is formed, preventing the further oxidation of molybdenite.

2. *Fluorination of Kaolinite by its Sintering with Ammonium Fluoride*

Fluorination of kaolinite by its sintering with ammonium fluoride was found to be complicated. Different products of sintering are obtained, depending on the temperature and amount of ammonium fluoride. The DTA curves (Fig. 3; ABDEL REHIM, 1979) indicate the formation of the ammonium aluminium fluoride complex and cryptohalite at 120-280°C and the appearance of aluminium fluoride, topaz and mullite at 640°C. The intensive formation of topaz takes place at 750°C which then dissociates at 940°C to form corundum. The very small endothermic peak at 1010°C represents the formation of mullite. The end product of sintering consists of corundum and mullite in the case of using kaolinite-ammonium fluoride mixes of ratios 1:1 and 1:1.3. When using mixes of ratios 1:1.7 aluminium fluoride constitutes the main composition end product (Fig. 4). The X-ray diffraction patterns of the products of sintering at 120°C, 290°C, 750°C and

940°C are shown in Fig. 5. Topaz constitutes the total composition of the product at 750°C, while the product at 940°C consists mainly of corundum and topaz with some mullite.

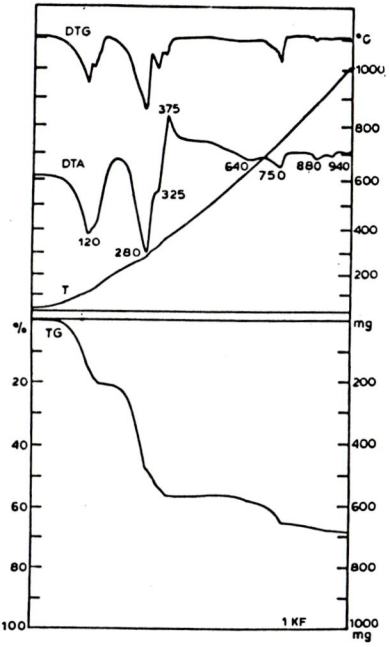
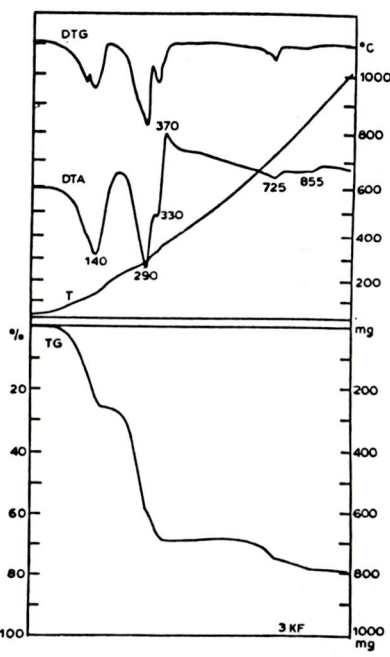

Fig. 3: Derivatogram of kaolinite sintering with ammonium fluoride of ratio 1:1 (ABDEL REHIM, 1979).

Fig. 4: Derivatogram of kaolinite sintering with ammonium fluoride of ratio 1:1.7.

Weight of sample 1000 mg, heating rate 10°C per minute.

The mechanism of the reaction was found to be a complicated one and can be considered as the following:

At 120°C: The reaction of kaolinite with ammonium fluoride takes place with the formation of ammonium hexafluorite and cryptohalite.

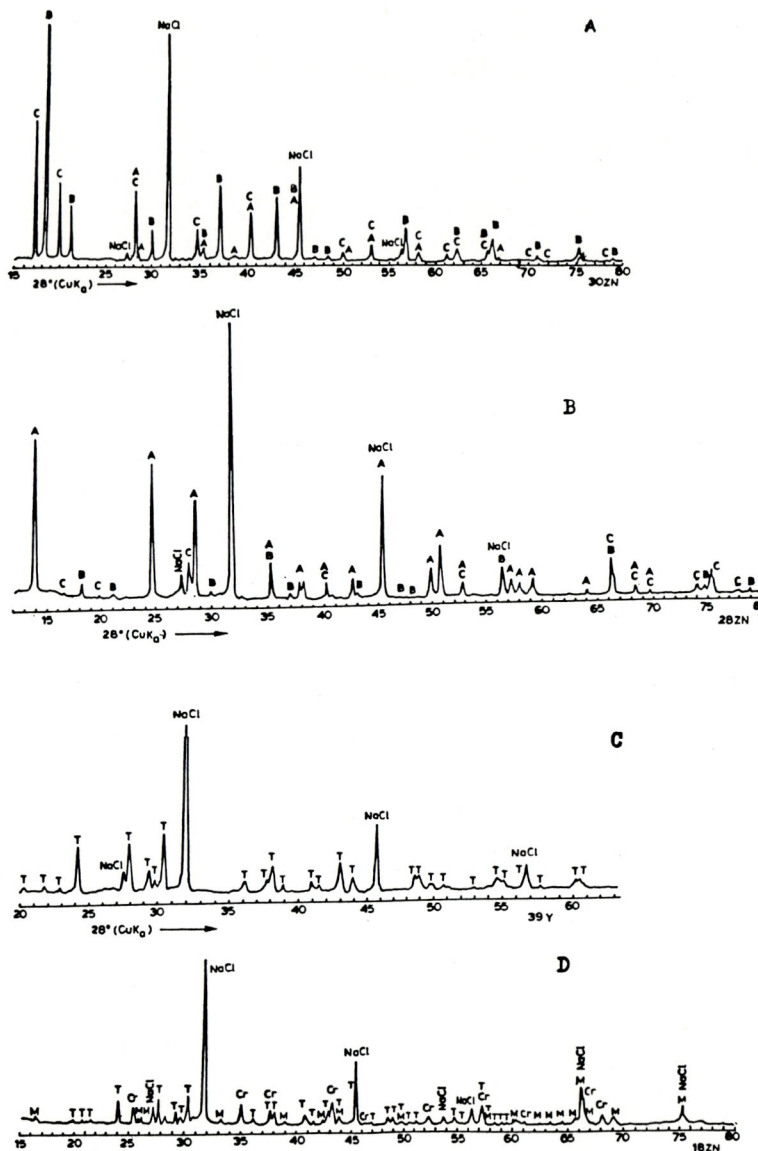

Fig. 5: X-ray powder diffraction patterns of the products of kaolinite sintering with ammonium fluoride. (A), (B), (C) and (D) at 120°C, 290°C, 750°C and 940°C respectively. A: NH_4AlF_4, B: Cryptohalite $(NH_4)_2SiF_6$, C: $(NH_4)_3AlF_6$, T: Topaz, M: Mullite and Cr: Corundum.

$$Al_2Si_2O_5(OH)_4 + 24\,NH_4F \rightarrow$$
$$2\,(NH_4)_3AlF_6 + 2\,(NH_4)_2SiF_6 + 14\,NH_3 + 9\,H_2O$$

At 140°C: Ammonium aluminium hexafluoride is unstable and begins to decompose at a higher temperature, yielding the more stable ammonium aluminium tetrafluoride, according to:

$$(NH_4)_3AlF_6 \rightarrow NH_4AlF_4 + 2\,NH_3 + 2\,HF$$

or the reaction of kaolinite with ammonium fluoride takes place at 140°C, according to:

$$Al_2Si_2O_5(OH)_4 + 20\,NH_4F \rightarrow$$
$$2\,NH_4AlF_4 + 2\,(NH_4)_2SiF_6 + 14\,NH_3 + 9\,H_2O$$

At 290°C: Cryptohalite is unstable and dissociates with liberation of ammonia, silicon tetrafluoride and hydrogen fluoride

$$(NH_4)_2SiF_6 \rightarrow 2\,SiF_4 + 2\,NH_3 + 2\,HF$$

or the reaction of kaolinite with ammonium fluoride takes place, as

$$Al_2Si_2O_5(OH)_4 + 16\,NH_4F \rightarrow$$
$$2\,NH_4AlF_4 + 2\,SiF_4 + 14\,NH_3 + 9\,H_2O$$

At 330°C: The dissociation of ammonium aluminium tetrafluoride takes place with the formation of aluminium fluoride

$$NH_4AlF_4 \rightarrow AlF_3 + NH_3 + HF$$

or

$$Al_2Si_2O_5(OH)_4 + 14\,NH_4F \rightarrow$$
$$2\,AlF_3 + 2\,SiF_4 + 14\,NH_3 + 9\,H_2O$$

Topaz formation: The resulted aluminium fluoride may react with kaolinite in a hydrolysis-like manner or with the water produced by dehydroxylation of kaolinite. Above 300°C, aluminium fluoride will hydrolyze even under the effect of atmospheric moisture.

$$2\,AlF_3 + 3\,H_2O \rightarrow Al_2O_3 + 6\,HF$$

$$Al_2Si_2O_5(OH)_4 + 6\,HF \rightarrow Al_2(SiO_4)(OH,F)_2 + 3\,H_2O + SiF_4 + H_2$$

or

$$Al_2Si_2O_5(OH)_4 + 2\,AlF_3 \rightarrow Al_2(SiO_4)(OH,F)_2 + Al_2O_3 + SiF_4 + H_2$$

The reaction of kaolinite with aluminium fluoride with the formation of topaz takes place at 725-750°C.

Mullite formation: Under dynamical conditions, mullite formation proceeds in two ways. One starts from kaolinite, due to the deficiency of fluoride ion:

$$2\,Al_2Si_2O_5(OH)_4 + 2\,AlF_3 \rightarrow 2\,Al_2O_3 \cdot SiO_2 + SiF_4 + 2\,HF + H_2O$$

The second, from the dissociation of topaz, as indicated by the very small endothermic peak at 1010°C.

$$Al_2(SiO_4)(OH,F)_2 \rightarrow 2\,Al_2O_3 \cdot SiO_2 + F_2 + 2\,OH$$

Corundum formation: Corundum may be formed in different ways, either by the hydrolysis of aluminium fluoride or the desilication of topaz or mullite with aluminium fluoride according to:

$$Al_2(SiO_4)(OH,F)_2 + 2\,AlF_3 \rightarrow 2\,Al_2O_3 + SiF_4 + 2\,F_2 + H_2$$

also,

$$3\,(2\,Al_2O_3 \cdot SiO_2) + 4\,AlF_3 \rightarrow 8\,Al_2O_3 + 3\,SiF_4$$

From this DTA study, it can be concluded that topaz dissociates at 1010°C with liberation of fluorine and hydroxyl groups with the formation of mullite. In the presence of aluminium fluoride, desilication of topaz takes place at 940°C with the formation of corundum.

3. Production of the Refractory Material of Corundum and Baddeleyite

For production of high zirconium refractory material of baddeleyite (ZrO_2) and corundum, zircon should be converted to the oxide and then mixed with aluminium oxide. The different chemical processing methods of zircon (MIRSON, 1965; ABDEL REHIM, 1974; ABDEL REHIM, 1976), include sintering of zircon with soda or alkali and leaching with alkali solutions. These methods are achieved in stages. One prospective variant for direct production of this refractory material is the sintering of zircon with aluminium fluoride in presence of graphite.

The DTA curves (Fig. 6; ABDEL REHIM, 1974) indicate that the desilication of zircon with aluminium fluoride takes place intensively at 820°C. The three endothermic peaks at 105°C, 315°C and 450°C agree with the thermal data of aluminium fluoride (Fig. 7). The exothermic peak at 580°C represents the burning of graphite and the slow reaction of desilication of zircon and combustion of volatiles. The desilication of zircon results in the production of baddeleyite and corundum (Fig. 8), which are suitable for refractory purposes.

4. Production of Corundum from Kaolinite

There are different chemical processing methods for the production of aluminium oxide from kaolinite. These include sintering with limestone or soda and acid leaching. One variant is its desilication with aluminium fluoride (LOCSEI, 1968; ABDEL REHIM, 1975) and volatilization of silicon in the form of silicon tetrafluoride, resulting in the formation of corundum or α-aluminium oxide.

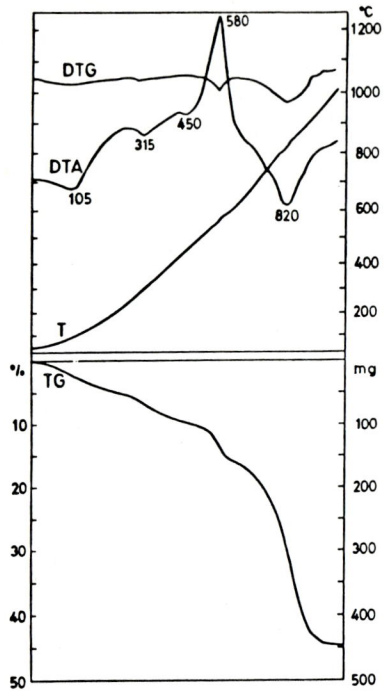

Fig. 6: Derivatogram of zircon sintering with aluminium fluoride in presence of graphite. Weight of sample 1000 mg, heating rate 10°C/min (ABDEL REHIM, 1974).

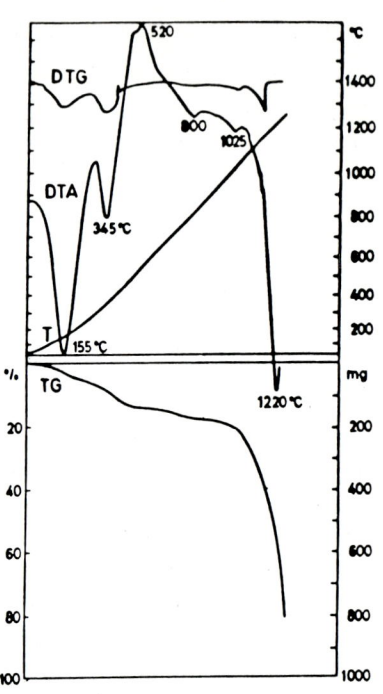

Fig. 7: Derivatogram of aluminium fluoride. Weight of sample 1000 mg, heating rate 10°C/min.

The DTA curves (Fig. 9) and X-ray diffraction patterns (Fig. 10; ABDEL REHIM, 1975) indicate the appearance of both topaz and mullite at 625°C. The desilication of kaolinite is observed to occur in two distinct steps. The first is marked by an endothermic peak at 750°C, representing the formation of topaz and the second by a sharp endothermic peak at 950°C, representing its subsequent dissociation. Desilication results in the production of corundum which is useful for many industrial purposes.

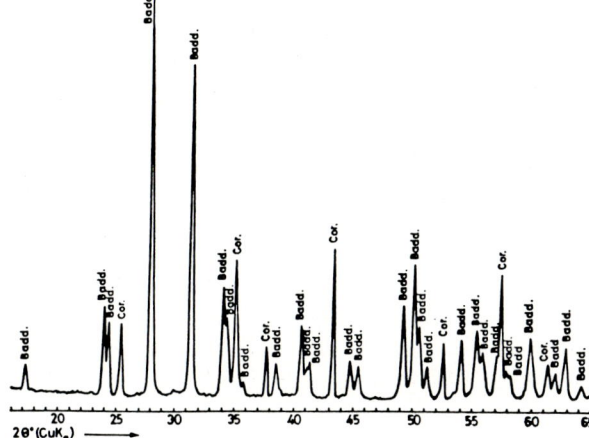

Fig. 8: X-ray powder diffraction pattern of the product of zircon sintering with aluminium fluoride. Badd.: Baddeleyite, Cor.: Corundum.

5. *Fluorination of Corundum with Ammonium Fluoride*

The thermal study of fluorination of corundum with ammonium fluoride (ABDEL REHIM, 1980) indicates that the reaction takes place in three distinct steps. The DTA curves (Fig. 11) shows the decomposition of ammonium fluoride and the formation of ammonium aluminium hexafluoride at 180°C. The endothermic peak at 225°C represents the dissociation of the resulted ammonium bifluoride and ammonium aluminium hexafluoride with the formation of ammonium aluminium tetrafluoride. The intensive formation of ammonium aluminium tetrafluoride takes place at 300°C by an exothermic reaction. The last product dissociates at 360°C into aluminium fluoride.

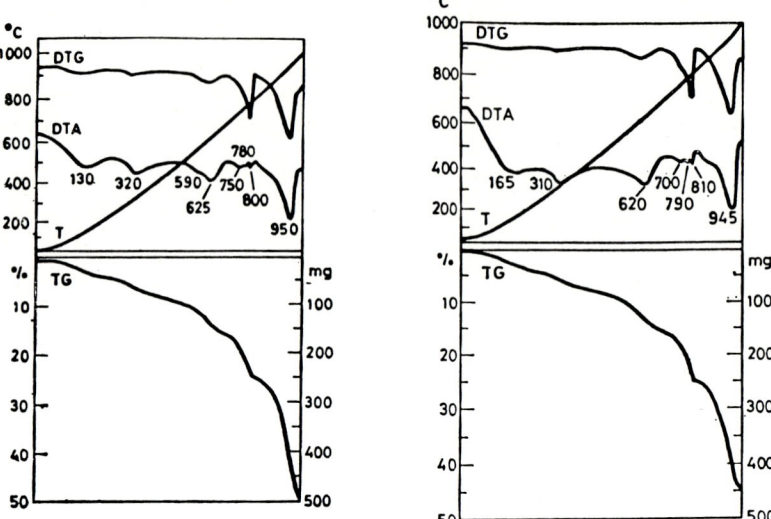

Fig. 9: Derivatograms of kaolinite sintering with aluminium fluoride of 150% (above left) and 125% (above right) of theoretical amount. Weight of sample 1000 mg, heating rate 10°C/min.

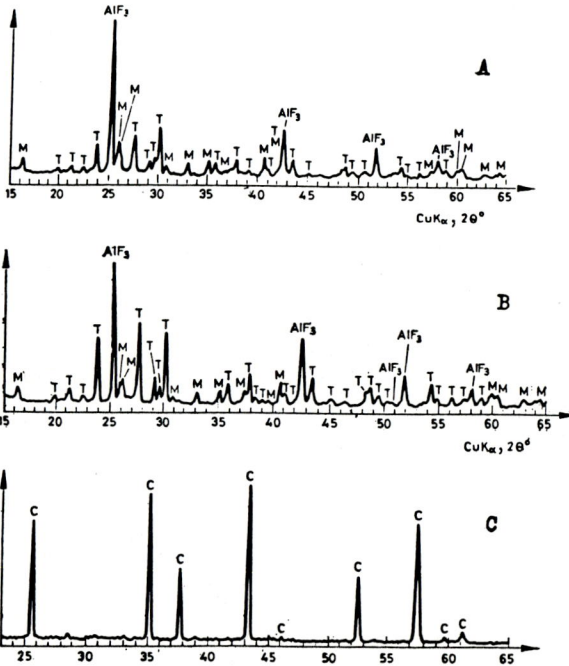

Fig. 10: X-ray powder diffraction patterns of the products of kaolinite sintering with aluminium fluoride of 150% amount. (A) at 625°C, (B) at 750°C and (C) at 950°C respectively. M: Mullite, T: Topaz, C: Corundum.

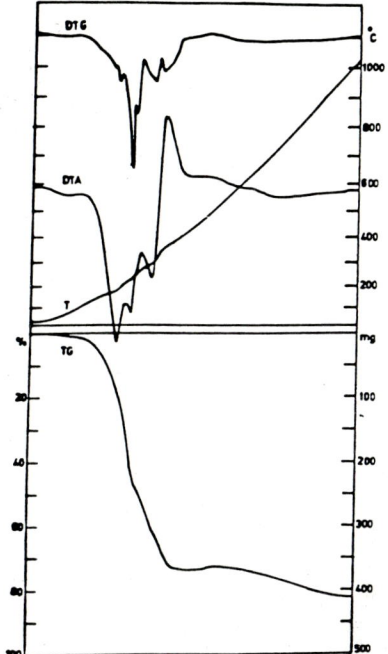
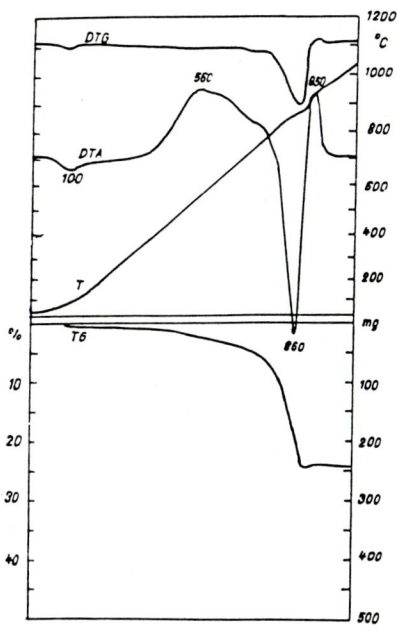

Fig. 11: Derivatogram of fluorination of corundum with ammonium fluoride (mix 1:1; ABDEL REHIM, 1980).

Fig. 12: Derivatogram of ilmenite sintering with calcite and graphite (ABDEL REHIM, 1976a).

6. Formation of Perovskite in Titanium Slags

Perovskite is one of the components of titanium slags, obtained from smelting of ilmenite (MIRSON, 1965; ABDEL REHIM, 1976a). Processing of ilmenite by smelting is used for the enrichment of titanium in the low grade titaniferrous iron ore. Smelting of ilmenite with coke is carried out at 1100-1150°C in the presence of fluxes (limestone or lime, magnesite and soda).

The thermal study (ABDEL REHIM, 1976a) of formation of perovskite by ilmenite sintering with calcite in presence of graphite (Fig. 12) shows the intensive formation of perovskite at 950°C by the exothermic reaction between ilmenite and dissociated calcite. The decomposition of calcite takes place at 860°C as represented by the large and sharp endothermic peak. Fig. 13 shows the intense and large peaks of perovskite and calcium ferrite of composition $Ca_2Fe_2O_5$. The ilmenite peaks disappear, indicating its complete conversion to perovskite.

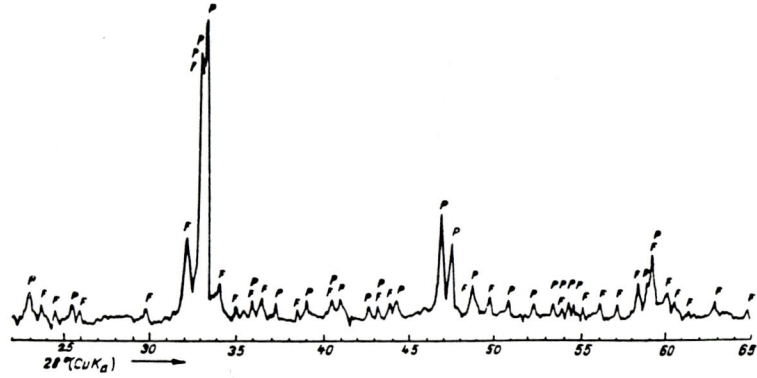

Fig. 13: X-ray diffraction pattern of the product of ilmenite sintering with calcite and graphite at 950°C. P: Perovskite, F: Calcium ferrite.

Processing of Refractories and Ceramics

Refractories are materials, usually non-metallic, used to withstand high temperatures. Refractoriness is the capability of maintaining the desired degree of chemical and physical identity at high temperature and in the environment and conditions of use. Fire clay refractories are composed essentially of fire clay and/or calcined fire clay in varying proportions. Fire clays include all clays having a fusion point above 1425°C. They are mainly composed of hydrated aluminium silicates with minor amounts of accessory minerals. The primary types of refractory bricks include fire clay and alumina bricks, basic bricks (magnesia and chrome ore), silica brick, insulating brick and special refractories (composed of carbon, SiC, baddeleyite, zircon, borides and nitrides). Processing of refractory bricks involves generally heat treatment at high temperature before use, but in some instances, they are unburned and/or chemically bonded (BUDNIKOV, 1964; NORTON, 1968; LEFOND, 1975; ABDEL REHIM, 1976).

A ceramic product is a solid composed of materials, which have been subjected to heat above 500°C (LEFOND, 1975). In ceramic production the clay is generally a refractory filler or skeletal material, having a plasticity derived from its colloidal grain size. On the other hand, silica is a glass former or non plastic filler. Clay is less of a glass former than pure silica and is more refractory. Ceramic processing involves heat treatment of clays and clay mixes, during which two processes can occur: Solid-state reactions and sintering (BUDNIKOV, 1964; DOLLIMORE, 1980).

Thermal analysis is a valuable tool in the study of refractory and ceramic processes. Detailed DTA and TG studies are used in the identification of clay and other minerals by their characteristic thermal behaviour of each mineral. Thermal analysis applied to the starting raw materials, dehydration and formation of new phases, production stages of ceramics and refractories and the end products. It can assist in controlling drying of ceramics and refractories by preventing cracking and thereby increasing the process efficiency.

When moist clay is dried at 100°C, it loses surface water and becomes hard and brittle. This process can be demonstrated by using DTA and TG techniques. The DTA data shows a strong endothermic peak at 500-600°C, representing the loss of combined water (constitutional OH radicals) and the structure breaks down and a hard porous mass results. At 900-1000°C, there is often a sharp exothermic peak, the clay becomes less porous and more strong as a consistence. At higher temperatures between 1400-1850°C, this solid phase will melt to a glassy mass (DOLLIMORE, 1980).

Clays are generally found mixed with fine grains of quartz, feldspar, mica, iron oxides, etc.. Kaolin or china clay is the purest form, and is a white earthy substance. Upgrading of kaolin is carried out by wet benefication processes, including degritting, followed by fine particle classification. A typical DTA curve of Kalabsha kaolinite is shown in Fig. 14. The dehydroxylation of kaolinite takes place at 575°C with the formation of metakaolinite. The sharp exothermic peak at 965°C represents the conversion of metakaolinite into an Al-spinel. An exothermic peak at 1200-1300°C can be attributed to rapid crystallization of mullite and/or cristobalite (DOLLIMORE, 1980).

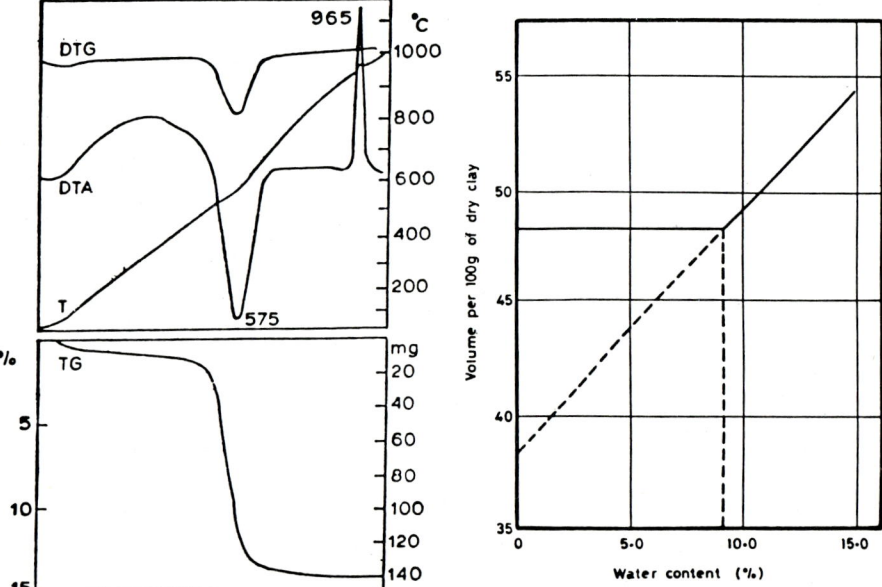

Fig. 14: Derivatogram of Kalabsha kaolinite. Weight of sample 1000 mg, heating rate 10°C/min.

Fig. 15: Shrinkage limit test.

The thermal behaviour of clays is very important in ceramic and refractory industries and soil mechanics. Clays swell when they adsorb water and shrink when they lose water. The shrinkage continues until the shrinkage limit is reached. The shrinkage limit is the water content, at which clay does not continue to shrink on further loss of moisture. The shrinkage limit (SL) is an important parameter, relating the shrinkage occurring in drying of clays with the water content. Fig. 15 shows the results of a shrinkage limit test. A thermomechanical analysis unit will be required with DTA and TG units.

Influence of Super-Fine Grinding on the Thermal Behaviour of Kaolinite

Super-fine grinding affects the thermal reactions of kaolinite considerably in the temperature ranges, covering both the endothermic and exothermic effects, as shown in Figs. 16 and 17 (KORNEVA & YUSUPOV, 1976).

For the dry-ground kaolinite (Fig. 16) the endothermic dehydration maximum at 560°C shifts to lower temperatures and new endothermic effects appear, indicating that several dehydration steps occur. The size of the 960°C exothermic peak depends essentially on the time of grinding. The reaction is most intense at grinding times of 10 minutes and 20 minutes and the formation of Al-spinel coincides with the increase of intensity of 960°C exothermic effect. Grinding of kaolinite in air leads to almost complete destruction of its structure and the product is essentially amorphous. Hence, the greater the disorder in the structure, the greater the amount of Al-spinel formed.

The thermal behaviour of kaolinite ground in aqueous medium is quite different (Fig. 17). Here, grinding even for a long time does not destroy the kaolinite structure completely, as represented by the endothermic effect remaining on the DTA curve, even after 30 minutes of grinding, although it becomes smaller. The exothermic effect at 960°C for 1 minute is less intense than for the original kaolinite and dry ground samples. This effect is caused by the formation of γ-aluminium oxide. Mullite was not observed in this temperature region. The size of the 960°C exothermic peak decreases gradually with increasing time of grinding. Simultaneously, an exothermic peak at 860°C appears and gradually increases in size.

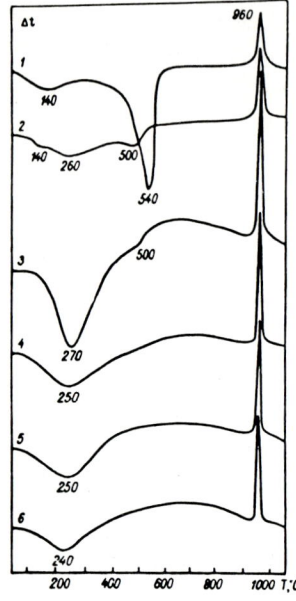 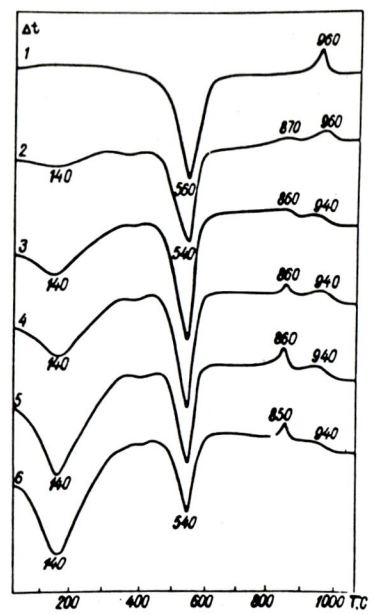

Fig. 16: DTA curves for kaolinite ground in air for:
1: 1 min, 2: 5 min, 3: 10 min, 4: 15 min, 5: 20 min, 6: 30 min. (KORNEVA & YUSUPOV, 1976).

Fig. 17: DTA curves for kaolinite ground in an aqueous medium for:
1: 1 min, 2: 5 min, 3: 10 min, 4: 15 min, 5: 20 min, 6: 30 min. (KORNEVA & YUSUPOV, 1976).

Effect of Ceramic Stabilizers on the Bonding of Free Lime in Refractory Portland Cement Concrete

The thermal analysis is a valuable tool in studying the refractory Portland cement concrete and the effect of ceramic stabilizers on the bonding of free lime (LACH, 1968). Ceramic stabilizers are used to modify the behaviour during heating and phase composition of refractory concrete and their contribution to bonding, i.e. to the change from hydraulic to ceramic bonding, which results in the stabilization of the product.

Stabilizers are chosen with a view to their ability to react with free calcium oxide, with simultaneous formation of new phases, which are resistant to humidity and prevent the formation of $Ca(OH)_2$. Also, they should improve the refractory properties of the binding agent, i.e. mainly the resistance to softening at high temperature. The material used as ceramic stabilizer includes dehydrated aluminium hydrosilicate (Fire-clay), kaolinite, chromite, sintered magnesite, slag, silica and diatomaceous earth.

The residual amount of free CaO (degree of stabilization) can be determined thermogravimetrically. DTA curves (Figs. 18 and 19; LACH, 1968) show that at least 10% of stabilizer, fire-clay, clay or chromite, calculated for the cement quantity is needed. When magnesite is used as a stabilizer, a considerable amount of CaO (up to 10.8%) always appears, even with percentage as high as 20% of the weight of cement.

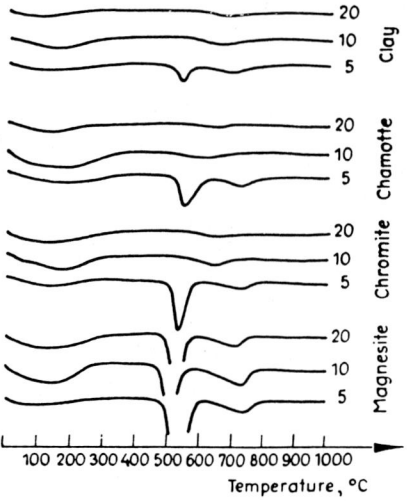

Fig. 18: DTA diagrams of PZ 450 after 7 days. Rel. humidity ~50% (LACH, 1968).

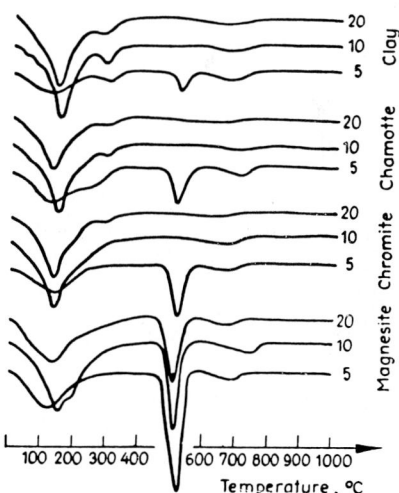

Fig. 19: DTA diagrams of PZ 450 after 7 days. Rel. humidity ~98% (LACH, 1968).

From this study, thermal analysis is shown to be a critical tool in ceramics and refractories processing for material and processes improvement.

Glass Industry

In the glass industry, considerable work has been done on the thermal behaviour of the starting materials of glass production and on the thermal behaviour of glass itself, which is characterized by a series of typical intervals within which its physical, chemical and structural properties change (BLAZEK, 1973).

The starting materials are silica (sand), alkali (sodium carbonate or hydroxide) and lime (chalk, limestone or dolomite), but there are many variations. These mixtures are heated beyond a certain temperature where the materials fuse and the liquid is cooled. The structure is a long-range disordered structure of a liquid, whilst retaining a short range structure. The starting materials are mixed into "batches" and used in powdered form (DOLLIMORE, 1980).

Silica, which is the glass forming oxide, needs a very high temperature (1800°C) for melting and working. Fluxes (sodium carbonate) are therefore added to reduce the melting temperature. The lowest melting mixture of soda and silica results from the reaction to produce sodium disilicate, which with silica melts at 783°C. This can be demonstrated by the use of DTA and TG runs on these mixes. WILBURN & DAWSON (1972) and DOLLIMORE (1980) identified several steps in a two phase mixture of soda and silica:

1- Endothermic peak at 100°C, representing loss of water,

2- Endothermic peak at 573°C, representing inversion of quartz,

3- Formation of sodium disilicate by endothermic reaction and weight loss,

4- Endothermic melting of sodium carbonate, sodium disilicate and silica,

5- Decomposition of melted sodium carbonate.

Reactions 3 and 4 can occur usually around 850°C, but highly dependent upon particle size. The glass produced by chilling the eutectic mixtures from above 783°C is readily dissolved in water. Lime is usually added as limestone, to increase durability of such a glass.

The combination of DTA and TG can be extended to investigate the behaviour of the other components of glass, both singly and in mixture. The DTA and DTG curves (Fig. 20; RICHARD OTT, 1980) of glass batch

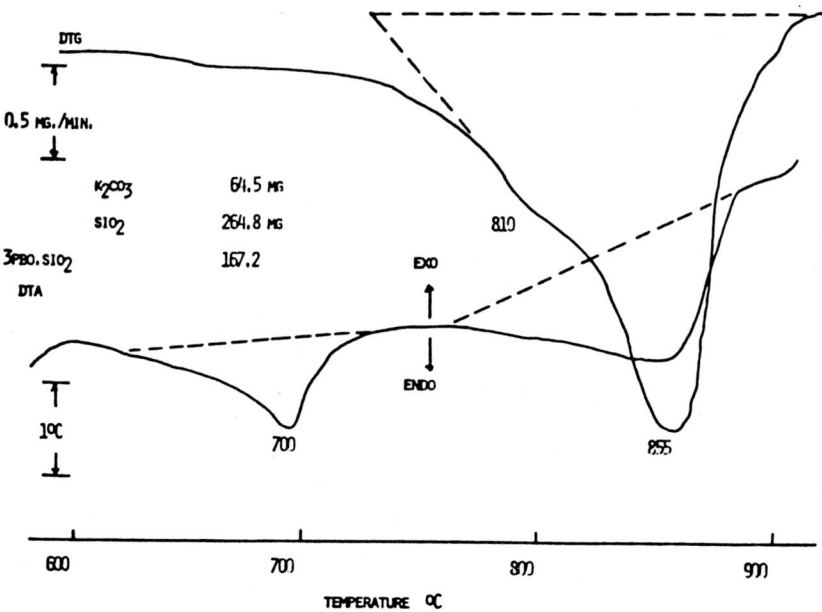

Fig. 20: DTA and DTG data of glass batch mixture (potassium carbonate, silica and ground tribasic lead silicate) (RICHARD OTT, 1980)

mixture (potassium carbonate - silica - ground tribasic lead silicate) show only the peaks characteristic of the raw materials up to 600°C. The broad DTA endotherm over the range 650-720°C was the result of the reaction between silica and ground tribasic lead silicate. This reaction precluded the reaction between potassium carbonate and ground tribasic lead silicate. The main portion of decomposition reaction was at 855°C.

Also, thermal methods may be used to study the behaviour of glass during heating (annealing peaks appear in the DTA) and the devitrifaction of glass. The transformation range can be recognized by a small endothermic effect as the temperature is raised. The temperature at which the effect occurs, depends on the glass composition and its magnitude on the previous thermal history. In general, the effect is small for poorly annealed glass, but greater for well annealed samples (Fig. 21; DOLLIMORE, 1980). DTA

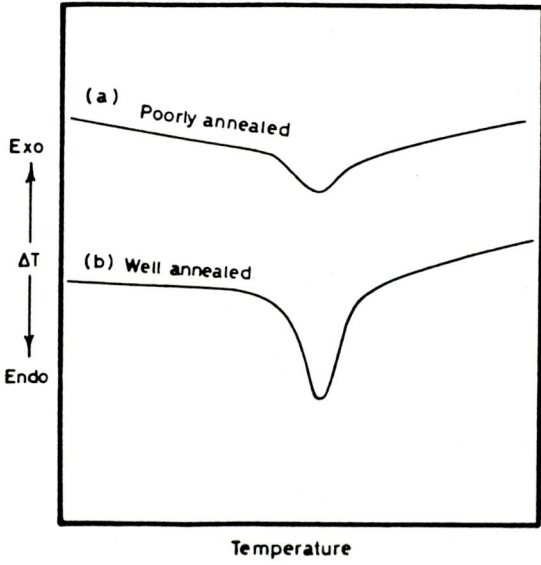

Fig. 21: DTA curves of a glass. (DOLLIMORE, 1980).

curves of some glasses may show an exothermic peak at higher temperature, which is due to the growth of solid phase crystals from glass, i.e. devitrifaction. The temperature of the peak and its magnitude depend on the rate of crystal growth.

Mineral Synthesis

Thermal analysis is widely used in the field of mineral synthesis, mechanism of reaction of synthesis and optimum temperature, and the thermal behaviour of minerals and starting materials. One example of this application was given on the thermal study of synthesis of topaz. It was reported, that topaz can be synthesized using different starting materials as (LOCSEI, 1968; ABDEL REHIM, 1975; ABDEL REHIM, 1975a; ABDEL REHIM, 1979; ABDEL REHIM, 1982):

1 - Quartz and aluminium fluoride,

2 - Corundum-quartz mix with ammonium fluoride,

3 - Kaolinite and alumiunium fluoride,

4 - Kaolinite and ammonium fluoride

1. Synthesis of Topaz Using Quartz-Aluminium Fluoride Mixtures

The DTA data of synthesis of topaz using quartz mix with 150% of theoretical amount of aluminium fluoride are shown in Fig. 22 (ABDEL REHIM, 1975a). The first two wide endothermic peaks at 140°C and 330°C represent the dehydration of aluminium fluoride. The small endothermic peak at 550°C represents the transition of β-quartz to the α-form and the beginning of the reaction between quartz and aluminium fluoride. The large and sharp endothermic peak at 780°C indicates the intensive formation of topaz. This process is connected with a remarkable decrease in weight (TG curve) due to the volatilization of silicon tetrafluoride. The resulting topaz dissociates at 960°C by an endothermic reaction in presence of excess of aluminium fluoride, giving rise to corundum or α-aluminium oxide. This is also accompanied by a sharp decrease in sample weight due to the removal of silicon tetrafluoride and the loss of constitutional OH radicals.

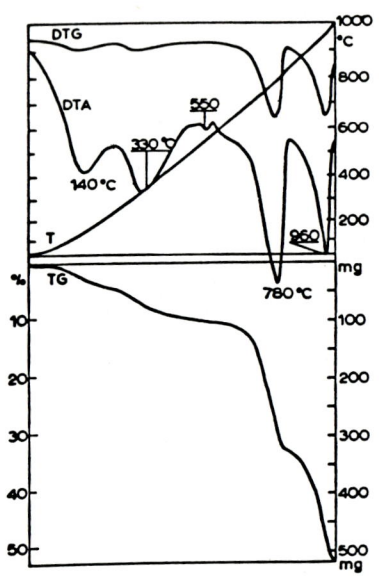 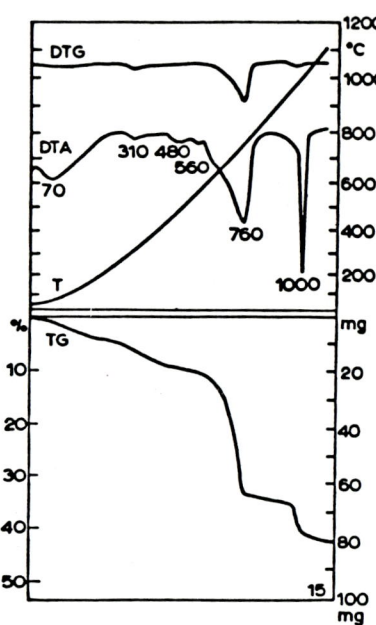

Figs. 22 & 23: Derivatograms of synthesis of topaz using quartz mixed with 150% of theoretical amount of aluminium fluoride, respectively. Weight of samples 1000 mg and 184 mg respectively, heating rate 10°C/min (ABDEL REHIM, 1975a).

The X-ray powder diffraction patterns of these products are shown in Fig. 24 (A at 550°C, B at 780°C and C at 960°C). Topaz is present in large amounts at 780°C and its peaks have completely disappeared in the run at 960°C.

Topaz synthesis using quartz mix with theoretical amount of aluminium fluoride, as shown by DTA data (Fig. 23) indicate that the product of the reaction at 760°C is composed of topaz and unreacted quartz, representing incompleteness of the reaction. The sharp endothermic peak at 1000°C represents the formation of mullite as the end product, due the deficiency of fluoride ion.

From this thermal investigation, topaz can be synthesized by sintering quartz with an excess of aluminium fluoride at 780°C.

2. *Synthesis of Topaz Using Corundum-Quartz Mixtures with Ammonium Fluoride*

The thermal analysis data (ABDEL REHIM, 1982), of synthesis of topaz using corundum-quartz-ammonium fluoride mixture of ratio 1:1:6.4 are

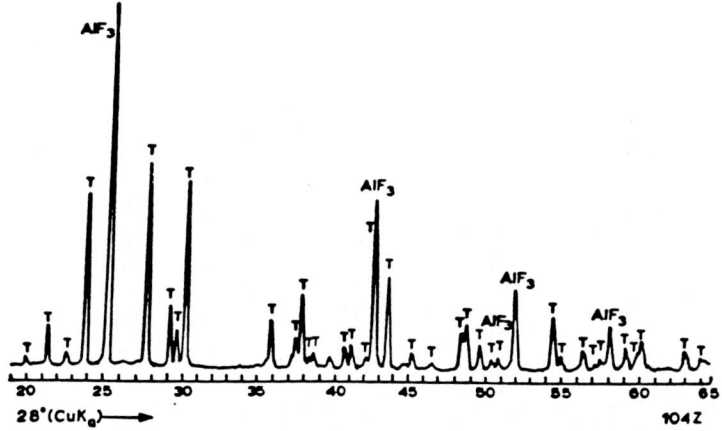

Fig. 24(a): X-ray powder diffraction patterns of topaz synthesis using quartz with aluminium fluoride (150% of theoretical value) at 550°C. T: Topaz, C: Corundum and Q: Quartz.

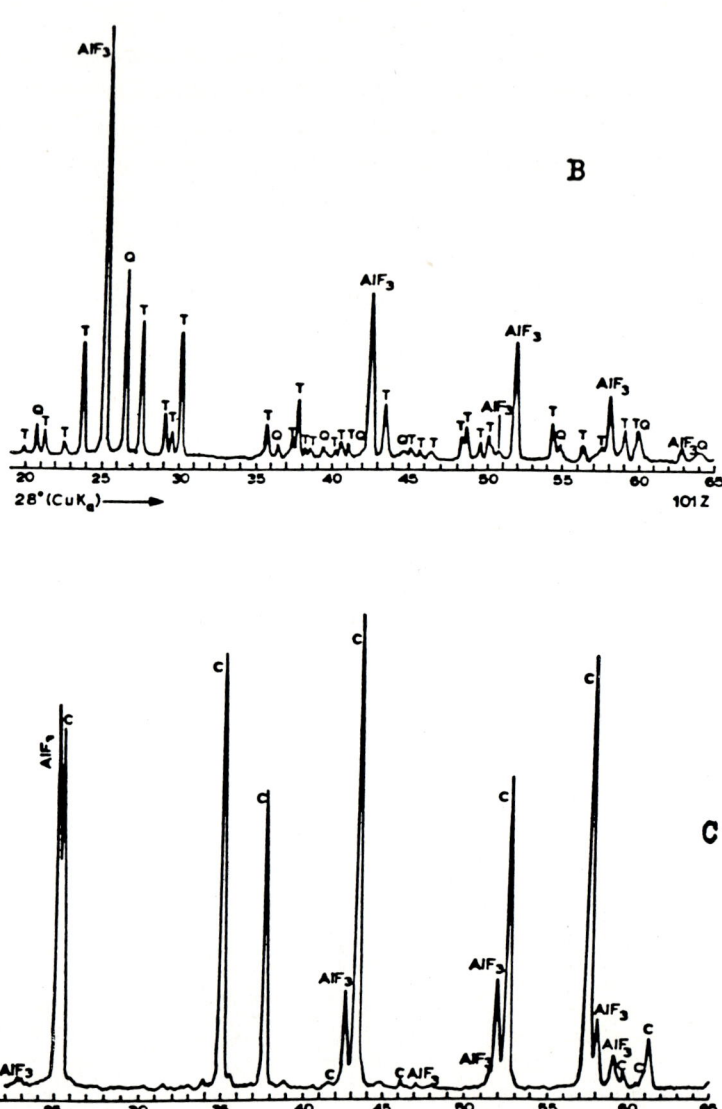

Fig. 24(b) & (c): X-ray powder diffraction patterns of topaz synthesis using quartz with aluminium fluoride (150% of theoretical value). T: Topaz, C: Corundum and Q: Quartz; (B) at 780°C and (C) at 960°C.

shown in Fig. 25. The first wide and large endothermic peak at 115°C represents the formation of ammonium aluminium hexafluoride and cryptohalite (ammonium silicon hexafluoride). The small endothermic peak at 240°C represents the dissociation of the resulted ammonium bifluoride and the unstable ammonium aluminium hexafluoride, yielding the more stable ammonium aluminium tetrafluoride. The large and sharp endothermic peak at 345°C represents the intensive dissociation of cryptohalite and ammonium aluminium tetrafluoride. The loss of weight (TG curve) is due to the volatilization of silicon tetrafluoride and removal of ammonia and water vapours.

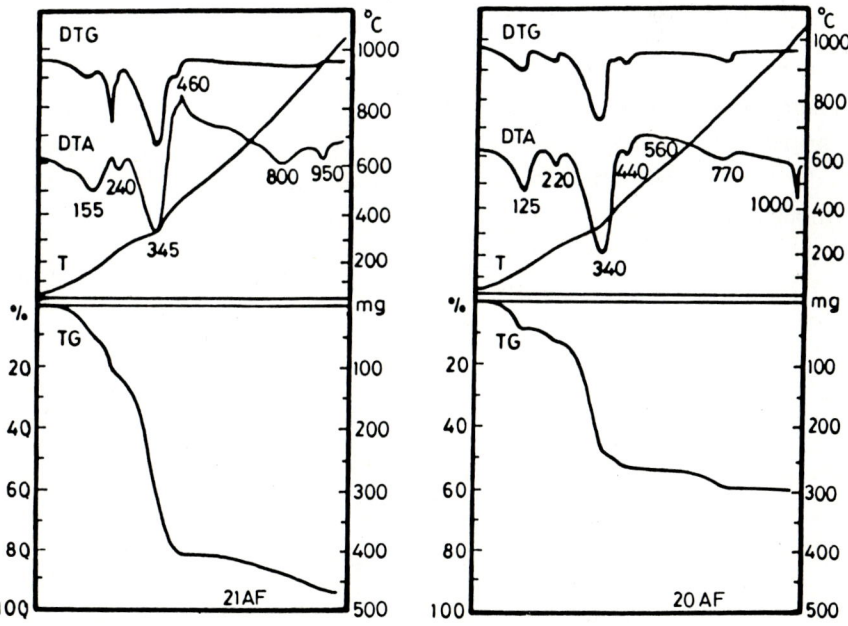

Figs. 25 & 26: DTA curves of synthesis of topaz using corundum-quartz-ammonium fluoride mixes of ratio 1:1:6.4 and 1:1:2.5 respectively. Weight of samples 500 mg each, heating rate 10°C/min (ABDEL REHIM, 1982).

The intensive formation of topaz takes place at 770-800°C by an endothermic reaction. Topaz dissociates at 950°C by an endothermic reaction and its simultaneously desilication, in the presence of an excess of aluminium fluoride, with the formation of corundum. Using mixes of ratio 1:1:2.5 topaz dissociates at 1000°C, as indicated by the sharp endothermic peak, with the formation of mullite due to the deficiency of fluoride ions (Fig. 26).

From this study, it is concluded that topaz can be synthesized by sintering of corundum-quartz mixtures with excess of ammonium fluoride at 770-800°C.

Synthesis of topaz by sintering of kaolinite with aluminium fluoride or ammonium fluoride was discussed before.

Thermal analysis is prooved to be an important and valuable tool in Mineral Technology for evaluating raw materials, the mechansims of their industrial processing, production stages, process improvement and end products. Several applications of thermal analysis in this field, are demonstrated.

References

ABDEL REHIM A. M. (1974) - Proceed. 4th Int. Conf. Therm. Anal., Vol. 1, 523, Budapest.

ABDEL REHIM A. M. (1975) - Acta Geol. Acad. Sci. Hung., Vol. 19, N.3-4, Budapest.

ABDEL REHIM A. M. (1975a) - Thermochimica Acta, 13, 231

ABDEL REHIM A. M. (1976) - Acta Mineralogica-Petrographica, Szeged, Hungary, Vol. 22, N.2, 353

ABDEL REHIM A. M. (1976a) - Acta Geol. Acad. Sci. Hung., Vol. 20, N.1-2, 19, Budapest.

ABDEL REHIM A. M. (1979) - Thermochimica Acta, 30, 127

ABDEL REHIM A. M. (1980) - Proceed. 6th Int. Conf. Therm. Anal., Bayreuth, Germany, Vol. 1, Birkhäuser Verlag.

ABDEL REHIM A. M. (1982) - Proceed. 7th Int. Conf. Therm. Anal., Kingston, Ontario, Canada.

BLAZEK A. (1973) - Thermal Analysis.- Van Nostrand Reinhold Co.

BUDNIKOV P. P. (1964) - Technology of Ceramic Refractories.- MIT Press, Cambridge, Mass.

DOLLIMORE D. (1980) - The State & Art of Thermal Analysis.- Proceed. Workshop, Nat. Bur. Stand., Maryland, (MENIS O., ROOK H. L. & GARN P. (eds.)), May 21-22, 1979, 1-31

IVANOVA V. P. (1961) - Thermograms of Minerals.- Zapisky Vsusausnovo- mineralog. Obshestva, 90, 1, 50-90, Izdat. AMUSSR.

JAYAWEERA S. A. & SLEEMAN P. (1976) - Proceed. First Europ. Sympos. Therm. Anal., Sept. 1976, Heyden, 287

KOPP O. C. & KERR P. F. (1957) - Amer. Mineralogist, 42, 445

KORNEVA T. A. & YUSUPOV T. S. (1976) - 1st Europ. Sympos. Therm. Anal., Proceed., Sept. 1976, Heyden, 336

LACH V. (1968) - Proceed. 9th Silicate Industry, Budapest, 675

LEFOND S. J. (ed.) (1975) - Industrial Minerals and Rocks (Nonmetallics other than Fuels).- Amer. Inst. Min. Metallurgy and Petrol. Engineers, Inc.

LOCSEI B. (1968) - Proceed. 9th Silicate Industry, Budapest, 103

LOSKOTOV F. M. (1965) - Metallurgy of Lead and Zinc.- Metallurg. Izdat.

MACKENZIE R. C. (1962) - "Scifax" Differential Thermal Analysis Data Index.- Cleaver-Hume Press, London.

MIDGLEY H. G. (1976) - Proceed. First Europ. Sym. Therm. Anal., Salford, Sept. 1976, (DOLLIMORE D. (ed.)), Heyden, 378

MIRSON G. A. & ZELIKMAN A. N. (1965) - Metallurgy of Rare Metals.- Metallurg. Izdat.

NORTON F. H. (1968) - Refractories.- 4th ed.., McGraw-Hill, New York.

PAULIK F., PAULIK J. & ERDEY L. (1966) - Talanta, $\underline{13}$, 1405

RICHARD OTT W. (1980) - State and Art of Thermal Analysis.- Proceed. Workshop Nat. Bur. Stand., Maryland, May 1979, 99

SMYKATZ-KLOSS W. (1974) - Differential Thermal Analysis. Application and Results in Mineralogy.- Minerals and Rocks, $\underline{11}$, Springer Verl., Berlin-Heidelberg-New York.

TSIDLER A. A. (1958) - Metallurgy of Copper and Nickel.- Metallurg. Izdat.

WATERS B. H. (1967) - Austr. Mineral Develop. Labs. Bull., $\underline{3}$, 31

WILBURN F. W. & DAWSON J. B. (1972) - Differential Thermal Analysis.- Vol. 2, (MACKENZIE R. C. (ed.)), Academic Press, London, 229

Application of Thermal Analysis in Mineral Technology

THERMAL INVESTIGATIONS IN TECHNICAL MINERALOGY

J. Schomburg

DURTEC - Ingenieur-, Beratungs- und Laborgesellschaft mbH
Neubrandenburg, Germany

Abstract

The lecture is concerned with the application of thermal analytic methods to problems in technical mineralogy. Examples for the characterization of clay raw materials (e.g. for ceramics), for the determination of sulphidic impurities in clays and for the hydrothermal synthesis of industrial minerals (zeolites, smectites) are given and discussed. The methods used were mainly differential thermal analysis, thermogravimetry and thermodilatometry.

Among the advanced techniques (Table 1) thermal analysis methods have a traditional position.

Table 1: Analytical methods related to technical mineralogy

X-ray diffraction (XRD)
light microscopy (> 20 μm)
transmission (TEM) and scanning (SEM)
electron microscopy
with micro probe analysis
infrared (IR) and raman spectroscopy
thermal methods, e.g. differential thermal analysis (DTA)
thermogravimetry (TG)
differential scanning calorimetry (DSC)
thermodilatometry (TD)
thermal optical analysis (TOA)
evolved gas analysis (EGA)

Studying the problems in the field of technical mineralogy we will find different objectives (Table 2).

Table 2: Examples for solving problems with thermal methods

	Method / use	examples
1.	Studying processes during heating and cooling (DTA, TG, TD)	ceramic firing cement clinker production glass manufacturing
2.	Quality control of products (DTA, TD)	moisture expansion of ceramic products glaze-body adaption of ceramics
3.	Investigation of dusts (DTA)	quartz content
4.	Investigation of ash and slags (DTA)	brown coal ash, different slags
5.	Determination of specific surface of clay minerals (TG)	three layer silicates
6.	Investigation of hydrothermal processes and products (DTA, TG)	pitchstone perlite
7.	Investigation of phasediagrams (DTA)	calcite-aragonite ferrite formation
8.	Investigation of building stone decay (DTA, TG)	kaolinite and gypsum formation in sandstones
9.	archaeometry (investigation of ancient ceramics) (DTA, TD)	ancient potteries, glazes and engobes

On the other hand the use of raw materials, for example clay materials for industrial application is strongly connected with heating processes (Table 3). In this way with the aid of thermal methods we have some

Table 3: Application of TA methods to clay raw materials

clay raw material	applications with thermal treatment
kaolins	ceramic industry binder for chemical products filler for rubber and plastics alumina production glass wool and glass fibre production production of polished products
bentonites	bleaching earth production foundry binder casting powder production binder for chemical products catalyst production production of expanded materials (porosite etc.) aid for deep drilling muds adsorber granulates ceramic industry production of pozzolana cements
micas	isolator material special papers lithium production
vermiculites	production of expanded materials isolator material nickel production
serpentines (antigorite, lizardite, chrysotile)	nickel production ceramic industry (special ceramics, cordierite) fire resistant materials

talc / pyrophyllite	fire resistant materials ceramic industry
chlorites	iron production (chamosites) nickel production ceramic industry
sepiolites	fire resistant materials pipes
palygorskites	analogous to bentonites
different clays	ceramic industry binder for chemical products production of expanded materials fire resistant materials alumina production sometimes analoguos to bentonites

possibilities for the reconstruction of transformation processes during industrial application.

Example 1

In the production of ceramics the following problems are of central interest:

- drying (= dehydration of clay minerals),
- sintering (= structure decomposition of clay minerals),
- firing atmosphere (= environmental protection).

The drying process (maximum temperature, evolved water vapour content) can be studied with differential thermal dilatometry, DTD. Different types of interlayer cations of bentonites cause different dehydration behaviour (Fig. 1).

Sulphur minerals, i.e. pyrite, are responsible for the development of corrosive gases during heating.

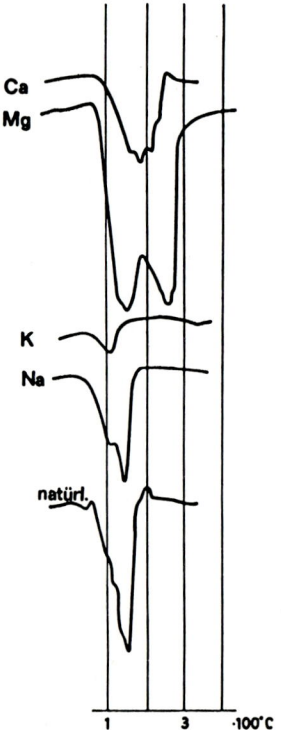
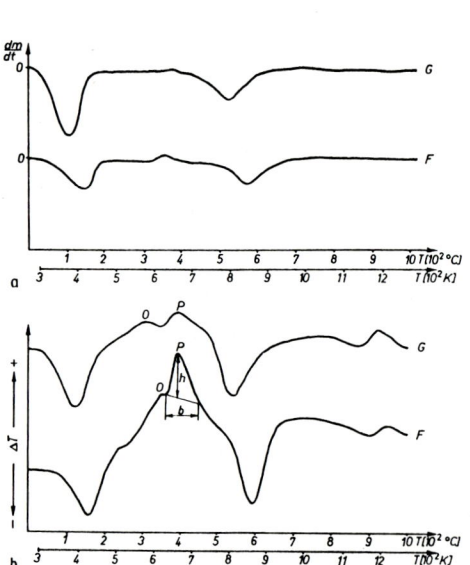

Fig. 1: DTD-curves of different cation-exchanged samples from montmorillonite (Belle Fourche, U.S.).

Fig. 2: DTA/TG-curves of pyrite-containing clays from Grimmen and Friedland, Mecklenburg-Vorpommern, Germany.

From this point of view ceramic producers are interested to know the mineralogical composition and dissociation interval (temperatures) of sulphur minerals. From Fig. 2 we may see DTA/TG measurements of pyrite containing clay from Friedland locality, Germany. Calculated data are shown in Table 4.

The beginning of the structural decomposition of clay minerals is of interest from the economic point of view for ceramic production. We found that DTD-measurements are very useful in this field (Figs. 3 and 4). Fe(III)-content in the octahedral position of three-layer silicates which is responsible for bloating processes ("secondary porosity") can be checked from dilatometrical curves.

Table 4: Results of pyrite determination of the Friedland clay by means of DTA-TG-measurements compared with chemical analysis

Sample	peak temp. (°C)	peak height (mm)	%-pyrite	peak area (cm^2)	%-pyrite	⌀-value pyrite DTA	%	pyrite content (%) by chemical analysis
F 1	380	56	3.5	5.99	3.7	(3.6)	3-4*	3.8
F 2	440	53	3.4	5.99	3.7	(3.5)	3-4	3.4
F 3	380	15	1.3	1.49	1.4	(1.3)	1	1.8
F 4	380	21	1.7	1.80	1.6	(1.6)	1-2	2.0
F 5	380	3.7	2.6	2.71	2.1	(2.3)	2	2.0
E	460	80	5.0	8.02	5	(5.0)	5	5

*: value for mineralogical data
F 1 - F 5: Friedland clay samples
E: internal standard sample

Example 2

Hydrothermal synthesis is used in different industrial production processes, i.e. for production of zeolites, synthetic smectites and others.

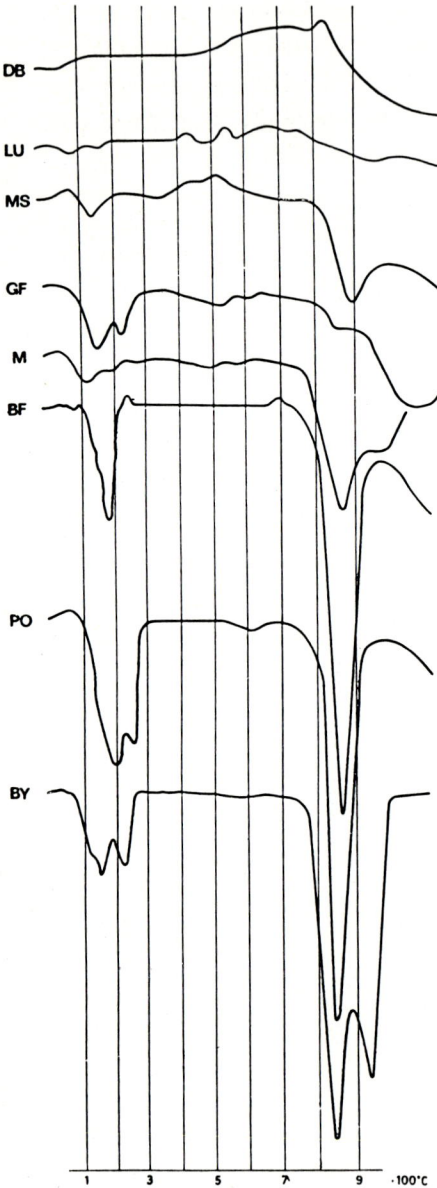

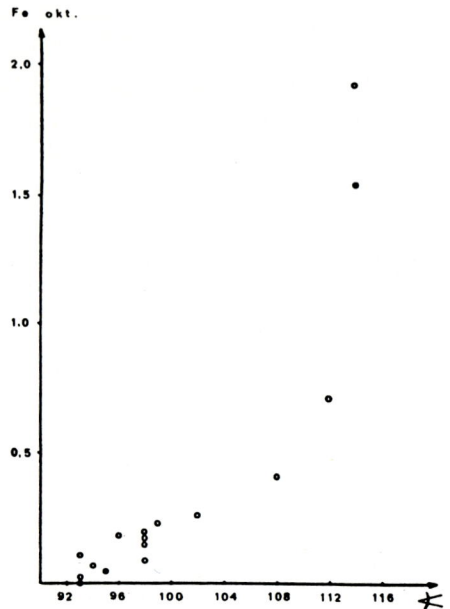

Fig. 4: Correlation between the structure decomposition angle of DTD-curves and octahedral fixed iron content for different smectites.

Fig. 3: DTD-curves of different three-layer-silicate minerals (DB: muscovite from Dolni Bory, CSFR; LU: hydromuscovite from Lukavice, CSFR; MS: illite from Morris, U.S.; GF: nontronite from Garfield, U.S.; M: beidellite from Mittelherwigsdorf, Germany; BF: montmorillonite from Belle Fourche, U.S.; PO: montmorillonite from Polkville, U.S.; BY: montmorillonite from Bayard, U.S.).

Different methods are used to obtain the reaction course. Using thermogravimetry we got more valid results on the relation between parent phase (pitchstone) and neoformed three-octahedral smectite than with XRD (the very small size of neoformed plase gives broad reflections) or with microscopy (the whole particle surface is coloured with benzidine or $FeCl_3$-solution).

After checking the characteristic weight loss between 500-1000°C for pitchstone (1.2%) and saponite (5.0%) we use the following formula:

$$WL_{mix} = C_{sm} \cdot WL_{sm} + C_p \cdot WL_p$$

$$C_{sm} + C_p = 1$$

WL_{mix} : weight loss of reaction product
WL_{sm} : weight loss of 100% smectite
WL_p : weight loss of 100% pitchstone
C_{sm} : content of neoformed smectite
C_p : content of relict pitchstone

Example 3

Clay minerals used as the main raw material for pottery production show some characteristic transformation effects which can be used for the determination of the firing conditions and former raw material composition of ancient potteries. Table 5 presents an overview of thermoanalytical criteria (> 700°C), so that clay minerals recommended as "archaeological thermometers" during studying ancient potteries may be known.

Table 5: Clay minerals, their use and thermal reactions (> 700°C) with relation to ceramics and DTA/TG criteria for determination.

clay mineral and use	process and temperature interval	DTA-method average peak (°C)	DTA-method endothermic(-) exothermic(+)	TG-method weight loss (mass-%)
kaolinite (glaze, engobe, body)	Al-Si-spinel formation (920-1000°C)	950	(+)	-
talc (engobe, body)	dehydroxylation (800-900°C)	880	(-)	4.0-5.0
muscovite (body)	dehydroxylation (700-950°C)	850	(-)	4.0
illite (engobe-"glanzton"-sheet, body)	spinel and mullite formation (850-950°C)	910	(+)	-
mixed-layer-minerals (engobe, body)	structure decomposition (820-980°C) cristobalite, mullite and spinel formation (880-980°C)	850-940 890-960	(-) (+)	0.1 -
montmorillonite (engobe, body)	structure decomposition (840-900°C) quartz, cristobalite and spinel formation (900°C)	880 910-1060	(-) (+)	0.1 -
Mg-chlorite (body, glaze)	dehydroxylation of the talc-like sheet layer (650-850°C) forsterite formation (800-830°C)	780 810	(-) (+)	2.3-3.0 -
palygorskite (body, engobe)	structure decomposition (800-830°C) enstatite and cristobalite formation (1000-1300°C)	800 900-1000	(-) (+)	- -

Thermal Investigations in Technical Mineralogy

APPLICATION OF THERMAL METHODS IN RAW MATERIAL CONTROL AND DURING THE PRODUCTION PROCESS

S. Starck

Oerlikon-Schweißtechnik
Eisenberg/Pfalz, Germany

Abstract

The paper deals with rapid water detection in welding fluxes by means of a CWA equipment (carbon/water analyzer). Water or carbon dioxide released by heating from minerals or technical products is detected by means of IR absorption spectroscopy. The influence of grain size on the water content of minerals is discussed. The given examples show that the water release temperatures are more important than the total water contents in respect to the possible use of the raw materials.

Introduction

Water incorporated in basic coated electrodes or welding flux has a negative influence on the welding properties of high strength fine-grained steel or stainless steel. In stainless steel water increases or generates porosity, in fine-grained steel water dissociates in the electric arc during the welding process and generates diffusible hydrogen (Hd) which is responsible for hydrogen induced embrittlement of the weld.

Water therefore is a dangerous factor in welding. Techniques detecting water in raw materials and products are therefore very important.

Determination of Hd is complicated, expensive and takes 72 hours. That is why Hd testing is unsuitable for immediate control of the production process because results can only be expected four days later! The amount of Hd in weld metal is about 4 ml per 100 g of deposited metal, which is about 1.6 ppm! Therefore we tried to establish a correlation between Hd and the water contents of welding flux.

Water detection

The instrument we use for detection is a RC-412 made by Leco. It heats the sample to a pre-determined temperature. Heating rate and steps during heating can be programmed. The maximum temperature which can be

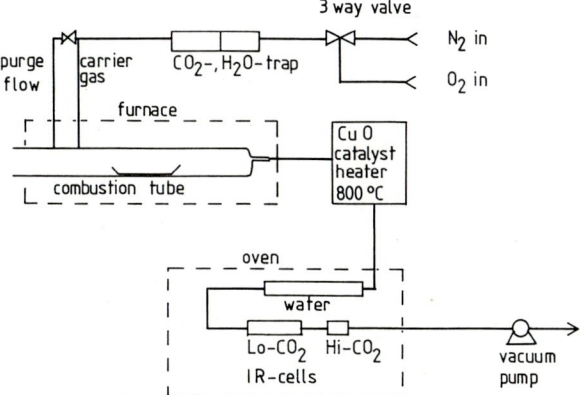

Fig. 1: Simplified flow-diagram of RC-412.

achieved is 1200°C; however, we usually work at 950°C or 1150°C depending on whether we check the water contents of flux or test a new raw material. The water or carbon dioxide contents of the evolved gas are detected by means of infrared absorption spectroscopy. Oxygen or nitrogen can be used as the carrier gas. This makes possible the distinction between inorganic carbon fixed to carbonates which is released when oxygen or nitrogen are used, and organic components and graphite which only decompose in an excess of oxygen. A simplified flow-diagram of this instrument is shown in Fig. 1. Figure 2 shows the water release plot of a calcined muscovite A which is suitable for welding electrodes. The water release plot of calcined muscovite B, given in Fig. 3, shows that this muscovite still contains water evolving at 950°C. Muscovite B is not suitable for electrode production although the total water contents of muscovite A or B are nearly the same.

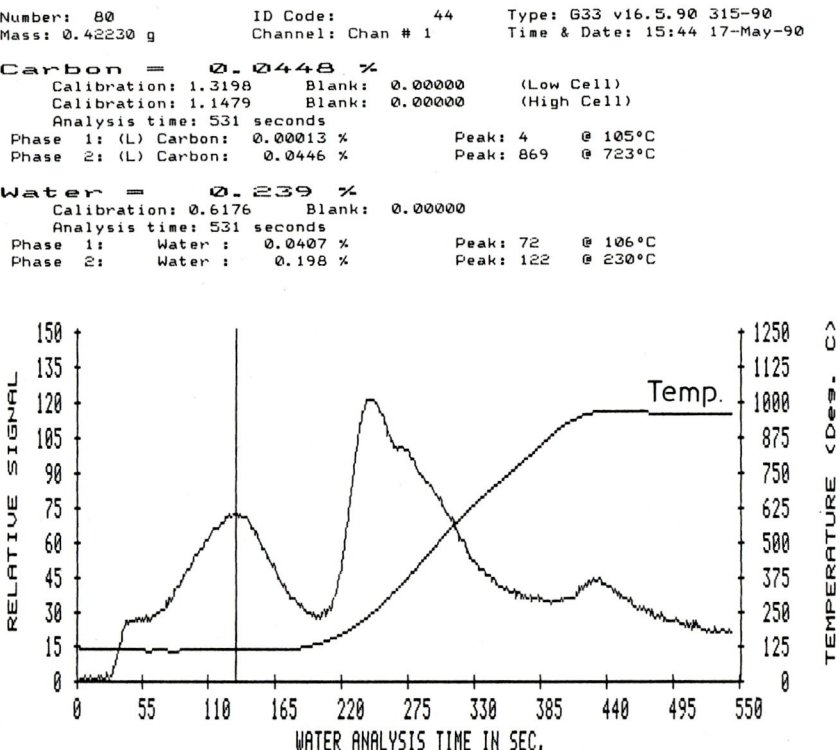

Fig. 2: Calcined muscovite A. Only small amounts of water are released above 900°C and rehydrated moisture can be seen from the water release plot. This material is suitable for electrodes which are affected by H_2O.

Application of Thermal Methods in Raw Material Control and During the Production ...

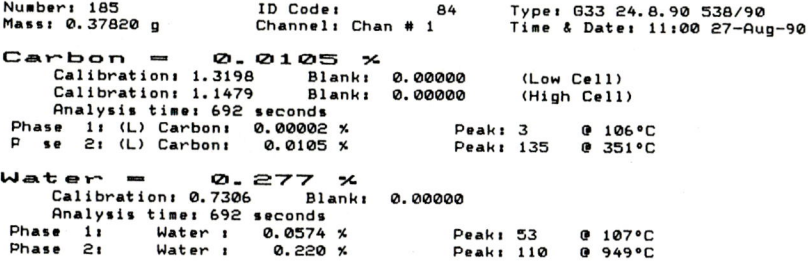

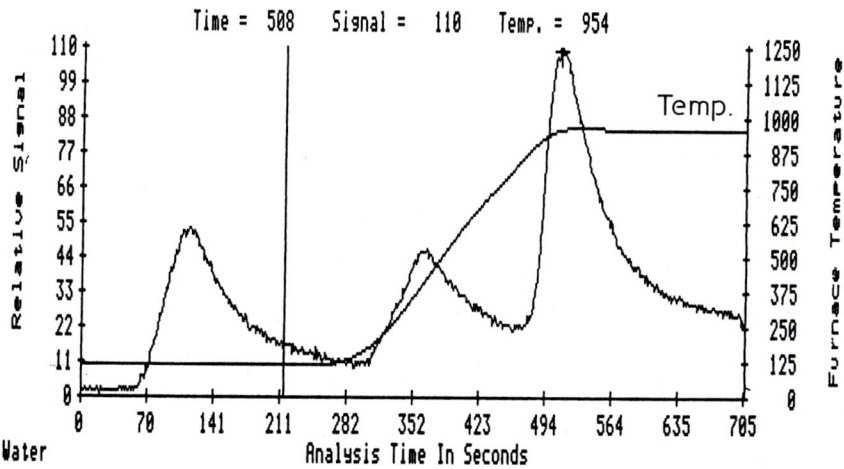

Fig. 3: Calcined muscovite B. This muscovite was calcined, too, but the temperature used was not sufficient. It still contains water evolving at 950°C.

Problems of Water Detection

Metals

Metals may cause reactions like Me + H_2O → MeO + H_2 if the amount of oxygen is not sufficient.

The IR-detectors are unable to detect the evolved H_2; therefore less water will be observed than actually is in the sample.

The same problem occurs if a thermo-gravimetric method for detection is used. Only the loss of hydrogen will be detected.

We had that problem with the former instrument we used which was unsuitable for oxygen (as a carrier gas) and had no step between furnace and IR-detector to convert H_2 back to H_2O by oxidation.

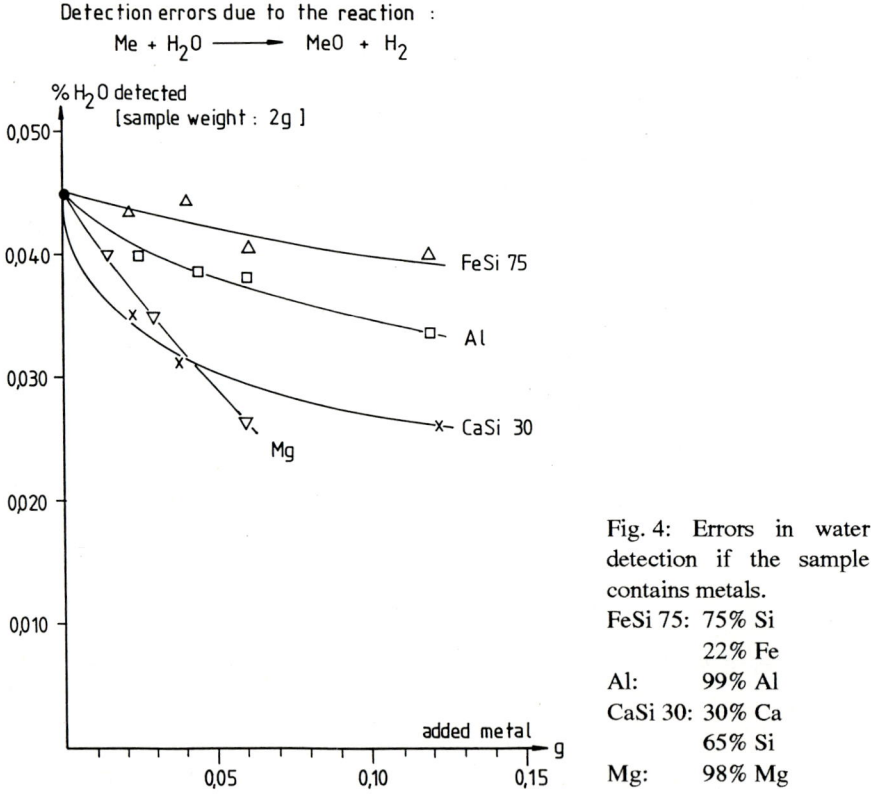

Fig. 4: Errors in water detection if the sample contains metals.
FeSi 75: 75% Si, 22% Fe
Al: 99% Al
CaSi 30: 30% Ca, 65% Si
Mg: 98% Mg

Figure 4 shows how different metals lead to the detection of less water than actually is in the sample, if there is not enough oxygen present or if there are no hydrogen-to-water conversion steps like those taking place in the catalyst heater of the RC-412 where copper-oxide is reduced to copper.

Grain size of the sample

The mean water contents of our welding flux is approximately 0.07% H_2O and the mean grain size range is 0.5-1.0 mm. The problem is to decide whether it is better to grind the sample and risk a moisture pick-up or not to grind it and risk an insufficient detection of water due to a temperature discrepancy inside and outside each grain and the short period of time available for water release.

We solve this problem by comparing the amount of water detected before and after grinding to the diffusible hydrogen (Hd) in the weld (Figs. 5 and 6).

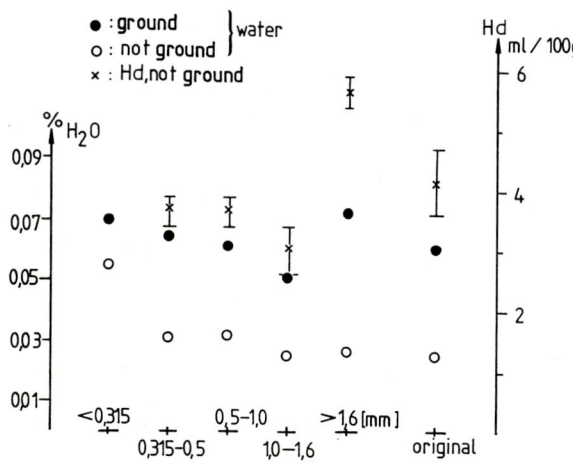

Fig. 5: Water and diffusible hydrogen detected in different grain sizes of flux.

Discussion of Figure 5:

1. Flux with a grain size smaller than 0.3 mm has relatively high water contents which can be detected if ground or not. (Here no hydrogen test was performed due to lack of sample).

2. High water contents of coarse flux can only be detected when the flux is ground. Higher hydrogen contents in the weld can be observed as well.

3. It is obvious that there should be a better correlation of hydrogen to water detected in ground flux.

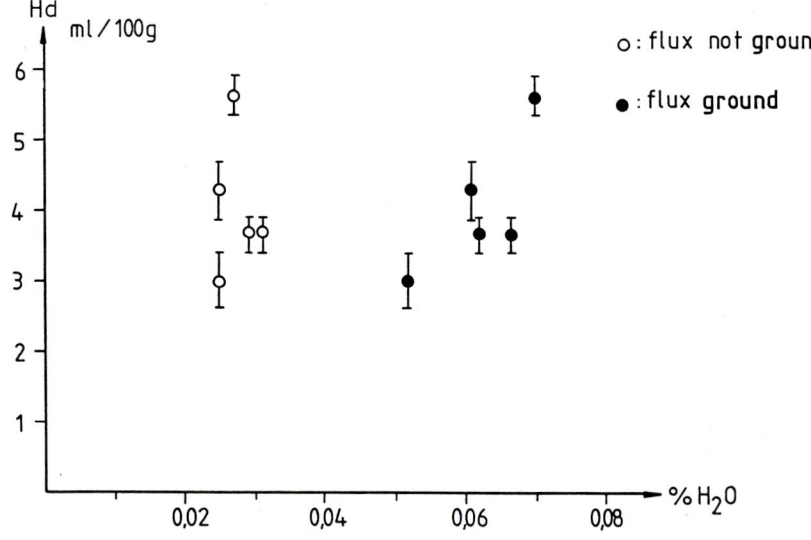

Fig. 6: Correlation of water detected in ground flux and not ground flux to diffusible hydrogen in the weld metal.

Before grinding less moisture is detected and there is no correlation, after grinding more water is detected and there is a good correlation. There should be an even better correlation when not using different grain sizes (Fig. 7).

Conclusion

1. Thermal control of raw materials helps to decide whether an ingredient is suitable for welding electrodes or fluxes. The water release temperatures of the raw material are even more important than the total water contents. These temperatures should be lower than the baking temperature of electrode or flux. If they are higher, water escapes during welding, which causes porosity or diffusible hydrogen. Both are very dangerous for the weld.

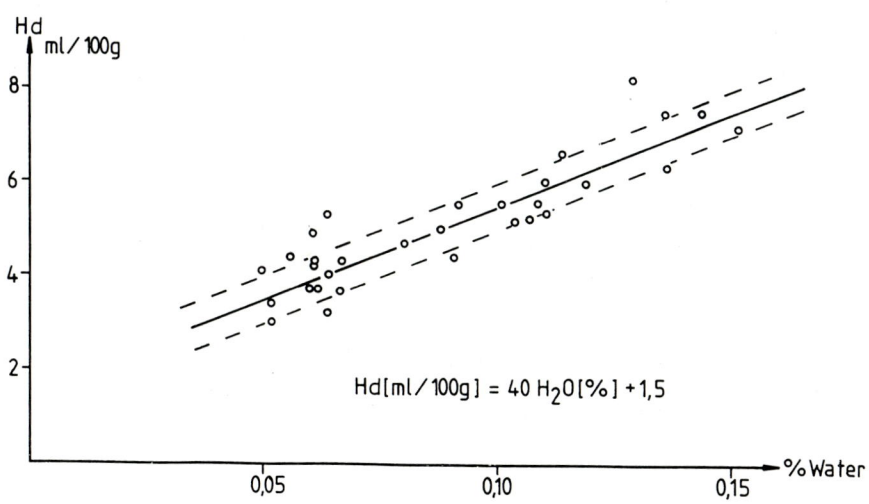

Fig. 7: Correlation of deposit weld metal to the water contents of a special welding flux. The dotted lines are drawn in a distance of ±0.5 ml diffusible hydrogen.

2. Knowing the correlation between water contents of a flux and diffusible hydrogen produced by welding with it, we now have the ability to adjust production parameters whilst production is still running.

Both steps help us to lower production costs by decreasing waste, and increasing quality control.

Clay Mineralogy and Applied Geology

A STUDY OF THE THERMAL BEHAVIOUR AND GEOTECHNICAL PROPERTIES OF A MARINE CLAY AND ITS COMPOSITES

E. T. Stepkowska[1], Z. Sułek[1], J. L. Perez-Rodriguez[2], C. Maqueda[3] & A. Justo[3]

[1] Institute of Hydroengineering, Polish Academy of Sciences
Gdansk, Poland

[2] Instituto de Ciencias de Materiales de Sevilla
Consejo Superior de Investigaciones Cientificas, CSIC
Sevilla, Spain

[3] Instituto de Recursos Naturales
Consejo Superior de Investigaciones Cientificas, CSIC
Sevilla, Spain

Abstract

A change in geotechnical properties with the addition of fly ash was observed in a marine clay, which was intended for use in earth embankment reconstruction. This was correlated with the drying rate and weight loss on heating. Composites with the optimum fly ash content (6% to 8%) also indicated the highest drying rate $\Delta G/\Delta t$ for $\Delta t = 4$ h to 1 d. Increase in composite strength was due to aggregation of the material and to the formation of a cementing compound from the amorphous matter present as shown by SEM. This resulted in an increase in weight loss, on heating, of the composite as compared to the parent material. Thus the study of the drying rate $\Delta G/\Delta t$ and weight loss on heating gives some information on microstructure and strength of the system, which were checked by XRD and SEM.

Introduction

This paper resulted from a study on the improvement of the geotechnical properties of a local marine clay, to be used for reconstruction of earth embankments. Such a material should have both restricted swelling, shrinkage and appropriate strength properties, which may be correlated with microstructure and thermal behaviour.

Various additives, indicated in the literature, were tried during this study, such as calcium hydroxide, cement, fly ash, water glass, sodium carbonate, calcium chloride, phosphogypsum, and the results will be presented in a separate paper. As the most suitable, the composite with added fly ash was selected for the detailed study and for design of the optimum mixture composition, which will be discussed below.

In previous research work on water sorption and thermogravimetry of a bentonite powder and of the drying rate of some bentonite suspensions it was observed, that these parameters were dependent on sample pretreatment and that from these data some information may be obtained concerning the microstructure of the system and thus its mechanical properties, i.e.:

1. Berkbent bentonite powder (and its parent material from Woburn, Poland) stored at increasing relative humidity (p/p_0 = 0.5; 0.95 and 1.0), during the water sorption measurement, indicated a smaller weight loss ΔG (110°C to 800°C), than the untreated powder, whereas, storage at decreasing relative humidity (p/p_0 = 1.0; 0.95 and 0.5), during the water retention measurement, resulted in a higher ΔG (110°C to 800°C), Fig. 1. Simultaneously in the first case (water sorption) the powder formed macroscopically a loose aggregated structure, whereas separate grains were composed mostly of parallel particles (SEM). In the second case (water retention) the macroscopic structure was quite dense, whereas in separate grains the particles formed a loose floc structure. This seemed to be the reason for an easy

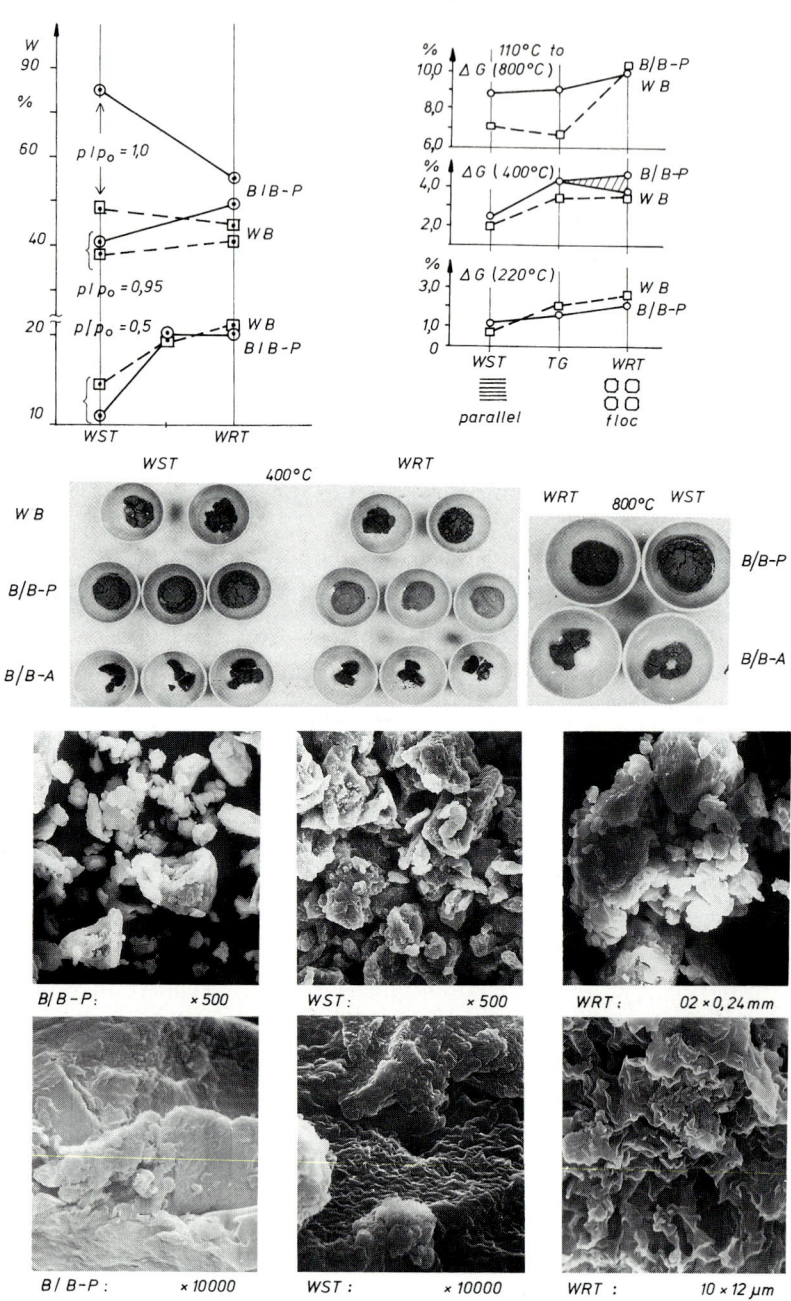

← Fig. 1: (a) Water sorption (WST) and water retention (WRT) of Berkbent bentonite powder (B/B-P), and of its parent material from Woburn (WB).
(b) Weight loss on static heating after these treatments and of the untreated powder (TG).
(c) Macrostructure and microstructure of B/B-P after the WS test and the WR test (STEPKOWSKA & JEFFERIS, 1987).

and high escape of water, whether in dehydration of interparticle and interlayer spaces, or in dehydroxylation of the crystal lattice (STEPKOWSKA & JEFFERIS, 1987);

2. change in weight loss, on static heating, may be due to phase transformation dependent on microstructure, thus caused by sample pretreatment (paragonite formation from parallel structure, feldspar formation from cluster and zeolite formation from floc structure, see STEPKOWSKA et al., 1988);

3. stirring of the Berkbent bentonite suspension at a high energy (5000 rpm, 15 min) resulted, after drying, in an increased water sorption, which is inversely proportional to the particle thickness δ, thus indicating particle delamination. This was also measured by XRD: $\delta = 6.2$ unit layers per particle in unstirred suspension and $\delta = 5.3$ unit layers per particle in stirred suspension (STEPKOWSKA & JEFFERIS, 1983). Also the weight loss on static heating was dependent on stirring energy (Table 1);

4. bentonite suspension formed stepwise aggregates, which increased in size and in shape regularity with the time of storage (ageing). The bigger the aggregates, the greater are the attractive forces between them and thus the higher is the strength of the system. This increased attraction also results in an increased drying rate of the suspension, which was clearly observed after prolonged storage (STEPKOWSKA & JEFFERIS, 1982; Fig. 2);

Table 1: Weight loss on static heating -ΔG(110°C to 800°C) of Berkbent bentonite (B/B) slurry stored for 4 months, dried at 45°C and heated.

Sample	Preparation of suspension	-ΔG(110°C to 800°C)
B/B-5	4%, 2000 rpm, 5 min	6.82%
B/B-6	5%, 2000 rpm, 5 min	7.16%
B/B-7	6%, 2000 rpm, 5 min	6.36%
B/B-8	5%, 7000 rpm, 5 min	7.31%
B/B powder	untreated	4.58%

5. storage of a clay at a high water content sometimes also results in particle delamination, i.e. in a decrease of the average particle thickness and an increase in water sorption of clay particles, but this process may be reversible. In a clay stored for one month, at a water content close to the liquid limit, the average particle thickness was changed from $\delta = 14 - 16$ to $\delta = 10 - 11$ unit layers per particle (Grimmen clay, Germany; $w_l = 80\%$, STEPKOWSKA & STÖRR, 1986), or from $\delta = 23$ to $\delta = 13$ unit layers per particle (Zebiec bentonite, Poland; $w_l = 220\%$, STEPKOWSKA & OLCHAWA, 1987), followed by a collapse to the initial value after prolonged storage;

6. consolidation pressure (overburden pressure) and negative pore water pressure (suction) may cause particle collapse (see STEPKOWSKA & POZZUOLI, 1990). The particle thickness is correlated with the size of aggregates.

Thus a marine clay, which has been suspended in water for a prolonged time, may indicate a small particle thickness (a high water sorption), which varies dependent on storage conditions and on the overburden pressure. Thus after prolonged storage in a dry state (one year) the water sorption of a marine clay dropped from $WS(p/p_0 = 0.5) = 8.3\%$ to 4.7% and from $WS(p/p_0 = 0.95) = 21.3\%$ to 13.7%, thus by ca. 0.4.

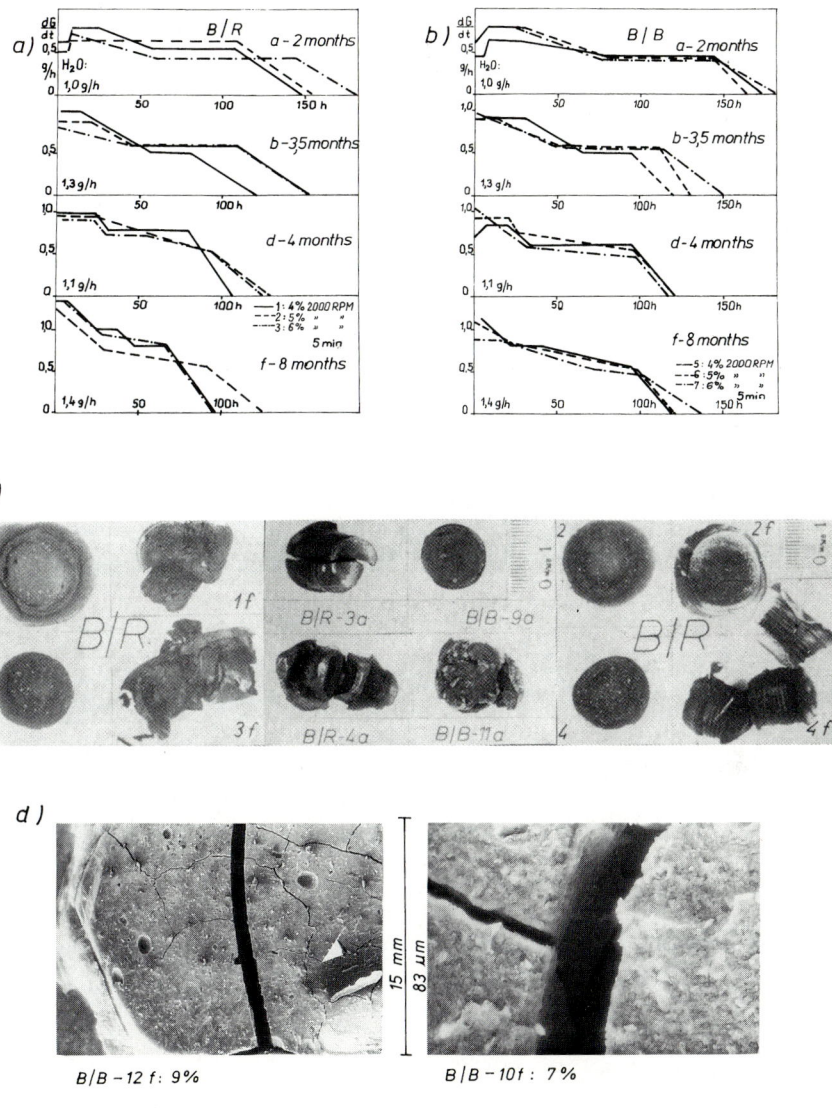

Fig. 2: Increase in drying rate with time of storage of Brebent (B/R) and Berkbent (B/B) bentonite suspensions (a) and (b), STEPKOWSKA & JEFFERIS (1982).

Remolding of a marine clay and subjecting it to freeze-thaw cycles caused some change in the water sorption, particle thickness and thermal behaviour (Table 2; STEPKOWSKA, 1988). Thus the water sorption decreased with the cyclic freezing (most probably due to particle collapse in parallel structure, which was formed by the influence of cryosuction) and this change was the higher, the smaller the sampling depth was, i.e. the less the clay was influenced by the overburden pressure and the more it was able to form a particle arrangement. At a greater depth, exceeding 5.4 m, the overburden pressure might have caused cluster and aggregate formation, which prevented the extensive development of the parallel particle arrangement.

Table 2: Water sorption and weight loss on static heating of the marine clay: undisturbed (U), remolded (R) and subject to ten freezing (-20°C) - unfreezing cycles (F).

Nr.	Sampling depth [m]	State	W [%] at p/p_0= 0.5	W [%] at p/p_0= 0.95	$-\Delta G$(110°C to 800°C)
1	3.6-3.7	U	2.9 ± 0.2	15.2 ± 0.5	10.8 ± 0.2
		R	2.5 ± 0.2	16.6 ± 1.3	11.6 ± 0.7
		F	1.6 ± 0.3	10.0 ± 1.6	15.4 ± 0.8
2	5.4-5.5	U	1.7 ± 0.1	12.8 ± 0.8	14.3 ± 0.4
		R	1.8 ± 0.2	11.8 ± 1.0	14.5 ± 0.5
		F	1.4 ± 0.2	10.5 ± 0.3	15.1 ± 1.5
3	5.5-5.6	U	1.8 ± 0.2	13.0 ± 1.5	15.0 ± 0.04
		R	2.0 ± 0.3	13.2 ± 0.5	15.0 ± 0.3
		F	1.9 ± 0.2	11.3 ± 1.9	11.6 ± 0.9

Also the weight loss ΔG (110°C to 800°C) increases with cyclic freezing, except in samples from the greatest depth. The comparison of the values, indicated in Table 2, gives the impression, that after the cyclic freezing the dehydration is removed towards higher temperatures. This may be due to parallel particle arrangement, formed around the ice lenses due to cryosuction and to the pressure exerted by the forming ice crystals of increased specific volume.

In this case the hydrating water has some difficulty in escaping both from the external surfaces of clay particles in parallel arrangement and from the interlayer spaces, due to a long diffusion path. Also the temperature of dehydroxylation may be increased because of the same reason.

The aggregated clay from a greater depth has a lower tendency to parallel particle arrangement, thus the diffusion path of the hydrating water is smaller and dehydration is not moved towards higher temperatures. The water sorption is lower, and the particle thickness is larger than that at the lesser depth (3.6 m). The decreased weight loss on heating of the 5.5 m specimen may be due to phase transformation, e.g. formation of feldspar during the heating of cluster structures, which was indeed observed in bentonite slurries mentioned above.

Thus the study of water sorption, thermal behaviour and drying rate may give interesting information on the microstructure and other properties, provided that the microstructure is not disturbed by sample preparation, as is usually the case in standard methods, derived to obtain reproducible results. Here rather the variation of test results is of interest.

Materials

Marine clay from Gronowo Dolne, Poland, taken from the depth 0.5-1.4 m, indicated:

    plastic limit, W_p = 30% (55.6% after half a year storage),

    liquid limit, W_1 = 81.2% (84.7% after half a year storage),

    water content, W_n = 68% (54% after half a year storage).

Water sorption and weight loss on static heating are shown in Tables 3 and 4.

Table 3: Water sorption and water retention of the marine clay from Gronowo Dolne, (mean of three specimens).

Test	W_0 %	W [%] at p/p_0= 0.5	0.95
June '88			
WS from 110°C	67.4	3.27	7.7
WR from W_0	65.1	4.48	12.4
January '89			
WS from 110°C	58.2	3.88	8.4
WS from W_0	55.9	4.8	9.2
WR from W_0	52.0	4.8	12.7

This marine clay (see below, Figs. 3a and 3b) was shown by XRD and SEM to be composed of clay minerals (illite, montmorillonite, chlorite, partly in mixed layers and kaolinite). Non-clay minerals were also detected (quartz, feldspar, calcite and gypsum), and organic matter is present.

Amorphous materials, most probably allophane, is present in appreciable amount, giving a high weight loss within the temperature range 110°C to 220°C and a background of an appreciable intensity on the diffractograms.

Fly ash, E, was taken from the local power plant in Elblag, Poland, and it indicated the following chemical composition: SiO_2 54.6%, Fe_2O_3 6.1%, Al_2O_3 20.7%, CaO 3.9%, MgO 2.8%, SO_3 0.5%, P_2O_5 0.22%, Na_2O 2.6%, K_2O 2.6%.

Its water content was 0.56%, content of water soluble material 2.04%, HCl soluble material 14.9%, organic matter 9.81%, and accessible CaO soluble in HCl 1.5% to 2.0%.

Table 4: Weight loss in static heating, ΔG, after the given treatment (mean of three specimens).

Test	W_0 %	$-\Delta G$ in % from 110°C to		
		220°C 8 h	400°C 4 h	800°C 1 h
June '88				
WS from 110°C	67.4	3.41	3.98	7.29
WR from W_0	65.1	3.34	4.61	7.77
TG untreated	68.5	2.88	5.13	8.69
January '89				
WS from 110°C	58.2	4.2	6.6	12.4
WS from W_0	55.9	5.2	7.0	12.2
WR from W_0	52.0	4.2	4.9	9.8
TG untreated	50.8	7.8	8.0	14.9
September '90				
TG air dry	3.47	3.69	5.84	10.69

Some DTA and SEM work was done on samples with the addition of the fly ash from Konin powerstation (K), which has a higher content of organic matter (16.9%) and CaO (15.9%).

Methods

Marine clay with the approximate initial water content ($W_0 = 64.5\%$) was mixed with the following amounts of fly ash E: (a) 2%, (b) 4%, (c) 6%, (d) 8%, and (e) 10%, related to the dry basic material.

Properties of the composites were determined either after 1 day or 28 days storage in plastic bags in a desiccator containing distilled water:

1. Strength, τ_f, was measured by a fall cone (angle $2\alpha = 30°$, weight $P = 80$ g or 100 g) calculating strength from the penetration depth, h, as:

$$\tau_f = K \cdot (P / h^2) \qquad \text{in kPa}$$

assuming $K = 1.00$

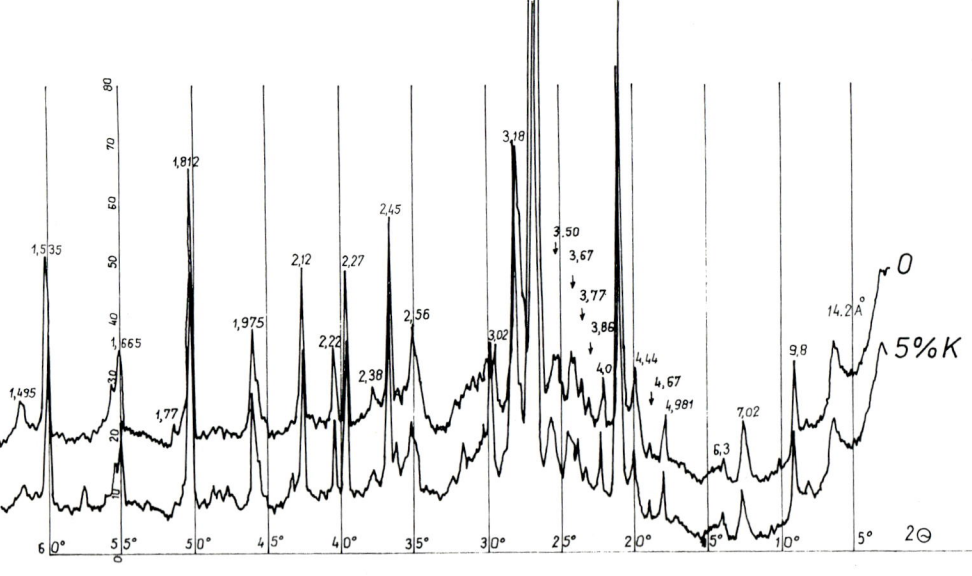

Fig. 3a: XRD patterns of the marine clay (O) and the fly ash K composite (5% K).

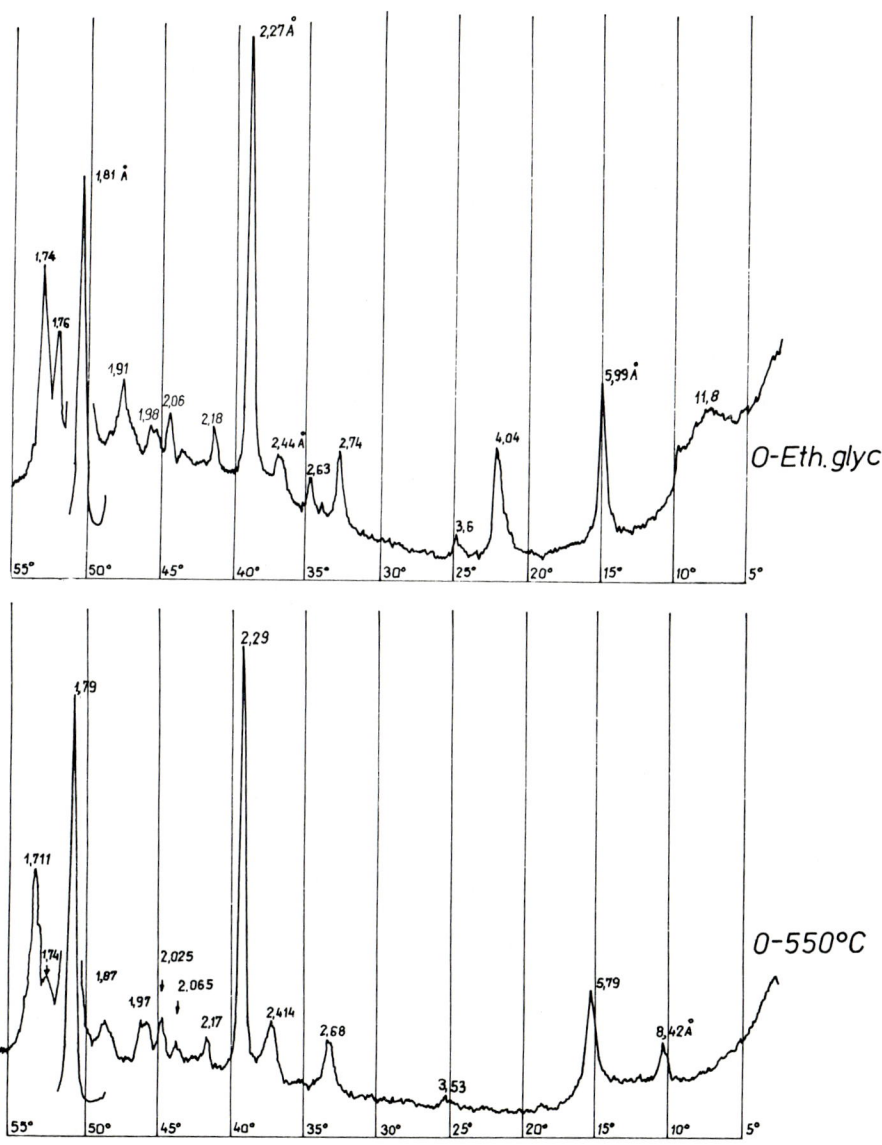

Fig. 3b: XRD patterns of the marine clay with ethylene glycol and the marine clay heated at 550°C.

2. Drying rate $\Delta G/\Delta t$ at 30°C of the cubes ca. $2.7 \cdot 2.7 \cdot 2.7$ cm^3 ≈ 20 cm^3 (series I) or ca. $3.0 \cdot 3.0 \cdot 3.0$ cm^3 ≈ 27 cm^3 (series II) whose dimension were measured exactly on particular specimens before and after the drying procedure. Sample weight decreased during 6 days of drying, followed by its oscillation ±20 mg about the equilibrium value. Series I was prepared at the indicated additions of fly ash E. In series II the influence of mixing was estimated, thus the specimens were prepared with the addition of 6% and 8% fly ash E, either by (a) thorough mixing or (b) by sandwiching the fly ash between three layers of marine clay.

3. Free swelling, shrinkage in drying and secondary swelling were determined in the Wasiliew apparatus (Fig. 4).

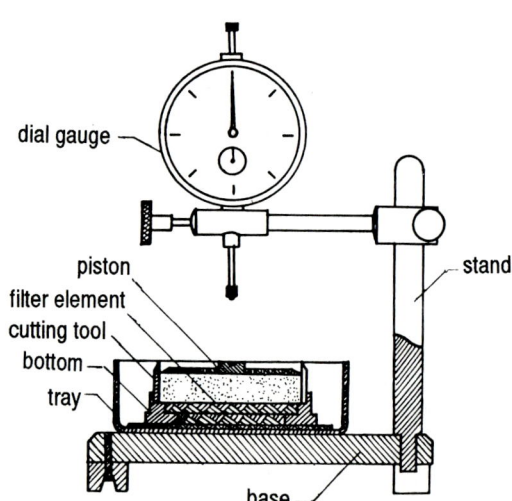

Fig. 4: Wasiliew apparatus

4. Disintegration in water was estimated on cylindrical samples, $\emptyset$ = 5.5 cm, h = 2.5 cm, stored for 12 days and placed in tap water.

5. Thermogravimetric behaviour was measured either on composites with added fly ash E or fly ash K, using either static heating or DTA

(1500 QD, MOM, Budapest, sample weight 250 mg, 10°C/min, sensitivity 100 mg).

6. XRD patterns were taken of the parent marine clay, before and after ethylene glycol addition and after heat treatment at 550°C. The 5% composite with fly ash K was studied as well.

7. SEM studies were made of the parent marine clay and the composite with 5% fly ash K.

Results

The results of the geotechnical tests and of drying rate will be discussed here in relation to the content of fly ash E.

The strength of the composites, after both 1 day and 28 days of storage, increased with the increase in fly ash up to 8% followed by a slight decrease at 10% fly ash (Fig. 5).

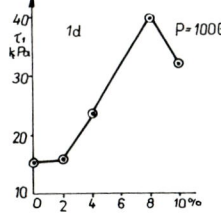
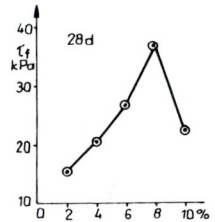

Fig. 5: Strength of the composites as a function of the fly ash E addition.

In the Figure 6 there is presented the drying rate within the given time interval Δt, as the function of fly ash addition. Initially the highest drying rate is indicated by the 4% or 6% composite. The most informative is $\Delta t = 4$ h to 1 d, when the highest drying rate, thus the highest attractive force is indicated by the 6% composite, the drying rate of which later decrease, as the water content becomes low.

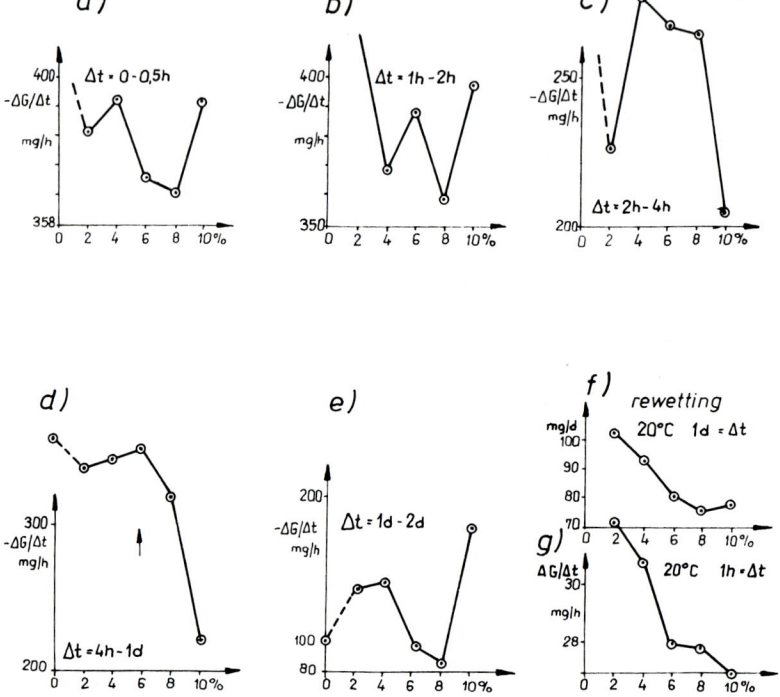

Fig. 6: The influence of fly ash E addition and of time on the drying rate at 30°C of the composites.

The presence of fly ash results in a decrease of the drying rate with respect to the parent material (Fig. 7a). The influence of mixing on the drying rate is of little importance, except at the beginning and at the end of the drying procedure (Figs. 7b and 7c).

Other parameters measured in the drying test indicate as well, that the 6% composite is the optimal one (Fig. 8). The shrinkage $-\Delta V/V_0$ (a), and the dry density γ_d (b) are the smallest. The shrinkage limit, W_s, is the highest (c) and the water content after 1 day drying is almost the smallest (d).

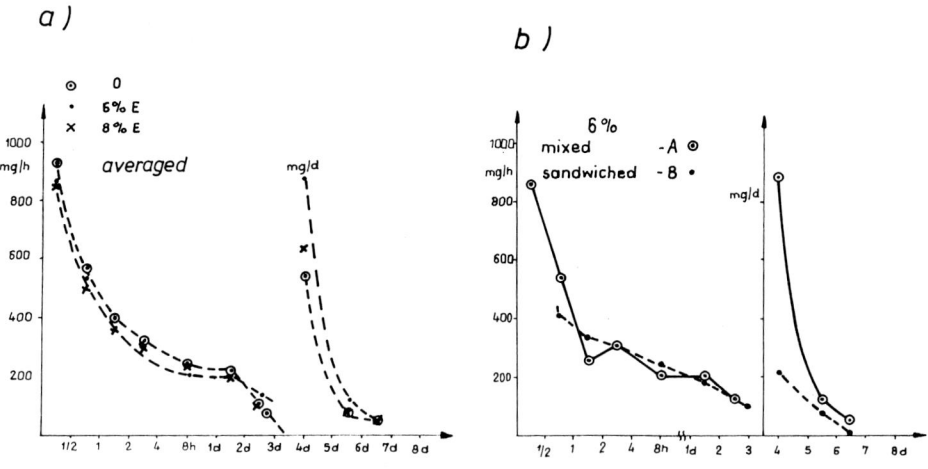

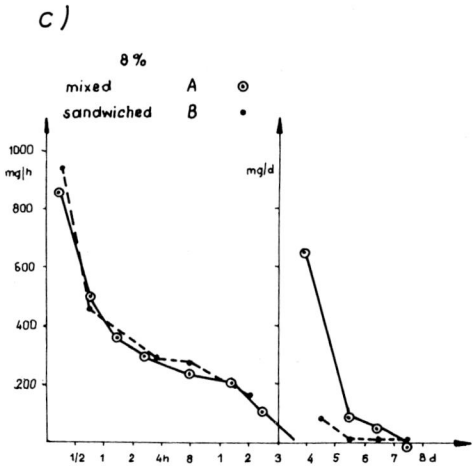

Fig. 7: The influence of the mixing procedure on the drying rate of the fly ash E composites (mean of three tests).

In Figure 9a, b and c there is presented the water content of the composites, dried at 30°C for various times, and rewetted at 20°C under laboratory conditions. Also the swelling, shrinkage and secondary swelling are presented in Fig. 9d, e and f which indicate some minima for the 6% composite.

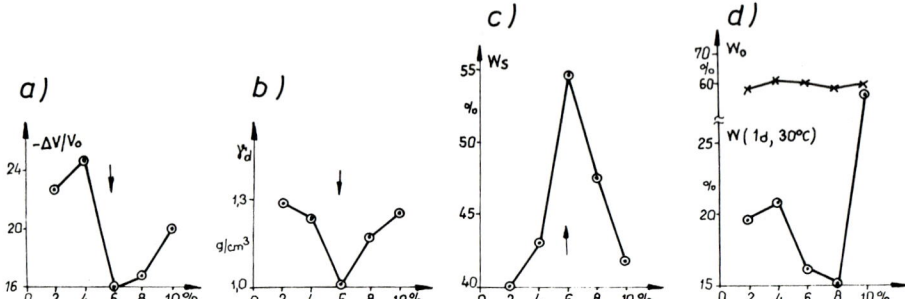

Fig. 8: The influence of the fly ash E addition on the shrinkage (a), dry density (b), shrinkage limit (c) and water content after one day drying (d).

Neither of the samples under study disintegrated in water, but the untreated material, the 2% and the 8% composites cracked after 1 day, 9 days and 1 day respectively.

In Figure 10 DTA curves and TG curves of the untreated material and of the 5% composite with fly ash K are compared. The addition of fly ash moved the first dehydration minimum from 115°C in the parent material to 105°C in the composite, which indicates a somewhat easier dehydration.

The exothermic maximum of organic matter disintegration was moved from 318°C to 325°C. Two other dehydration minima are slightly discernible at ca. 380°C and ca. 450°C. An endothermic dehydroxylation minimum is present at ca. 560°C and the endothermic minimum at ca. 770°C corresponds most probably to carbonate decomposition.

From the TG curve it may be seen, that the total weight loss ΔG (20°C to 1000°C) is higher in the 5% composite (13.4%) than in the untreated air dry parent material (11.6%). The difference of 1.8% occurs mainly within the temperature range 230°C to 700°C and it cannot be attributed to the organic matter content in the fly ash (16.9% · 0.05 = 0.84%).

The comparison of the XRD patterns of the parent marine clay (O) and of the 5% composite with fly ash K addition, indicates the increase in crystallinity of the system: the intensity of the background is much higher in the O-specimen, whereas the relative intensity of some peaks is increased in the 5% composite. Formation of new phases could not be detected by XRD.

Drying test.

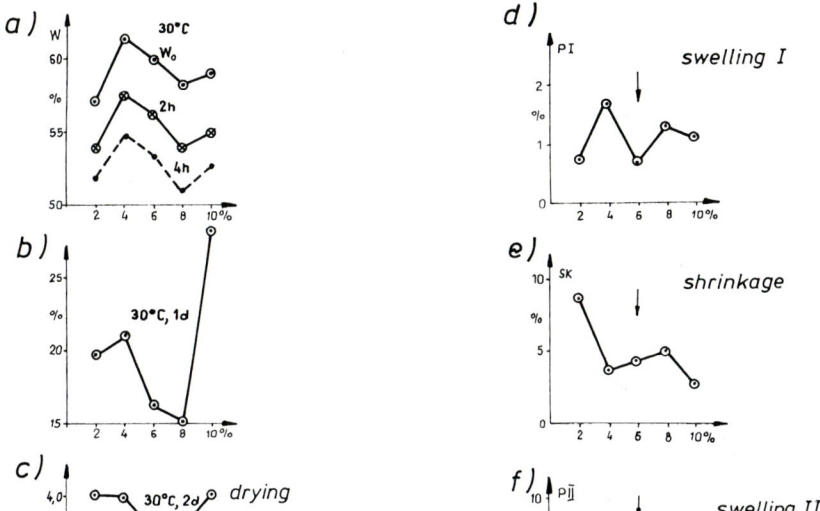

Fig. 9: Water content of the composites during drying (a, b, c) and the influence of the fly ash E addition on the swelling on wetting (d), shrinkage on drying (e), and secondary swelling on wetting (f).

Ethylene glycol caused the displacement of smectite peaks and the formation of a broad band at $2\Theta = 5°$ to $10°$. Some peaks disappeared or their intensity decreased (e.g. quartz at 4.25 Å), and some of the accessory minerals showed an increased peak intensity. Heating of the marine clay at

550°C caused the disappearance of the kaolinite 7 Å peak, but the higher order 3.53 Å chlorite peak remained at a low intensity. The peak at 5.8 Å indicates the possibility of interstratification of montmorillonite/chlorite.

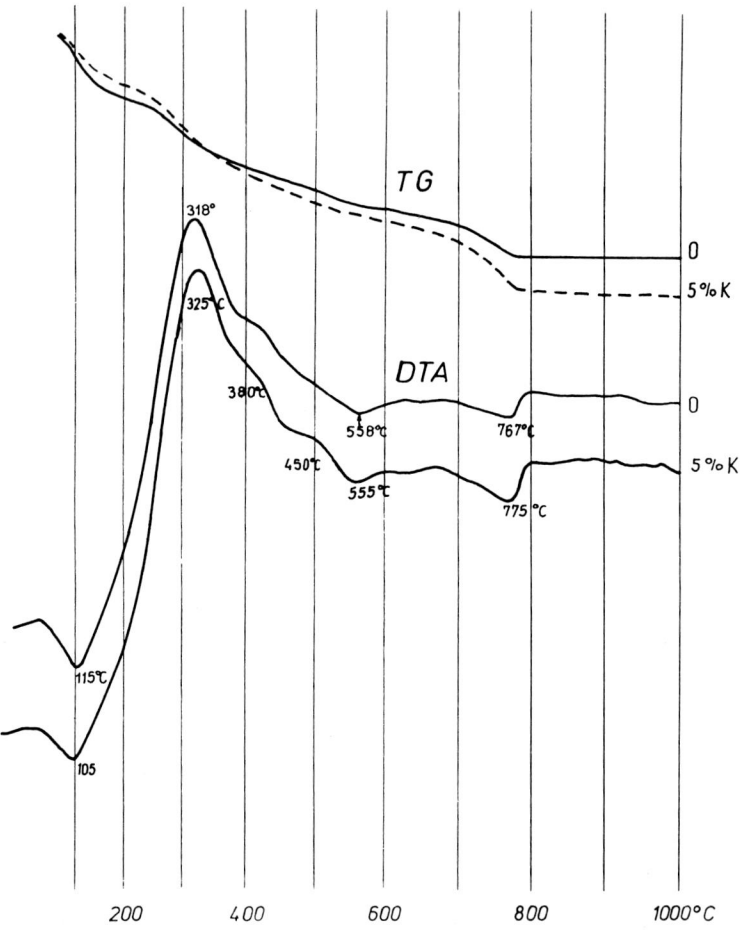

Fig. 10: DTA and TG curves of the parent marine clay and the composite with 5% fly ash K added.

SEM studies indicate the presence of clay particles, accessory minerals and some flora (diatoms) in the parent material (Figs. 11a to c). The addition of fly ash caused a visible aggregation (Fig. 11d) and an increase in size, thickness and shape regularity of the crystals in the form of thin rigid plates (Fig. 11e). The grains of fly ash of a regular round shape but indicating some delamination, are visible in the clay matrix (Fig. 11f). The change in strength properties is most probably due to the active CaO and the formation of plate-like or foil-like cementing material, similar to that observed by WILD et al. (1986) in a marl + 10% lime composite, cured for 24 weeks at 75°C (their Fig. 11). This cementing component forms most probably from the amorphous material, increasing the strength of the composite, while its weight loss on heating and results in the decreased background intensity of the XRD patterns. This cementation causes also the aggregation of the material, which increases its strength (Fig. 11d).

Conclusion

The geotechnical properties of the marine clay are improved by the addition of 5% to 8% of the fly ash, 6% being the optimum: the strength is higher and the swelling and shrinkage are lower than those of the parent material.

Weight loss on heating at 220°C to 700°C increases due to fly ash addition. This is most probably caused by the formation of some compounds from amorphous matter, present in the marine clay, with the accessible CaO in the fly ash. WILD et al. (1986) suggest this to be an amorphous calcium silicate aluminate hydrate gel. This process in the marine clay under study causes a decrease in background intensity of the XRD patterns thus the new compound formed may be not amorphous. The presence of this cementing material results in aggregation of the composite, an improvement in its strength and other geotechnical properties.

The change in properties may be studied by the measurement of the drying rate: it is decreased in composites as compared to the parent marine clay, as

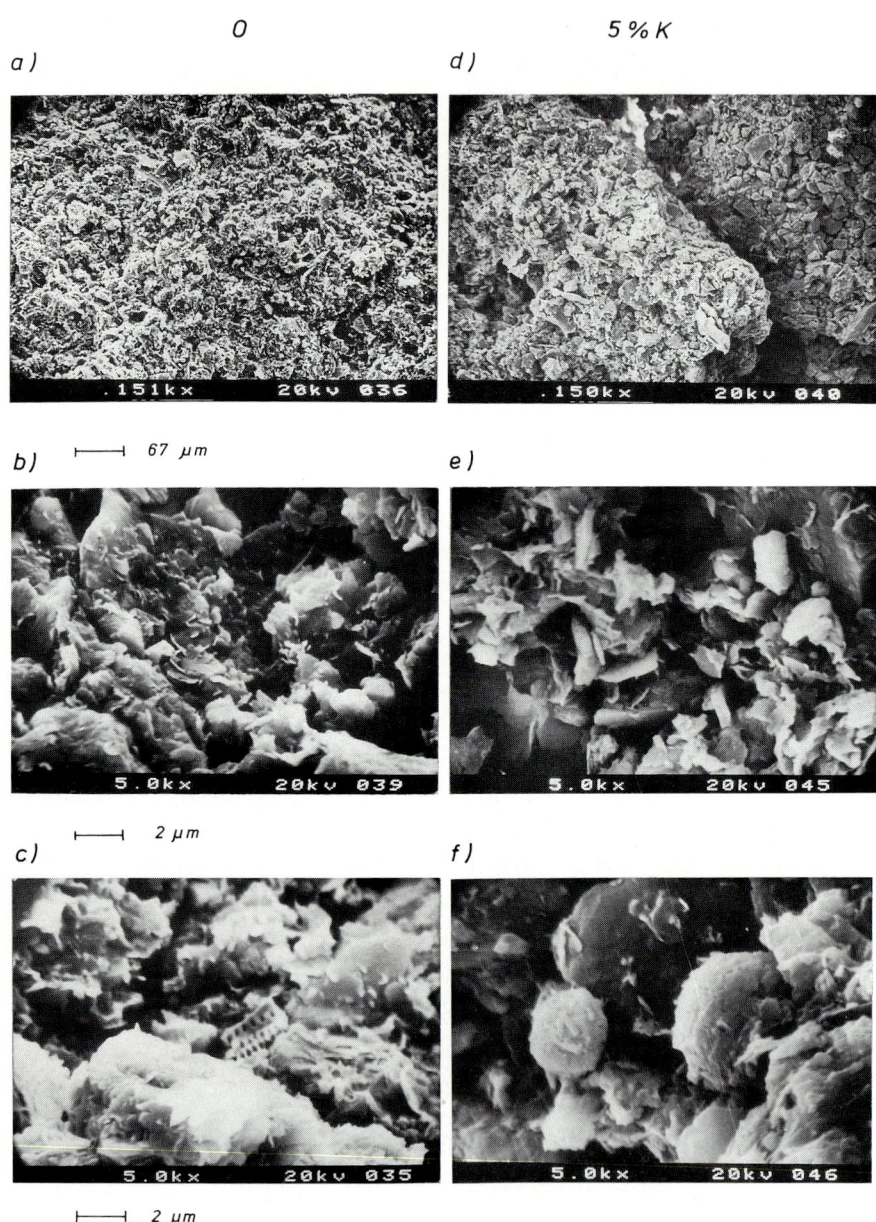

← Fig. 11: SEM photographs of the marine clay (a), (b) and (c) and of the composite with 5% fly ash K added (d), (e) and (f).

their aggregates are stronger and less deformable than those in the parent material. Drying rate of the composites increased with the size of aggregates and with the magnitude of attractive forces between them. Thus from the drying rate the quality of the composite may be inferred.

References

STEPKOWSKA E. T. (1988) - Influence of the freezing-unfreezing cycles on the thermal behaviour of a marine clay.- Thermochimica Acta, 135, 313-318

STEPKOWSKA E. T. & JEFFERIS S. A. (1982) - Various types of microstructure in smectite and their influence on drying behaviour.- 9th Conference on Clay Mineralogy and Petrology, Zvolen (Czechoslovakia), (KONTA J. (ed.)), 35-42

STEPKOWSKA E. T. & JEFFERIS S. A. (1983) - Study in microstructures of clay slurries.- Archiwum Hydrotechniki, 30, 193-211

STEPKOWSKA E. T. & JEFFERIS S. A. (1987) - Variability in water sorption and in thermogravimetry of bentonite.- Thermochimica Acta, 114, 179-186

STEPKOWSKA E. T. & OLCHAWA A. (1987) - Wpływ wilgotnosci, temperatury i czasu na grubosc czastki w bentonicie.- Prace Naukowe Instytutu Geotechniki Pol. Wrocławskiej, Nr 52, 137-142

STEPKOWSKA E. T., PEREZ-RODRIGUEZ J. L., JUSTO A., SANCHEZ-SOTO P. J. & JEFFERIS S. A. (1988) - Possibility of feldspar formation in bentonite suspensions during storage, drying and/or heating.- Thermochimica Acta, 135, 319-334

STEPKOWSKA E. T. & POZZUOLI A. (1990) - Microstructure and properties of clays.- Meeting of the Societa Italiana di Mineralogia e Petrologia, Bologna, (MORANDI N. (ed.)), in print.

STEPKOWSKA E. T. & STÖRR M. (1986) - Variability of water sorption by mechanical treatment in the clay from Grimmen, GDR.- 10th Conf. on Clay Mineralogy and Petrology, Ostrava, (KONTA J. (ed.)), 1988, 197-207

WILD S., ARABI M. & LENG-WARD G. (1986) - Soil-lime reaction and microstructural development at elevated temperature.- Clay Minerals, 21, 279-292

THERMAL ANALYSIS OF SELECTED ILLITE AND SMECTITE CLAY MINERALS. PART I.
ILLITE CLAY SPECIMENS

C. M. Earnest

Department of Chemistry, Berry College
Rome, Georgia (USA)

Abstract

The characterization of illite clay minerals by the use of the technique of differential thermal analysis (DTA), thermogravimetry (TG) and derivative thermogravimetry (DTG) is presented. This presentation is offered not only as a review of the thermal characteristics of this important group of clay materials but suggestions relative to the application of the thermal analysis techniques to contaminated illitic specimens; i.e., mineral mixtures, are included. Two commonly referenced illitic clay specimens, which have been widely distributed, were studied here. These were the American Petroleum Institute Reference Clay Specimen from Fithian, Illinois (API #35) and the Clay Mineral Society's Source Clay Specimen from Silver Hill, Montana (CMS-IMt).

These clay specimens were studied using a modern computerized differential thermal analyzer which also contained a "DSC" mode of operation for peak energy assignment. Representative DTA thermal curves using both DTA and computerized DSC modes are given for both clay specimens. The effect of the variation of heating rate and sample size on the observed peak temperatures and resolution is demonstrated for both illite specimens.

This study also demonstrates the use of carbon dioxide purge atmospheres for both shifting and enhancing the DTA peak signal observed for small carbonate contaminants in such clay materials. Finally, the actual inorganic carbon content of the Fithian (API #35) specimen is determined by acid decarboxylation of the carbonate component and subsequent measurement of the carbon dioxide which is evolved using a commercial element analyzer.

Illite Clay Minerals

The term illite was proposed by the American geologists GRIM, BRAY & BRADLEY (1937) as a general term, not as a specific clay mineral name, for the mica-like clay minerals. The name was derived from the abbreviation for the state of Illinois (USA). The term has now been widely used for a mica-type clay mineral with a 10 Å c-axis spacing which shows no expanding lattice characteristics.

GRIM et al. (1937) gave the general formula

$$(OH)_4 K_y (Si_{8-y} \cdot Al_y)(Al_4 \cdot Fe_4 \cdot Mg_4 \cdot Mg_6) O_{20}$$

for illites. In muscovite y is equal to 2, whereas in illite y is less than 2 and frequently equal to 1.0 to 1.5. However, fifty years later there are no definitions that enable the compositions of this important group of clay minerals to be closely specified. It is agreed that illites contain less potassium and more water than true mica. It is also believed that many materials that have been designated as illite in the literature are actually mixed layer clays (NEWMAN, 1987).

Thermal Analysis of Illite Clays

The scientific literature contains many DTA thermal curves for illite clay mineral. Most of the early thermal analysis data was produced by GRIM and co-workers (1940, 1942a, 1942b, 1947 and 1948a). SPEIL et al. (1945) also published curves for illite. Much of the more recently published thermal analysis data of illite clays relates to the abundance of the clay in coals as coal mineral. Illite is also a major component of shales and roof rock materials in the coal mining industry.

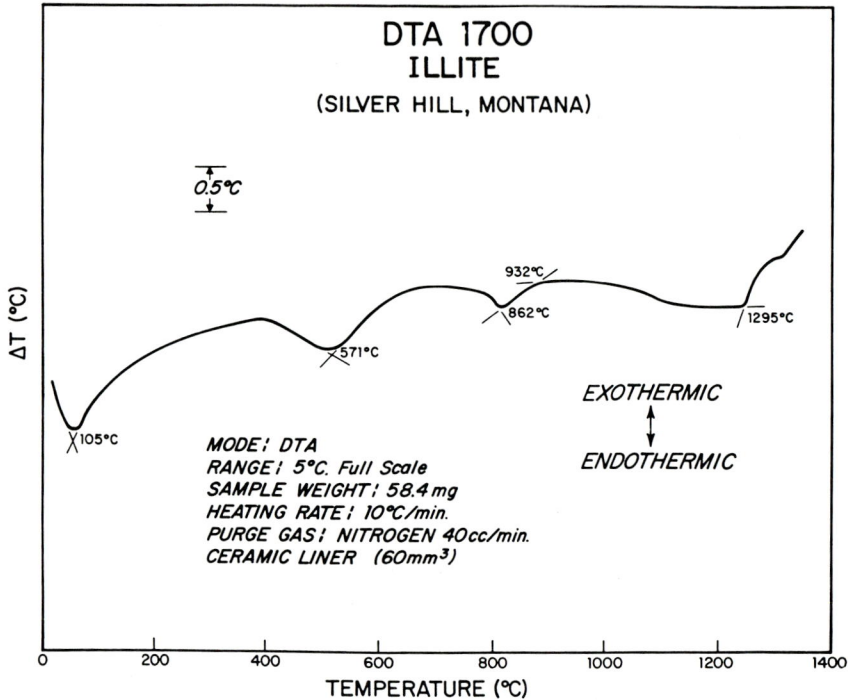

Fig. 1: DTA thermal curve for CMS-IMt illite specimen obtained in dynamic nitrogen atmosphere at a heating rate of ten degrees per minute.

Since it is the belief of the author that modern computerized thermal analyzers are much better than most of the instruments used in the older literature, thermal curves of some clay materials which have already been published by other investigators were repeated in his own laboratory. These have here-to-fore been used only for personal reference and will be given in the discussions which follow. The DTA and TG instrumentation used in these studies has been previously described (EARNEST, 1983a, 1983b and 1984).

Figure 1 shows the DTA thermal curve obtained for the Clay Mineral Society (CMS-IMt) illite specimen from Silver Hill, Montana. This thermal curve was obtained using a 58.4 milligram sample in a 60 mm^3 ceramic liner as the sample cup. The sample was heated at 10°C/minute in flowing nitrogen.

The first two endothermic events lead to the evolution of water from the clay material. The lower temperature event, with peak temperature 105°C, corresponds to the heat absorption necessary to remove molecular water from the clay. There is some uncertainty as to exactly how the molecular water is held in the clay itself (i.e., structural, sorbed, or both?). The breadth of this endothermic DTA peak indicates loss of molecular water over a wide range of temperatures which should correspond to a similar distribution of intermolecular or binding forces between the polar water molecule and the clay composition. One will note that the first endothermic event corresponds to the loss of 2.9% of the original sample as is shown in Figure 2.

The second endothermic DTA peak (T_{max}=571°C) corresponds to the dehydroxylation of the clay lattice which also leads to the evolution of water molecules. It can be observed in the TG thermal curve, shown in Figure 2, that this corresponds to a loss of 4.5% of the original sample weight of the illite specimen. This dehydroxylation event is also notably broad which, in this case, is often related to particle size distribution in the clay material itself. The derivative thermogravimetric (DTG) curve shown in Figure 2 describes the rate of water evolution as a function of temperature in a very analogous manner to that described by the DTA thermal curve in Figure 1.

GRIM (1948b) studied the dehydration and rehydration of illite specimens. He found that the first endothermic peak is reversible if given enough time in the presence of an atmosphere containing water vapor. Even the hydroxyl group which is lost near 500°C could be regenerated by the silicate lattice provided the clay had not been heated past ca. 800°C.

The third endothermic peak in the DTA thermal curve of illite clay specimens corresponds to the destruction of the silicate lattice. As can be seen in Figure 1, the matrix destruction is complete at 932°C when the Silver Hill, Montana, specimen is heated at ten degrees per minute in flowing nitrogen. This endothermic peak is immediately followed by an exothermic spinel crystallization peak in the DTA thermal curve for illites. GRIM & BRADLEY (1940) concluded that the middle sheet carrying the alumina, magnesium and iron lead to the spinel formation while the alkali metal and silica associated with the outer sheets lead to the formation of an amorphous glass. Furthermore, they found that if the glass forms concurrently with the lattice destruction then the amplitude (ΔT) of the third endothermic DTA peak may serve as an indication of the amount of potassium present in the illite.

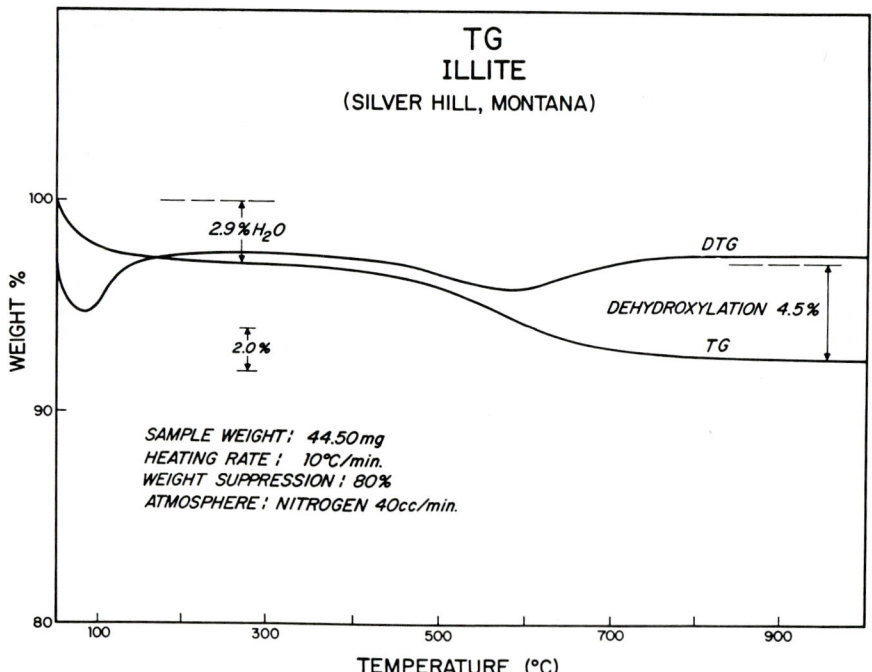

Fig. 2: TG and DTG thermal curves for CMS-IMt illite specimen (Nitrogen atmosphere, heating rate 10°C/min.).

One will note that the exothermic spinel crystallization is difficult to observe in the DTA thermal curve for the Silver Hill specimen of Figure 1. This crystallization peak may be intensified by increasing the heating rate to 20 degree/minute as is demonstrated by the DTA thermal curve in Figure 3. The exothermic effect is easily observed immediately after the endotherm before the DTA signal returns to the baseline level. One will also note that all three of the endothermic peaks are sharpened and intensified by an increase in heating rate when compared to the DTA thermal curve shown in Figure 1. The DTA peak minima are shifted to higher temperatures as a result of the use of the faster heating rate as well as the use of a larger amount of clay sample in the DTA.

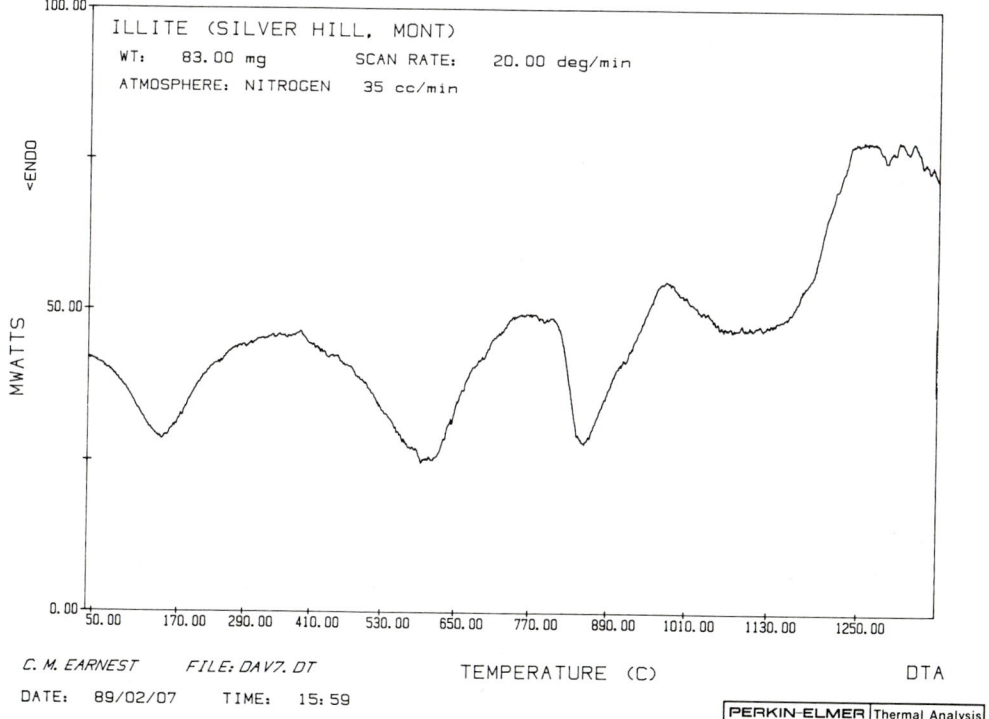

Fig. 3: DSC thermal curve for CMS-IMt illite specimen obtained in dynamic nitrogen atmosphere at a heating rate of twenty degrees per minute.

As the illite specimen is heated to temperatures in excess of 1200°C, further exothermic ordering of the material is observed. It is here at these elevated temperatures that the actual elemental ratios (chemical composition) become extremely important. The chemical composition at this point determines exactly what high temperature phases result. In the manufacture of brick and ceramic tile from such illitic clays this represents the firing temperature range, that is ca. 1200°C.

Thermal Analysis of the Fithian, Illinois Specimen (API #35) of Illite Clay Material

One of the most readily available, and hence well studied, specimens of illitic clay material is that of the American Petroleum Institute Specimen (API #35) from Fithian, Illinois (USA). Unfortunately, this specimen contains a good deal of other minerals. KERR, KULP & HAMILTON (1949) report that even the best material from Fithian includes carbonate, sulfides, carbonaceous matter and a certain amount of montmorillonite. Thus, the thermal curves given for the CMS-IMt (Silver Hill, Montana) specimen are probably more representative of the clay materials described as *illite*.

Figures 4 and 5 give the DTA and the TG-DTG thermal curves obtained in the author's laboratory for the Fithian, Illinois specimen. These thermal curves were obtained using a heating rate of 10°C/min and a flowing nitrogen atmosphere. One can observe by comparing these DTA and TG thermal curves that the first three endothermic peaks (124°C, 507°C and 729°C) correspond to weight loss events. The third endothermic DTA peak, in this case, is due to the decomposition of a carbonate contaminant. The broad endothermic event above 820°C corresponds to the matrix destruction of the silicate lattice. The broadness of this endothermic activity often implies the presence of a mixed-layer (illite-smectite) component in the clay material.

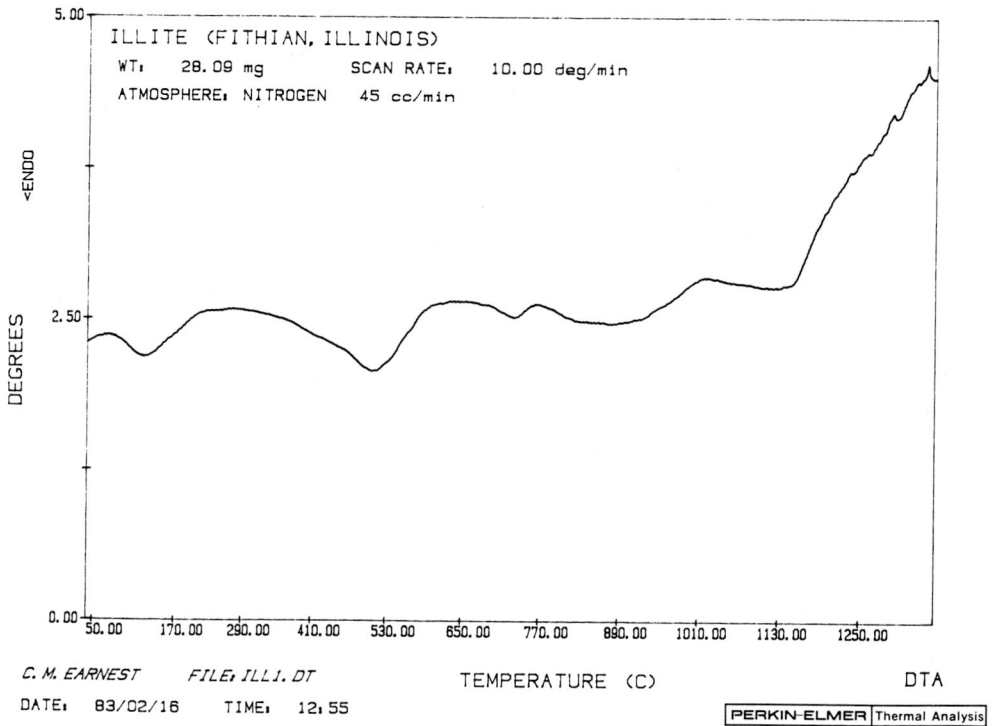

Fig. 4: DTA thermal curve for API #35 illite specimen obtained in dynamic nitrogen atmosphere at a heating rate of ten degrees per minute.

For the small sample size (28.09 milligrams) used for this DTA analysis, the exothermic ordering peak for illite clay mineral is only weakly observed at 1010°C. This is followed by the onset of the usual strong exothermic ordering actively near 1130°C. It should be noted that heating such illitic material past this temperature in ceramic DTA liners (sample holders) leads to the contamination of the DTA liner and a new sample cup is required for the next DTA study.

As was the case for the Silver Hill, Montana, specimen, a larger sample of the Fithian Specimen was also studied at a heating rate of 20°C/min. Figure 6 shows the thermal curve which resulted from the DTA analysis of a

86.60 milligram specimen using a large (100 mm^3) ceramic liner as the DTA sample cup. This study was also conducted using the DSC mode of the instrument so that quantitative peak area (energy) assignments were also possible. As one would expect, when compared to the DTA thermal curve in Figure 4, the peak temperatures in Figure 6 are observed at higher temperatures and are intensified for the larger sample and more rapid heating rate. One will also note that the third endothermic peak (carbonate decomposition) is less well resolved from the endothermic matrix destruction peak as a result of doubling the heating rate and greatly increasing the sample size.

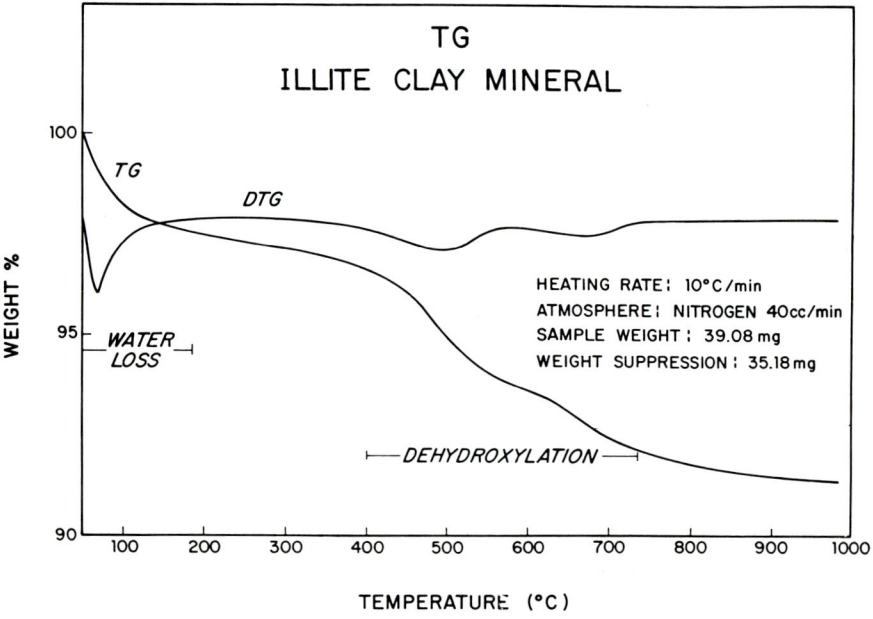

Fig. 5: TG and DTG thermal curves for API #35 illite specimen (Nitrogen atmosphere, heating rate 10°C/min.).

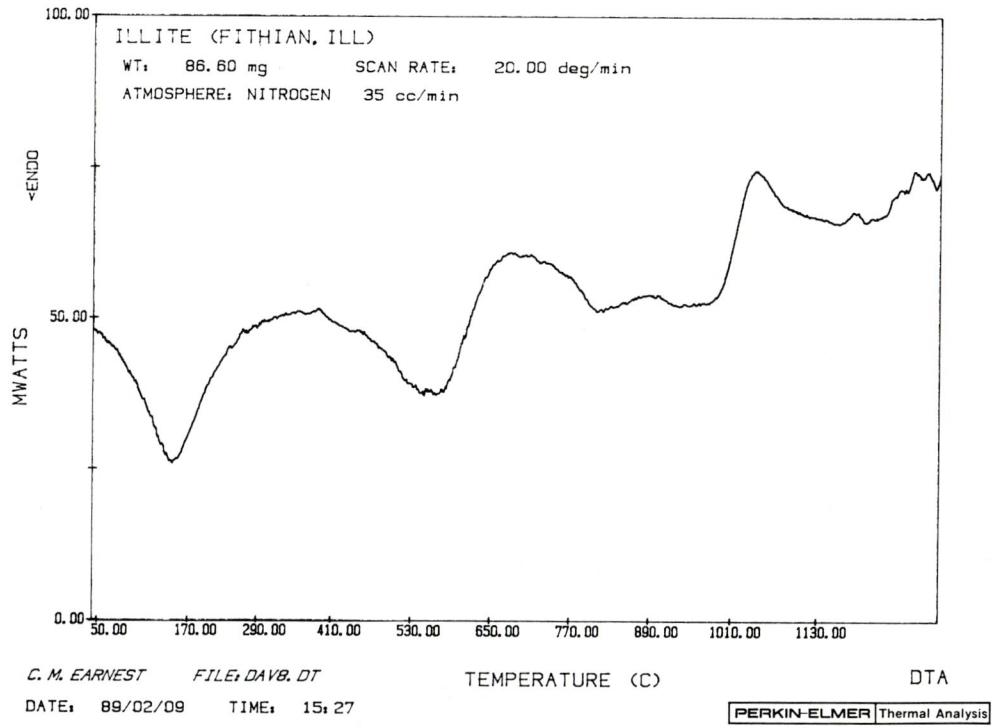

Fig. 6: DSC thermal curve for API #35 illite specimen (Nitrogen atmosphere, heating rate 20°C/min.).

Since the 20°C/min studies of both the CMS-IMt and API #35 specimens were conducted using the DSC Mode of the instrument, the peak temperatures and peak energies for these two thermal curves were tabulated for comparative purposes. These are given in Table 1.

Although the first and second peak temperatures (T_{min}) for the two illite specimens are given in Table 1, the contaminants present in the Fithian (API #35) specimen prevent quantitative comparison of identical thermal events. Furthermore, the overlap of the endothermic carbonate decomposition and the endothermic matrix destruction of the Fithian Specimen made it difficult to state a temperature range for the later when the study was conducted at 20°C/min. This activity is more clearly distinguished in the previous study conducted at 10°C per minute. Other than the obvious presence of the carbonate peak in the Fithian Specimen, one can also observe that the first endothermic peak (144°C) for the Fithian Specimen is sharper than that obtained for the CMS-IMt specimen. This is believed to be primarily the result of the presence of a certain amount of smectite clay component in the Fithian Specimen. The presence of smectite clay was originally suggested by KERR, KULP & HAMILTON (1949). Another observable difference in the two DTA studies, given in Figures 4 and 6, is that the exothermic ordering associated with spinel formation (GRIM & BRADLEY, 1940) occurs at a higher temperature for the Fithian Specimen. The reasoning behind this is most probably related to the difference in metal oxide content of the two specimens.

Table 1: DATA Summary for DTA/DSC of Illite Specimens at 20°C/min in Flowing Nitrogen Atmosphere

		CMS-IMt		API #35	
		T[°C]	ΔH[J/g]	T[°C]	ΔH[J/g]
1st Peak	(minimum)	141	84.0	144	116.0
2nd Peak	(minimum)	604	176.0	567	127.4
3rd Peak	(minimum)	852	92.8	750-990*	N/A
Exotherm	(maximum) (onset)	998 954	28.8	1042 1004	N/A

* Carbonate decomposition and matrix destruction overlap
N/A = Not assigned

It has been observed (ROWLAND & LEWIS, 1951) and further demonstrated (SMYKATZ-KLOSS, 1974; WARNE, 1975, 1978, 1979 and 1986; EARNEST, 1984) that the observed DTA peak temperature for many carbonate minerals are dependent upon the partial pressure of carbon dioxide in the DTA analyzer cell. Furthermore, the carbonate decomposition peaks obtained by DTA studies in CO_2 rich atmospheres are both sharpened and intensified. This behavior offers the thermal analyst a tool for diagnostic enhancement of carbonate peaks in natural specimens such as clay minerals. Thus, it is possible to observe carbonate components at levels which are not detectable in inert atmosphere such as flowing nitrogen or argon.

Figure 7 shows the DTA thermal curve obtained when the Fithian Specimen is studied in a flowing carbon dioxide atmosphere, $P_{CO_2}=1$ atm, using a heating rate of ten degrees Celsius per minute. By comparing the DTA curve of Figure 4 with that of Figure 7, one will see that the carbonate decomposition peak observed in flowing nitrogen, $T_{onset}=693°C$ and $T_{min}=729°C$, is shifted above 900°C in the CO_2 atmosphere. The expected sharpening and intensification of the peak is also observed. This behavior in this temperature range is generally associated with the decomposition of calcite ($CaCO_3$). The strong endothermic energy associated with this decomposition, ca. 1.7 kJ/g, coupled with the peak sharpening observed in CO_2 purge makes it possible to detect relatively low levels of calcite.

It should also be pointed out that by shifting the carbonate decomposition peak to higher temperatures, the thermal event associated with the destruction of the silicate lattice is clearly observed at 799.5°C. The exothermic ordering peak was observed at 1002°C and the dehydroxylation endotherm gave a DTA peak minimum at 542°C for the study carried out in carbon dioxide atmosphere.

Thus, the use of carbon dioxide as the purge atmosphere in this case not only leads to diagnostic tool for the carbonate component but also frees the adjacent endotherm from overlap with the carbonate decomposition. WARNE (1979) has similarly demonstrated the separation of overlapping decomposition peaks of pyrite and magnesite. EARNEST (1981) achieved separation of a small amount of calcite from a magnesite ore using the same change of atmosphere technique.

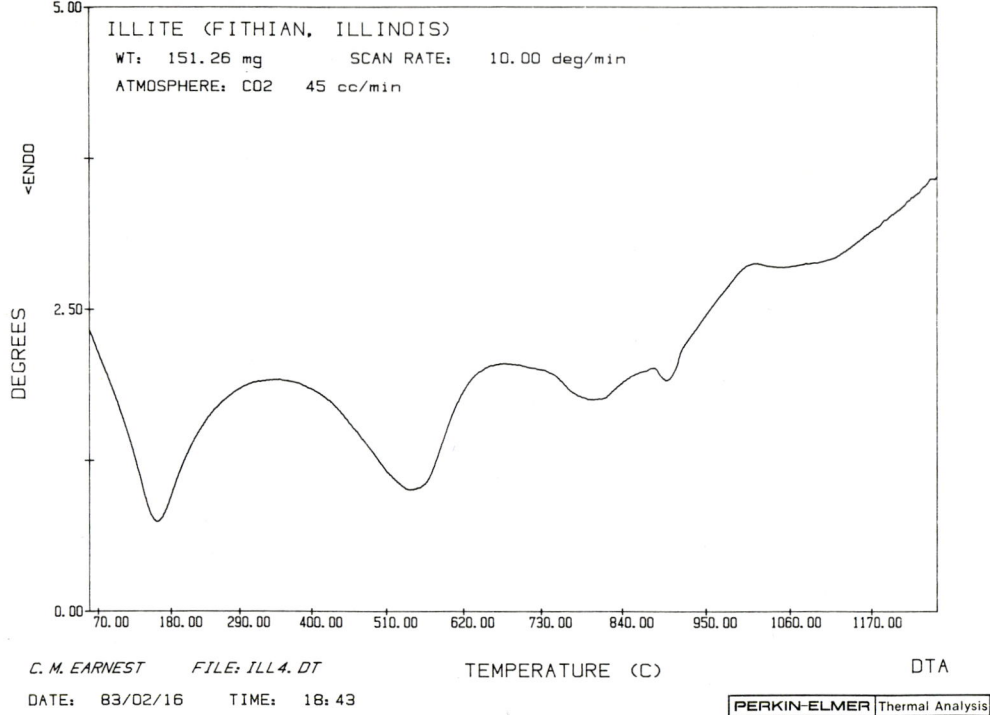

Fig. 7: DTA thermal curve for API #35 illite specimen (Carbon dioxide atmosphere, heating rate 10°C/min.).

Further experimentation with the Fithian Specimen to assess the carbonate level was conducted. In this case, an automated acid decarboxylation method associated with a commercial elemental analyzer which has previously been described (EARNEST, 1983) was used. In this experiment a small amount of the Fithian Specimen was accurately weighed and then dropped into a thermostated bath of diluted phosphoric acid. The carbon dioxide, thus evolved, was swept into the normal analytical train consisting of scrubber materials and differential thermal conductivity cells of an automated C, H and N analyzer. The amount of carbon dioxide evolved from the acid bath was measured by the usual carbon thermal conductivity bridge of the elemental analyzer.

The carbon bridge of the instrument was calibrated in this case by a series of weight samples of primary standard calcium carbonate. The percentage inorganic carbon content of the Fithian Specimen was determined by the analyzer to be 0.26%. By using stoichiometric mass ratios, this corresponds to the evolution of 0.95% carbon dioxide and a total carbonate anion content of 1.3% by weight in the Fithian illite specimen. This is believed by the author to be mostly calcite but there have been reports of small amounts of siderite (KERR, KULP & HAMILTON, 1949) in some specimens obtained from the Fithian, Illinois site.

Conclusion

As one could easily determine from the examples which were given in this work, the thermal analysis techniques of DTA, TG and DTG are valuable tools for the characterization of illite clay specimens. It can also be recognized that the lower temperature thermal events in these clays, which give rise to the evolution of water vapor, occur over broad temperature ranges for each of the events. Thus the corresponding DTA and DTG thermal curves arising from these activities are broad and are of relatively weak intensity. For this reason, many DTA thermal curves which have been published in the scientific literature do not describe these thermal events very well. As a matter of fact, many of these resemble wavy lines or are "snakelike" in appearance.

As was shown by the work presented here, the thermal events which are present in illite clays may be better observed in both DTA and DTG by the use of heating rates of at least twenty degrees Celcius per minute. This increase in peak amplitude can be at the expense of resolution when certain contaminant (e.g., calcite) minerals are to be separated and recognized using thermal analysis techniques. The use of more than one purge gas atmosphere in thermal analysis can often assist in both the separation and identification of contaminant minerals.

Illite clay materials also possess the special problem of being poorly defined with respect to composition and are thus not consistent in this from specimen to specimen. It is, therefore, not surprising that considerable variation in the thermal curves can be obtained for different illite specimens. The magnitude and temperature of the exothermic DTA peak associated with the spinel crystallization in such clays is an excellent example of this.

In closing, several reference sources which contain published thermal analysis data, as well as other discussion of illite clay specimens, should be mentioned. First, one of the earliest and most valuable publications by MACKENZIE (1957) should be consulted for examples of DTA thermal curves of any clay mineral. Mica-type clay materials are also included in a later book by MACKENZIE (1970) under the discussion of interstratified phyllosilicates. A book by DR. RALPH GRIM (1968), originator of the term *illite* gives much detail of illite characteristics as well as DTA thermal curves. More recently, a data handbook of clay materials has been published (VAN OLPHEN & FRIPIAT, 1979) which includes thermal curves for a reference clay specimen which originates in Le Puy en Velay, France.

References

EARNEST C. M. (1981) - Thermal Analysis of Mineral Matter in Coals and Coal Ash.- Pittsburgh Conf. Paper #493, Atlantic City, NJ.

EARNEST C. M. (1983a) - Thermal analysis of hectorite. Part I. Thermogravimetry.- Thermochim. Acta, 63, 277-289

EARNEST C. M. (1983b) - Thermal analysis of hectorite. Part II. Differential thermal analysis.- Thermochim. Acta, 63, 291-306

EARNEST C. M. (1983) - in Advances in Materials Characterization.- Mat. Sci. Res. Series, Vol. 15 (ROSSINGTON D. R. & SNYDER R. L. (eds.)), Plenum Press, New York, p. 515

EARNEST C. M. (1984) - Thermal Analysis of Clays, Minerals and Coal.- Perkin Elmer Corporation, Norwalk, CT.

GRIM R. E. (1947) - Differential thermal curves of prepared mixtures of clay minerals.- Am. Mineral., 32, 493-501

GRIM R. E. (1948a) - The illite clays.- Vol. Titels and Abstracts - Internat. Geological Congress, London, 127-128

GRIM R. E. (1948b) - Rehydration and dehydration of clay minerals.- Am. Mineral., 33, 50-59

GRIM R. E. (1968) - Clay Mineralogy.- Second Edition, McGraw Hill, York.

GRIM R. E. & BRADLEY W. F. (1940) - Effect of heat on illite and montmorillonite.- J. Am. Ceram. Soc., 23, 242-248

GRIM R. E., BRAY R. H. & BRADLEY W. F. (1937) - Mica in argillaceous sediments.- Am. Mineral., 22, 813-829

GRIM R. E. & ROWLAND R. A. (1942a) - Differential thermal analysis of clay minerals and other hydrous materials. Part I.- Am. Mineral., 27, 756-761

GRIM R. E. & ROWLAND R. A. (1942b) - Differential thermal analysis of clay minerals and other hydrous materials. Part II.- Am. Mineral., 27, 801-808

Kerr P. F., Kulp J. L. & Hamilton P. K. (1949) - Differential Thermal Analysis of Reference Clay Specimens.- API Project #49, Columbia University, New York, p. 40

Mackenzie R. C. (1957) - The Differential Thermal Investigation of Clays.- Mineralogical Society, London.

Mackenzie R. C. (1970) - Differential Thermal Analysis.- Vol. 1. Academic Press, London, p. 539

Newman A. C. D. (ed.) (1987) - Chemistry of Clays and Clay Minerals.- Wiley-Interscience, New York, 69

Rowland R. A. & Lewis D. R. (1951) - Furnace atmosphere control in differential thermo analysis.- Am. Mineral., 36, 80

Smykatz-Kloss W. (1974) - Differential Thermal Analysis: Application and Results in Mineralogy.- Springer-Verlag, Berlin.

Speil S., Berkelhamer L. H., Pask J. & Davies B. (1945) - Differential thermal analysis of clays and aluminous minerals.- U.S. Bureau of Mines Tech. Paper #664

Van Olphen H. & Fripiat J. J. (eds.) (1979) - DATA Handbook for Clay Minerals and Other Non-Metallic Minerals.- Pergamon Press, Oxford.

Warne S. St. J. (1975) - An improved differential thermal analysis method for the identification and evaluation of calcite, dolomite and ankerite in coal.- J. Inst. Fuel, 48, 142-145

Warne S. St. J. (1978) - Proben-Abhängigkeit (PA) curves of simple anhydrous carbonate minerals.- J. Thermal Anal., 14, 325-330

Warne S. St. J. (1979) - Identification of magnesite form pyrite in coal by DTA.- J. Inst. Energy, 52, 21-22

Warne S. St. J. (1986) - Application of variable atmosphere DTA (in CO_2) to improved detection and content evaluation of anhydrous carbonates in mixtures.- Thermochim. Acta, 109, 243-252

THERMAL ANALYSIS OF SELECTED ILLITE AND SMECTITE CLAY MINERALS. PART II. SMECTITE CLAY MINERALS

C. M. Earnest

Department of Chemistry, Berry College
Rome, Georgia (USA)

Abstract

In this paper, the application of the techniques of differential thermal analysis (DTA), thermogravimetry (TG) and derivative thermogravimetry (DTG) to the characterization of smectite clay minerals is presented. Several specimens including both dioctahedral and trioctahedral types of smectite clay mineral are included here. The use of DTA to separate the mineral montmorillonite into two subtypes is demonstrated. A unique industrial example of the use of TG to follow the cleaning of a raw bentonite clay and its subsequent conversion into an organo-clay product is demonstrated. The thermal analysis of both raw and processed hectorite clay specimens which was originally performed by the author is reviewed.

Introduction

Montmorillonite Clay Minerals (Smectites)

The montmorillonite group (smectites) consists of a number of clay minerals composed of tetrahedral-octahedral-tetrahedral silicate layers of both dioctahedral and triooctahedral types. These belong to the group of minerals termed "argillaceous minerals" along with the micas. Argillaceous minerals are hydrated aluminum silicates which derive from the $[Si_2O_5]^{2-}$ anion. The dioctahedral members of the group are montmorillonite, beidellite and nontronite. The trioctahedral members are hectorite and saponite.

Montmorillonite is the dominant clay mineral in bentonite (altered volcanic ash). Bentonite has the unusual property of expanding several times its original volume when placed in water. Due to this behavior, it has become industrially important as an oil drilling mud in which the montmorillonite is used to give the fluid a viscosity several times that of water. It is also used for stopping leaks in dams, soils and rocks. Another recent use is as a catalyst support material.

Unlike the kaolinite group of clay minerals, the smectites undergo extensive cation exchange or replacement in the crystal lattice. Thus, one will note that location of the origin of the clay is often listed with it. The degree of displacement and type of metal in the lattice can eventually govern the shape and position of many thermal analysis peaks.

Montmorillonite

Montmorillonite is by far the most abundant dioctahedral smectite. The theoretical formula is $Al_2(Si_4O_{10})(OH)_2$. In montmorillonite, the silicate layer charge is due to the replacement of Al^{3+} by Mg^{2+} in the octahedral sheet. Most montmorillonite specimens have additional substitution of Si^{4+} by Al^{3+} in the tetrahedral sheet as well as substitution of Al^{3+} by Fe^{3+} in the octahedral sheet.

Because of this substitution, montmorillonite always differ from the theoretical formula. In the tetrahedral plane, the substitution may proceed up to about 15 percent, however, in the octahedral plane, it may extend to completion (TODOR, 1976). This extensive substitution in the octahedral plane leads to a number of diverse minerals classified in this group.

Structurally montmorillonite is composed of units made of two silica tetrahedral sheets with a central alumina octahedral sheet. All of the tips of the tetrahedrons point in the same direction and toward the center of the unit. The tetrahedral and octahedral sheets are combined so that the tips of the tetrahedrons of each silica sheet and one of the hydroxyl layers of the octahedral sheet form a common layer. The atoms common to both the tetrahedral and octahedral layers became oxygens instead of hydroxyls. The layers are continous in the a and b directions and are stacked one above the other in the c direction.

In the stacking of the silica-alumina-silica units, oxygen layers of each unit are adjacent to oxygens of the neighboring units with a consequence that there is a very weak bond and an excellent cleavage between them. The outstanding feature of the montmorillonite structure is that water and other polar molecules, including polar organic molecules, such as amines and alcohols, can enter between the unit layers causing the lattice to expand in the c direction. Interlayer sheets of polar molecules many molecular layers thick may develop.

Exchangeable (charge balancing) cations exist between the silicate layers. The thickness of the water layers between the silicate units depends on the nature of the exchangeable cations at a given water vapor pressure. For example, it is reported (GRIM, 1962) that a montmorillonite with sodium as the exchangeable cation frequently has one molecular water layer while calcium will generally have two molecular water layers. This interlayer water, as well as the hydroxyl groups positioned in the clay structure, will be seen to play an important role in the observed thermal behavior of smectite clay minerals in general.

Hectorite

Hectorite is a trioctahedral member of the smectite clay minerals group. Unlike the dioctahedral members of this group, hectorite contains no aluminum ions in the silicate structure. In hectorite all possible octahedral positions, originally containing Al^{3+}, have been substituted by Mg^{2+} and Li^+

ions (GRIM, 1968). The formula suggested for hectorite by ROSS & HENDRICKS (1945) is

$$(OH)_4Si_8(Mg_{5.34}Li_{0.66})O_{20} \cdot n\ H_2O$$
$$Na_{0.66}$$

where n H_2O represents interlayer water. As is shown by the formula, the structural lattice charge results from a substitution of Li^+ for Mg^{2+}. The charge balancing cation, located external to the silicate layer, is sodium. Although not indicated by the formula of ROSS & HENDRICKS (1945), hectorite is known to contain fluorine atoms. In another formula calculated from chemical analysis data by KERR et al. (1949), the atoms of fluorine are equal in number to that of the -OH groups in the silicate structure. In a more recent compilation (GABIS, 1979), the chemical analysis of the SHCa-1 hectorite specimen (carbonate-contaminated) of the Clay Mineral Society Source Clays Collection was found to contain 2.75% fluorine in the raw hectorite ore.

Hectorite finds important uses in the ceramic industry as a suspending agent in glazes, as a plasticizing agent for non-plastic formulations such as high alumina or zirconium bodies, and as a non-migrating binder in extruded bodies. Due to its ability to gel water (MOLL, 1979), it finds many uses in both the paints and cosmetics industries. It is also used as a clarifier for beverages. Organo-clays derived from hectorite are used as oil drilling fluids and more recently (ANON., 1980) as stereo-specific catalysts.

Most hectorite ore specimens in the United States are limited to those found in southeastern California (near Hector, CA), and are always found associated with large proportions of calcite and, in some cases, varying amounts of dolomite. Thus, the hectorite ores are commercially processed to remove most of the carbonate component prior to use.

Experimental

All thermal curves presented in this paper were originally obtained by the author and have appeared elsewhere (EARNEST, 1980, 1983a, 1983b and 1988) at one time or another. Since the thermal curves are a result of original work of the author, the following experimental information is included.

All DTA work, shown by the thermal curves presented here, was performed with a Perkin-Elmer DTA 1700 Differential Thermal Analysis System. All results of thermogravimetry were obtained using a Perkin-Elmer Model TGS-2 Thermogravimetric Analysis System. The temperature axes of the TGS-2 System was calibrated using Curie point standards (Nickel, Perkalloy and Iron) which were supplied by the vendor with the instrument. The temperature axis of the DTA System was calibrated using metal fusion standards (aluminum and gold). All clay samples used in the DTA studies were accurately weighed using a Perkin-Elmer Model AD-2Z Autobalance.

The DTA thermal curves were obtained using 60 cubic millimeter ceramic liners in the DTA sample holders. The reference material used in all DTA experiments was powdered alumina (Al_2O_3) which had been previously heated to 1400°C. Both the conventional DTA mode as well as the DSC mode of the DTA system were employed in this work.

The "Cheto" type montmorillonite sample used in this study was the Chambers, Arizona reference clay (API #23) which was included in the API Project #49 (KERR et al., 1949). The Black Hills bentonite samples and organo-clay type drilling mud were obtained from Southern Clay Products (Gonzalez, Texas).

The SHCa-1 hectorite specimen was obtained from the Clay Mineral Society Source Clay Collection. The raw hectorite ore and processed hectorite specimen was obtained from the R. T. Vanderbilt Company (Norwalk, CT, USA).

Thermal Analysis of Montmorillonite

Classification into Subtypes by DTA

As was mentioned earlier, the mineral montmorillonite has been well studied over a period of many years. A number of different specimens have been studied from varying locations and reported in the literature. After extensive X-ray and thermal analysis studies, BRADLEY & GRIM (1951), GRIM and co-workers (1957, 1961) have classified the various montmorillonite specimens into two subtypes or mixtures of the two. These are the "Cheto" type and "Wyoming" type montmorillonites. Although there are those (LUCAS & TRAUTH, 1965) who subdivide the "Wyoming" type into "Wyoming" types I and II, the DTA thermal curves show the same pattern with the difference being primarily the peak temperatures. An "intermediate" type is also described (LUCAS & TRAUTH, 1965) which varies both in peak temperature and intensity from other "Wyoming" type species. Nonetheless, it too gives the same "Wyoming" type DTA pattern. While the montmorillonites also vary in their thermal behavior in the high temperature region, one would tend to agree with Grim from the standpoint of thermal analysis patterns observed from ambient temperature to 1000°C. The most obvious difference in this temperature region is simply "those that do" (Cheto) and "those that do not" (Wyoming I, II or Intermediate) exhibit significant temperature separation (typically 130-150°C) between the third (endothermic) and fourth (exothermic) peak in the DTA thermal curve.

An example of both the "Cheto" type and "Wyoming" type of montmorillonite is presented here. Figure 1 shows the DTA thermal curve of a "Cheto" type montmorillonite obtained in this study. This clay is the Chambers, Arizona reference clay (API #23). Figure 3 shows the thermal curve obtained in this work for Black Hills bentonite, a "Wyoming" type montmorillonite. Although "Wyoming" type reference clays are available and are present in our laboratories, this specimen obtained from a commercial vendor of clay products was chosen for demonstration purposes. The Black Hills bentonite is recieved at the commercial site as a bulk raw clay which is subsequently screened, washed and further processed to remove organic impurities. Figures 4 and 5 show the TG thermal curves obtained using the thermogravimetry for both the "raw" and "processed" Black Hills bentonite. By using air as the purge gas for oxidative purposes, thermogravimetry offers a rapid and reliable means of monitoring such a cleaning process.

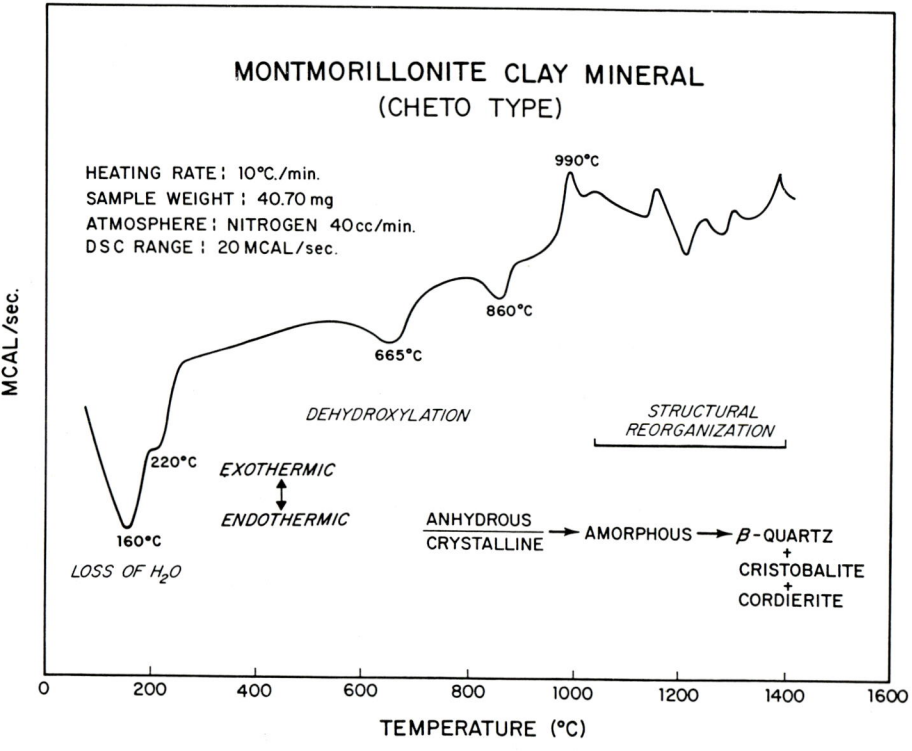

Fig. 1: DSC thermal curve for "Cheto" type montmorillonite (Chambers, Arizona), (from EARNEST, 1980; reprinted with permission of Prkin-Elmer).

Characterization by DTA and TG

The first three peaks in the DTA thermal curves are common to all types of montmorillonites. The *first peak* is a strong endotherm which corresponds to the loss of interlayer water. The shape and temperature at which this peak is observed will vary with the type of cation predominating in the

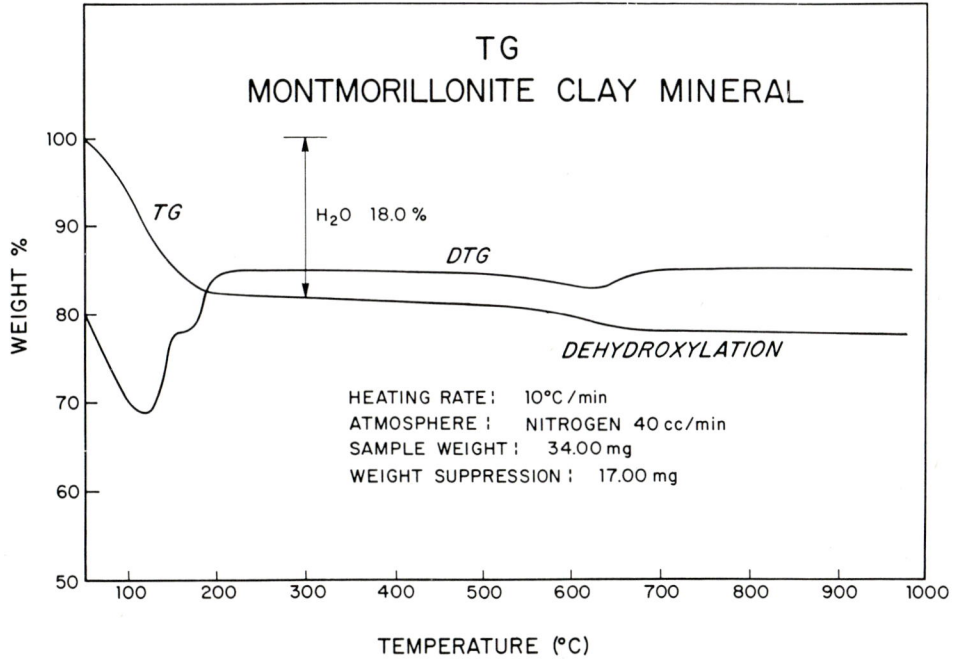

Fig. 2: TG and DTG thermal curves for "Cheto" type montmorillonite, (from EARNEST, 1980; reprinted with permission of Perkin-Elmer).

exchange layer, as well as the sample pretreatment. In the particular "Cheto" type studied here, the predominant exchange cation is the divalent ion Ca^{2+}, KERR et al. (1949), and is distinguished as a shouldered (double) peak appearing at 160°C and 220°C (Figure 1). Both Mg^{2+} and Ba^{2+} divalent ions also give a characteristic double peak. When the predominant exchange cation is one of the univalent ions Na^+, K^+, Li^+, H^+ or $[NH_4]^+$, the endotherm is singular and is generally sharper. When the exchange cation is $[NH_4]^+$, an additional weak endotherm is observed at approximately 400°C and corresponds to the evolution of NH_3. Lithium is an anomaly here in that it gives a double peak for the dehydration event. Figure 2 shows the thermogravimetric (TG) analysis along with the

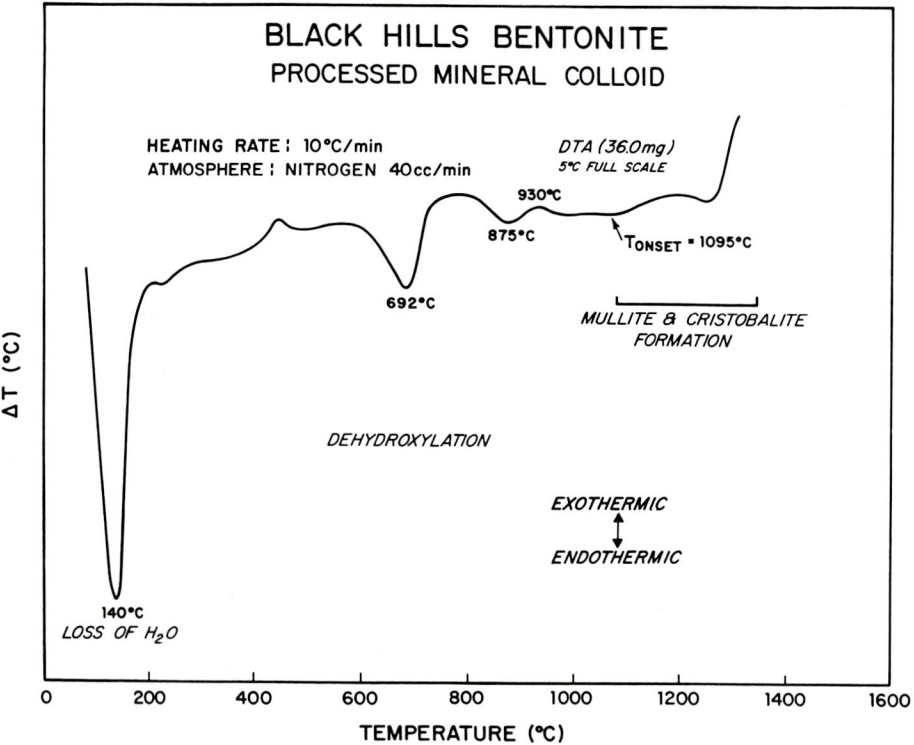

Fig. 3: DTA thermal curve for "Wyoming" type montmorillonite (Black Hills Bentonite), (from EARNEST, 1980; reprinted with permission of Perkin-Elmer).

derivative thermogravimetric curve (DTG) of the "Cheto" type montmorillonite. One will observe that the DTG thermal curve also gives the same shouldered dehydration profile as that observed in the DTA thermal curve.

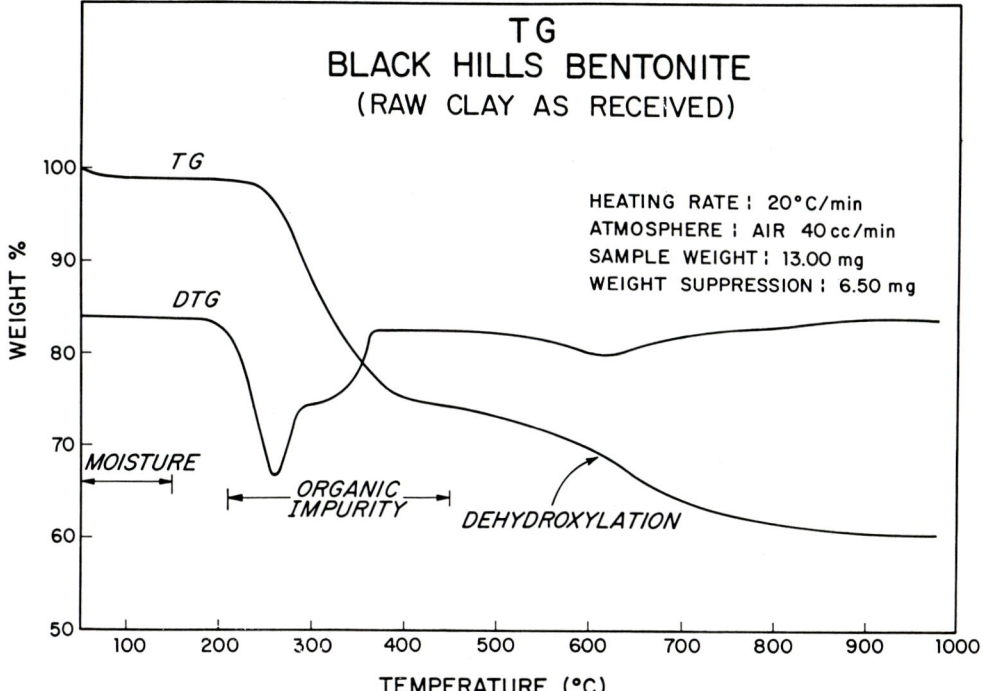

Fig. 4: TG and DTG thermal curves for "Wyoming" type montmorillonite (Black Hills Bentonite, raw clay as recieved), (from EARNEST, 1980; reprinted with permission of Perkin-Elmer).

The DTA thermal curve for the bentonite specimen (Figure 3) shows a much sharper dehydration endotherm with a peak maximum at 140°C. This dehydration endotherm is characteristic of a sodium montmorillonite in which the peak maximum generally occurs between 130°C and 150°C. Figure 5 shows the thermogravimetry of the bentonite along with the derivative thermogravimetric curve obtained in this study. By comparing Figures 2 and 5, one may see that for these particular samples, the "Cheto" type undergoes an 18.0% weight loss due to the loss of interlayer water, while the "Wyoming" type lost only 8.5% water. The dehydration process is reported to be completely reversible provided that the montmorillonite samples are not heated above 250°C (TODOR, 1976). The amount of water uptake depends on the relative humidity of the environment. For this reason, many workers expose the clay minerals to a 56% relative humidity prior to thermal analysis.

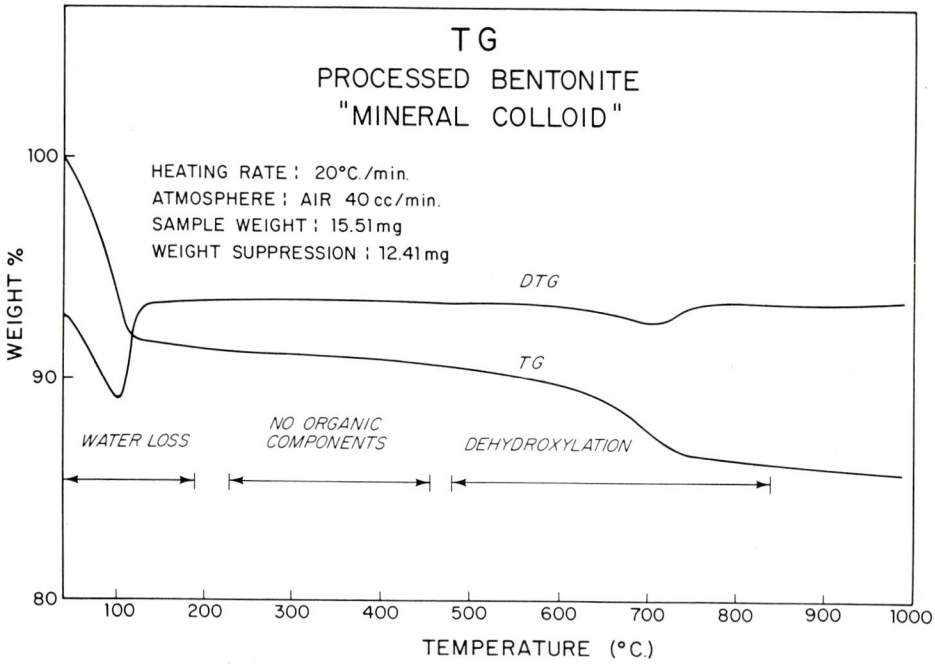

Fig. 5: TG and DTG thermal curves for "Wyoming" type montmorillonite (Processed Black Hills Bentonite), (from EARNEST, 1980; reprinted with permission of Perkin-Elmer).

The *second* major DTA peak observed in both the "Cheto" and "Wyoming" montmorillonites is an endothermic event corresponding to the dehydroxylation of the silicate lattice. This peak generally occurs in the 670°C to 710°C temperature range. This endotherm also involves a loss of mass by the clay minerals as water vapor is lost by the dehydroxylation mechanism. The TG thermal curves (Figures 2 and 5) show these mass losses as a percentage of the original sample weight. In the two specimens studied here, the "Cheto" type shows a 4.0% weight loss and the "Wyoming" type a 5.0% loss in weight. Since these two samples contained different amounts of interlayer water, they must be compared in this respect on a dry weight basis. When converted to a dry weight basis, these values become 4.9% and 5.45% respectively. It may appear that we are comparing one TG thermal curve obtained using a nitrogen purge (Figure 2) and one using an air purge (Figure 5). The bentonite (processed mineral colloid) was also studied under the same conditions as the "Cheto" type; N_2 purge and 10°C/min, the same percentage values were obtained as that obtained in air. It should be mentioned that a small amount of oxidizable contaminant is seen to be present in the processed bentonite sample. This is evidenced by the small exothermic oxidation peak in the DTA thermal curve (Figure 3).

The dehydroxylation peak observed in DTA represents most of the dehydroxylation, however some activity continues as the temperature is increased to 1000°C. This may be observed in the TG thermal curves as a continuing (slightly), negative slope of the weight percent thermal curve versus temperature. The dehydroxylation peak minima for the "Cheto" and "Wyoming" montmorillonite samples of this study were observed in the DTA thermal curve at 665°C and 692°C respectively when using a heating rate of 10°C/minute. Thus, for the two specimens studied here, the dehydroxylation of the "Wyoming" type occurs at a higher temperature than the "Cheto" type. This may also be observed in the DTG peak positions in Figures 2 and 5. The use of a faster heating rate will increase the T_{max} while slower heating rates will lower its temperatures. This peak is also influenced by the isomorphic substitution of cations which has taken place in the structure of the montmorillonite. A large amount of Fe^{3+} substitution for Al^{3+} will tend to shift the dehydroxylation to lower temperatures. On the other hand, Mg^{2+} substitution for Al^{3+} will cause an increase of the peak to higher temperatures (TODOR, 1976). The smectite clays may rehydroxylate (after dehydroxylation) to some extent if exposed to humidity but total recovery is often extremely slow (GRIM, 1968).

The *third* peak for both the "Cheto" and "Wyoming" montmorillonite is also endothermic. Unlike the two previous thermal events, this absorption of thermal energy does not lead to a loss of mass. The energy absorbed in this case is an order-disorder transition corresponding to a collapse of the dehydrated silicate matrix and an amorphous specie results. This transition is easily observed in the thermal curves of the "Cheto" type (860°C) and in the "Wyoming" type (875°C) given in Figures 1 and 3.

From this point on, the thermal behavior of the two montmorillonites differ. The "Wyoming" type undergoes a weak exothermic ordering immediately after the amorphous state has completely developed. This has been described as a shouldering effect directly above the point where the third endotherm achieves an apparent return to baseline. This is the result of spinel crystallization. The magnitude of this effect is an indication of the amount of spinel formed. For the bentonite sample used here, this very weak effect is at 930°C. The "Cheto" type, on the other hand, exhibits a distinct exotherm at approximately 990°C. Therefore, this ordering exotherm is separated from the previous disorder endotherm by approximately 130°C. This temperature separation is specific for, and is a characteristic of, the "Cheto" type only. X-ray data have shown that in the "Cheto" type, amorphous silica crystallizes into β-quartz and the β-quartz subsequently inverts to cristobalite from 1000-1200°C. The major difference between these two montmorillonites has been described as more structural than chemical when comparing their thermal behavior up to 1000°C (GRIM, 1961).

GRIM (1968) points out that the mineral montmorillonite tends to develop quartz if there is little substitution in the tetrahedral sheet and spinel if there is substitution of Al^{3+} in the tetrahedral positions. Also, montmorillonites exhibiting the delayed exotherm, as in the "Cheto" type, are relatively low in iron content. In the low iron containing "Cheto" type, enstatite, mullite or anorthite may also appear in the X-ray data at about 1000°C. This may, however, depend on the amount of MgO present and the amount of exchangeable Ca^{2+}.

As can be seen in Figures 1 and 3, additional structural reorganization peaks occur in temperatures above 1000°C. These peaks are predominantly exothermic and depend on which oxides are present. Thus, differences in chemical composition become important at these temperatures. Clays of either the "Wyoming" or "Cheto" type may show a fairly strong exotherm in the vicinity of 1200°C which represent the formation of either mullite or cordierite respectively. The onset of mullite nucleation is observed in Figure 3 as an exothermic plateau effect in the thermal curve for Black Hills bentonite at approximately 1100°C. This is followed by a stronger exothermic effect at about 1270°C which according to the X-ray studies of BRADLEY & GRIM (1951) and GRIM and co-workers (1957 and 1961) could correspond to the development of cristobalite.

The thermal curve of the "Cheto" type montmorillonite from 1000°C to 1500°C is more difficult to interpret than that of the "Wyoming" type. According to the continous X-ray data mentioned earlier (BRADLEY & GRIM, 1951; GRIM et al., 1957 and 1961), cristobalite continues to develop from 1000°C up to 1200°C and a small amount of anathorite is present from 1000°C to 1150°C. Cordierite appears at about 1260°C in the X-ray data. Both cordierite and cristobalite disappear at about 1450°C. Above 1450°C no crystalline forms exist.

TG Analysis of Organo-Clay Material

One of the major commercial uses of Wyoming Type montmorillonites, such as the Black Hills bentonite specimen of this study, is in the preparation of oil drilling fluids. The Black Hills bentonite has been found to be especially suitable for this purpose due to its yield strength, gel strength and low filter cake permeability. Even when used in small amounts, this bentonite produces an impervious clay layer on the wall of the drilling hole (JONES, 1983).

As has been previously published (EARNEST, 1980 and 1988), Figure 4 shows the TG thermal curve for the raw Black Hills bentonite which was dried and ground at a commercial plant prior to arriving at the author's laboratory. This material was subsequently subjected to a wet-refining (cleaning) process for removal of coarse non-clay material and other impurities. By comparing the TG-DTG thermal curves given in Figures 4 and 5 one can follow the effectiveness of the commercial cleaning process. In each case, the bentonite was subjected to thermogravimetry in dynamic air atmosphere using a heating rate of 20°C/min. One can observe the removal of the impurity component (which shows mass loss between 200°C and 400°C) by the cleaning process as well as the recovering of the interlayer water by the bentonite clay specimen.

The exchangeable cations present in montmorillonites can be replaced by long chain quaternary ammonium cations and the resulting organo-clays find considerable commercial application. These organo-clays have the important property of being able to swell and disperse in a variety of organic solvents. These organophilic clays are important constituents of oil well fluids, paints, plastics and greases. They are also used as rheological control agents in a variety of solvent systems (GRIM, 1962).

Figure 6 gives the TG and DTG thermal curves obtained for a commercial organo-clay material sold primarily for use as a component of oil drilling fluids. This organo-clay product was prepared by an ion exchange process using long chain quaternary ammonium cations and the wet-refined Black Hills bentonite clay which we have studied in both Figures 3 and 5 by DTA and TG techniques respectively. The TG and DTG thermal curves for the organo-clay product, Figure 6, show that most of the organo-cation component is oxidized away in the 200°C to 400°C temperature range.

Some of this component, however, does not leave the clay mineral until temperatures in and above the dehydroxylation temperature. Thus, more than one volatile decomposition product is emerging from the sample at elevated temperatures of this study. Since both water from the dehydroxylation event and oxidation products of the organo-cation are present, the best approach to this analysis would be through evolved gas analysis using either Fourier transform infrared spectroscopy or mass spectroscopic techniques.

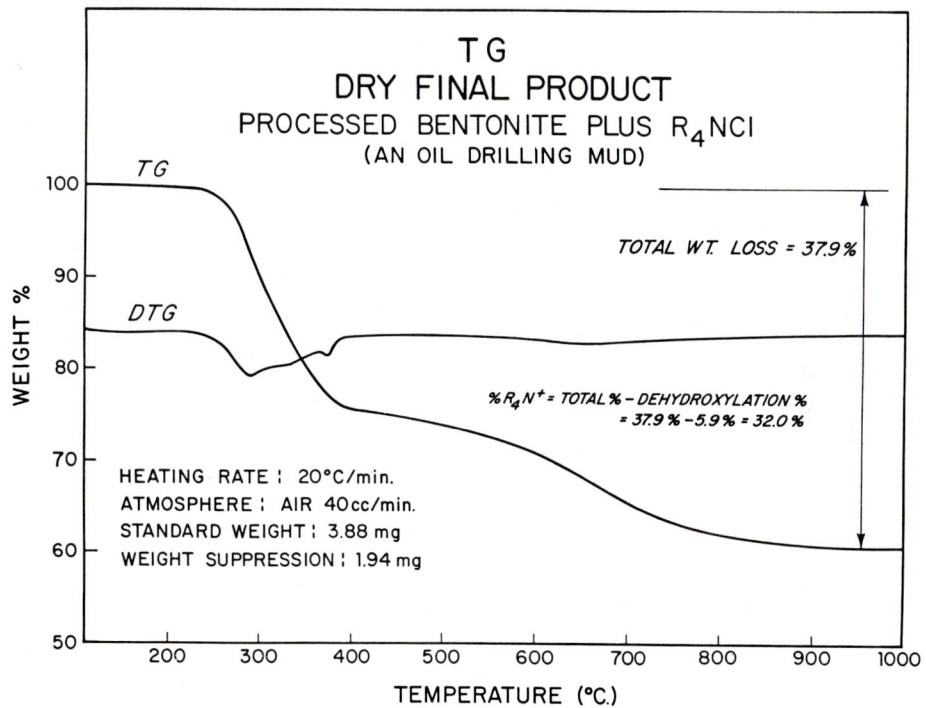

Fig. 6: TG and DTG thermal curves for organo-clay product prepared from processed Black Hills Bentonite clay, (from EARNEST, 1980; reprinted with permission of Perkin-Elmer).

DTA may also be used for such studies of organo-clay complexes. The observation of the organic additive in this case is based on the exothermic peak corresponding to the oxidation of the quaternary ammonium specie when the DTA thermal curve is obtained using a dynamic air atmosphere.

Thermal Analysis of Hectorite

The thermal behavior of most smectite clay minerals has been established by BRADLEY & GRIM (1951) and GRIM and co-workers (1957 and 1961) using both differential thermal analysis and X-ray diffraction techniques. Generally speaking, in DTA one expects to observe thermal events arising from the loss of interlayer water, dehydroxylation of the silicate structure, matrix collapse of the anhydride silicate structure, and an ordering exotherm as smectite clay mineral species are heated from ambient to ca. 1020°C. However, BRADLEY & GRIM (1951) have found that the trioctahedral members of this group do not exhibit an anhydride structure after dehydroxylation as do the dioctahedral members. Thus, a simultaneous matrix collapse is believed to occur during the dehydroxylation of hectorite and a separate endothermic event is not observed in the DTA thermal curve.

The trioctahedral smectites, saponite and hectorite, are considered to dehydroxylate at higher temperatures than those exhibited by the dioctahedral members. Due to the carbonate contamination of natural hectorite specimens, a problem exists in distinguishing the dehydroxylation event from the decomposition of the carbonate component.

EARNEST (1983a and 1983b) has studied several hectorite specimens by computerized DTA, TG and DTG. He also included acid evolved CO_2 studies, as well as plazma emission spectroscopy, in the analysis of the hectorite specimens. The work included both raw and carbonate cleaned specimens of hectorite.

Figure 7 shows the TG-DTG thermal curve as originally published by
EARNEST (1983a) for the SHCa-1 source clay hectorite specimen which is
distributed as a part of the Source Clays Collection by the Clay Mineral
Society. This specimen is fairly typical of those specimens obtained from

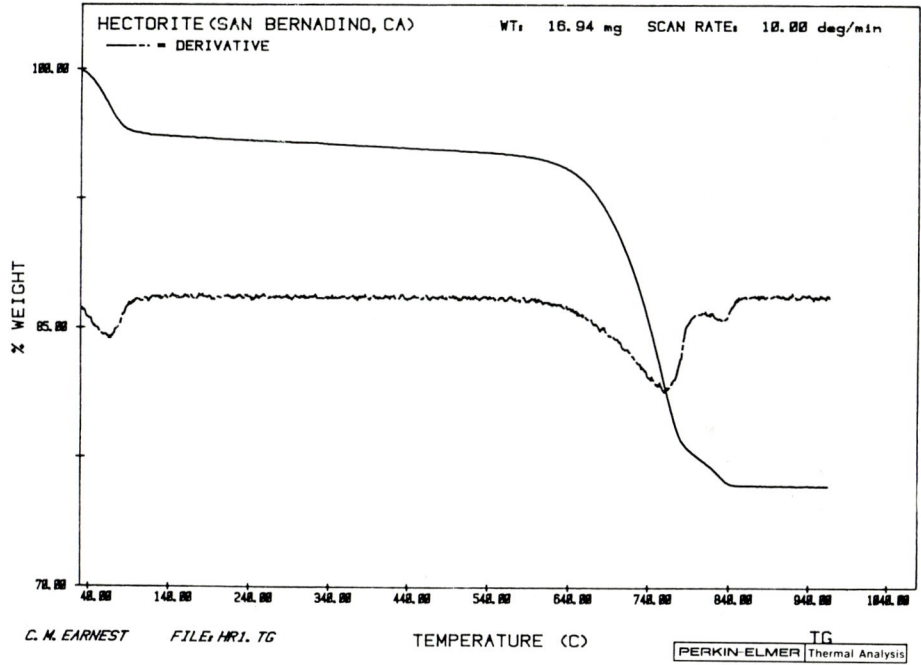

Fig. 7:   TG and DTG thermal curves for the SHCa-1 Hectorite Source Clay
Specimen (from EARNEST, 1983a; reprinted with permission of
Thermochimica Acta).

San Bernardino County, California. The thermal curve is best interpreted
by dividing the TG weight loss profile into three regions, these are
30-240°C, 240-540°C and 540-940°C. The first of these (30-240°C)
corresponds to the loss of 4.12% by weight of the original sample as

interlayer water. The second represents a slight but continuous bleed from the clay material which in this case corresponds to only 0.68% of the original sample mass. The third (540-940°C) corresponds to two overlapping thermal events. This is readily discerned by the dual peak nature of the DTG curve in this region. This corresponds to the dehydroxylation of the silicate lattice as well as the decomposition of a rather large carbonate component in the clay material and leads to a weight loss of 19.26%.

The total calcium carbonate level of the SHCa-1 hectorite source clay specimen was determined to be 40.0% based on the concentration of the Ca^{2+} ion in the dilute acid soluble portion of the raw clay specimen. On the other hand, the total carbonate anion $[CO_3]^{2-}$ content in this specimen was determined to be 25.15% using acid evolved CO_2 determination of inorganic carbon. By combining the TG weight loss data and acid evolved CO_2 data, it was concluded that the major decomposition observed in the ca. 500-850°C temperature region by hectorite ores is primarily due to the decomposition of carbonate component(s).

In support of this conclusion, EARNEST (1983a and 1988) reported the comparative TG thermal curves for both a raw hectorite ore and the resultant "carbonate cleaned" hectorite specimen. These thermal curves are given in Figure 8. One will note the drastic reduction in weight loss in the 540°C to 940°C region due to the removal of 92.4% of the carbonate component by the commercial cleaning process. Another outstanding feature of the TG thermal curve for the processed hectorite ore is that the percentage by weight of the interlayer water increases to 7.78% as a result of the removal of the carbonate contaminant. This is almost twice as much as that found in the raw hectorite ore.

Figure 8 demonstrates how the TG technique may be used to follow the industrial cleaning process to remove the carbonate component from raw hectorite ores. For many commercial applications of the trioctahedral smectite clay, the carbonate component must be minimized. In his studies of this clay mineral, EARNEST (1983a) determined the efficiency of the industrial batch cleaning process to be 92.4%. This efficiency was calculated based on inorganic carbon determinations with an automated elemental analyzer which has been previously described (EARNEST, 1983a).

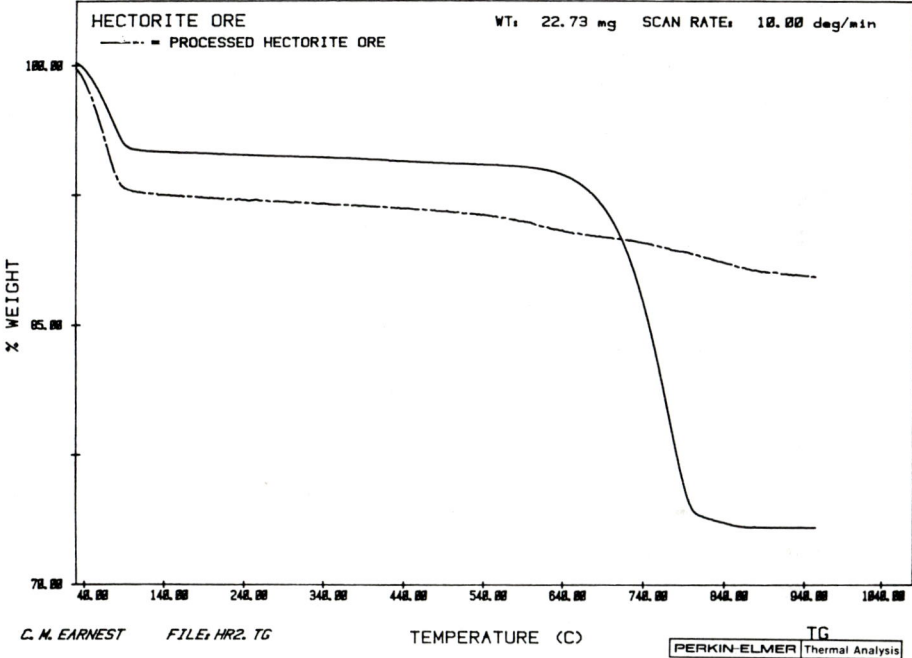

Fig. 8: TG thermal curves for a raw hectorite ore (Hector, CA) before and after processing for carbonate removal (from EARNEST, 1983a; reprinted with permission of Thermochimica Acta).

In the DTA study (EARNEST, 1983b) of the SHCa-1 hectorite source clay specimen, three different purge atmospheres were employed in the DTA sample tube. X-ray diffraction studies were also included to ascertain the identify of the high temperature phases which were formed above 1050°C.

Figure 9 shows the DTA thermal curve obtained by the author in the study of the SHCa-1 hectorite specimen in flowing nitrogen atmosphere. Essentially the same DTA thermal curve is obtained in dynamic air purge for this specimen. The endothermic peak at 121°C is due to the loss of interlayer water while the strong endothermic activity from ca. 600°C to 900°C is due to both dehydroxylation of the clay mineral and decomposition of the carbonate contaminant. The total enthalpy change associated with this endothermic activity was determined to be 524 Joules per gram of hectorite ore as is shown in Figure 10.

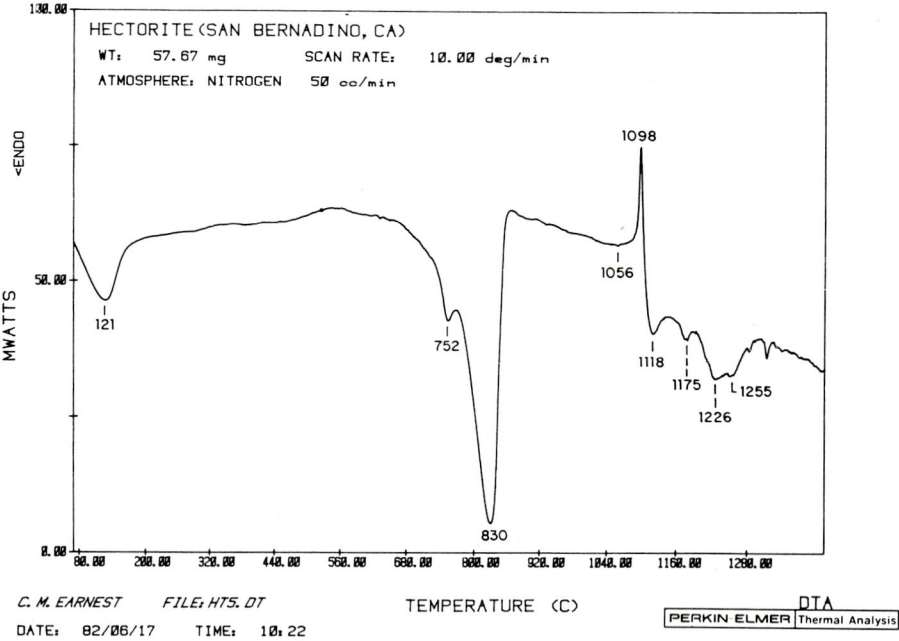

Fig. 9: DTA thermal curve for SHCa-1 Hectorite Specimen (from EARNEST, 1983b; reprinted with permission of Thermochimica Acta).

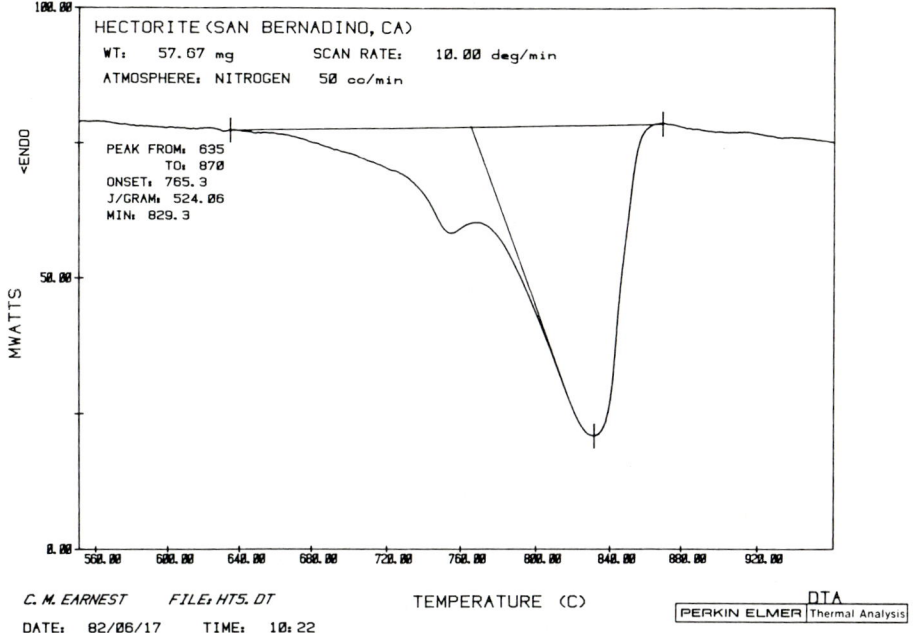

Fig. 10: Assignment of endothermic energy of transition for SCHa-1 Hectorite Specimen (from EARNEST, 1983b; reprinted with permission of Thermochimica Acta).

A sharp exothermic event (T_{max}=1098°C) followed by an immediate endothermic event and a marked endothermic shift in the baseline are also observed in the DTA thermal curve. This sharp exo-endo event was found to be very reproducible. However, as the sample size was increased to 80 mg, the exothermic portion on this event became less pronounced and a rounding effect was noted. On increasing the sample size to 100 mg, the exothermic portion of this event was completely eliminated. This suggests an atmospheric dependence due to the gaseous decomposition products arising from the preceding dehydroxylation/carbonate decomposition event.

Also, this may explain why this exothermic event has not been reported by most previous workers using either much larger samples or static air environment. Prior to this work, only SMYKATZ-KLOSS (1974) had reported an ordering exotherm in the DTA thermal curve of hectorite.

Following the exo-endo event, additional endothermic activity is observed in the DTA thermal curve. BRADLEY & GRIM (1951), SCHMIDT & HEYSTEK (1953) and MACKENZIE & CAILLERE (1979) have previously reported endothermic events in this temperature region. The sum of the endothermic areas observed between 1150°C and 1300°C in Figure 2 represents about 89 J·g^{-1} of the SHCa-1 hectorite source clay specimen (EARNEST, 1983b). BRADLEY & GRIM (1951) suggest that the major endothermic activity observed at these temperatures in their study was due to the formation of clinoenstatite.

In an attempt to further assign the phases associated with these thermal events observed at elevated temperatures (EARNEST, 1983b), numerous samples were heated in dynamic nitrogen purge in the DTA cell and stopped at various points while observing the DTA thermal curve on a XY_1Y_2 recorder. The XRD patterns showed that diopside [$CaMg(Si_2O_6)$], forsterite (Mg_2SiO_4) and clinoenstatite ($Mg_2Si_2O_6$) began to form as early as 1020°C. These species all increased in amount as the temperature was raised to 1115°C. The XRD patterns showed that these three were the major crystalline forms up to 1154°C. The difference between small sample specimens showing the sharp exotherm and large specimens which did not exhibit the exotherm in the DTA thermal curve, was determined using samples which were heated to 1115°C in the DTA cell. Those specimens exhibiting the exothermic event showed much stronger XRD patterns for all three crystalline forms than those not exhibiting the event. Hence, the difference is simply one of quantity (or crystallization rate) of the three phases.

References

ANON. (1980) - Chem Eng. News, 58, # 14, 42

BRADLEY W. F. & GRIM R. E. (1951) - High temperature thermal effects of clay and related materials.- Am. Mineral, 36, 182-201

EARNEST C. M. (1980) - Characterization of smectite clay minerals by differential thermal analysis and thermogravimetry. Part I. Montmorillonite.- Perkin-Elmer Thermal Application Study #31, 1-8, Norwalk, CT.

EARNEST C. M. (1983a) - Thermal analysis of hectorite. Part I. Thermogravimetry.- Thermochim. Acta, 63, 277-289

EARNEST C. M. (1983b) - Thermal analysis of hectorite. Part II. Differential thermal analysis.- Thermochim. Acta, 63, 291-306

EARNEST C. M. (1988) - in Compositional Analysis by Thermogravimetry ASTM STP 997.- (EARNEST C. M. (editor)), Amer. Soc. for Testing and Materials, Philadelphia, PA, 272

GABIS V. (1979) - in Data Handbook for Clay Minerals and Other Non-Metallic Minerals.- (VAN OLPHEN H. & FRIPIAT J. J. (editors)), Pergamon Press, Oxford, 128

GRIM R. E. (1957) - Etude aux rayons des réactions des mineraux argileux á hautes températures.- Bull Soc Franc. Ceram., 36, 21-28

GRIM R. E. (1962) - Applied Clay Mineralogy.- McGraw Hill, New York, 280

GRIM R. E. (1968) - Clay Mineralogy.- 2nd Ed., McGraw Hill, New York.

GRIM R. E. & KULBICKI G. (1961) - Montmorillonite: high temperature reactions and classification.- Am. Mineral, 46, 1329-1369

JONES T. R. (1983) - The properties and uses of clays which swell in organic solvents.- Clay Minerals, 18, 399-410

KERR P. F., KULP J. L. & HAMILTON P. K. (1949) - Differential thermal analysis of reference clay specimens.- A.P.I. Project 49, Columbia University, New York, 36

LUCAS J. & TRAUTH N. (1965) - Study of high temperature behavior of montmorillonite.- Bull. Serv. Cart. Geologique de l'Alsace Lorraine, 18, 4

MACKENZIE R. C. & CAILLERE S. (1979) - in Data Handbook for Clay Minerals and Other Non-Metallic Minerals.- (VAN OLPHEN H. & FRIPIAT J. J. (editors)), Pergamon Press, Oxford, 244

MOLL W. F. (1979) - in Data Handbook for Clay Minerals and Other Non-Metallic Minerals.- (VAN OLPHEN H. & FRIPIAT J. J. (editors)), Pergamon Press, Oxford, 109

ROSS C. S. & HENDRICKS S. B. (1945) - Minerals of the montmorillonite group.- U.S. Geol. Surv. Prof. Paper 205B, 23-79

SCHMIDT E. R. & HEYSTEK H. (1953) - A saponite from Krugersdorp district, Transvaal.- Min. Mag., 30, 201-210

SMYKATZ-KLOSS W. (1974) - Differential Thermal Analysis. Application and Results in Mineralogy.- Springer Verlag, Berlin.

TODOR D. N. (1976) - Thermal Analysis of Minerals.- Abacus Press, Kent, UK.

REMARKS ON THE APPLICABILITY OF THERMAL ANALYSIS FOR THE INVESTIGATIONS OF CLAYS AND RELATED MATERIALS

Anna Langier-Kuzniarowa

Geological Institute
Warsaw, Poland

Abstract

The paper reviews the application of thermal methods in clay mineralogy. It stresses the different improved methods and equipments, deals with the identification and differentiation of clay mineral species and with the determination of special properties by means of thermal and calorimetric methods. Particularly, the interactions of swelling clay minerals with organic molecules (stearic acid, rhodamine, pesticides) and their thermal characterization is treated. Finally, the possibilities and limitations of thermal characterization and determination of clay minerals are briefly summarized.

Introduction

The history of thermal examinations of clays has been long and closely connected with the development of theory and instrumentation. The rapid development of thermal investigations applied to geosciences, and then to clay mineralogy and petrology, began in the 1950's, when thermal analysis, especially DTA, seemed to be a perfect tool for the identification of clay minerals, which occur mostly in natural associations: clay rocks and soils, and sometimes in essentially monomineral geological occurrences. For this purpose the excellent book of MACKENZIE (1957), has been an essential, and still commonly used comprehensive source of DTA data for clay and accompanying minerals. But the significance of the separate DTA method has decreased simultaneously with the development of the knowledge and the recognition of very complicated structures of differentiated but generally similar clay minerals.

In the 1960's combined thermal methods came into wide-spread use with development of the first instrument for simultaneous DTA-TG-DTG methods, also with the possibility of joining dilatometry (TD-DTD), using the "derivatograph" developed by PAULIK F., PAULIK J. & ERDEY (1958). Successive generations of this instrument have appeared and have been further updated by the addition of equipment for the titration of the gaseous products of thermal reactions, and other attachments including a computer. The derivatograph has been commonly used for the studying of clay minerals, rocks and related materials and the atlas of derivatographs of clay minerals has been published already by LANGIER-KUZNIAROWA (1967).

In addition by the early 1960's, analogous instruments have been produced by several world-known firms as Mettler Instrumente AG, Stanton-Redcroft Ltd., Netzsch-Gerätebau GmbH, Linseis Messgeräte GmbH, DuPont Instruments, Rigaku Co., Setaram S.A., Deltatherm and others. These demonstrate a very high technological level, variety of experimental possibilities and elegance of workmanship. Most of them subsequently have joined these TA units to MS, EGA, and other analytical instruments.

Nowadays usually other techniques have also been used in combination with thermal treatments, e.g. X-ray diffraction, TEM, SEM, IR, NMR, and various analytical methods for the determination of the composition. Also calorimetric methods have been employed for examinations of clays, sometimes using new types of calorimeters. Simultaneously with the development of thermal methods, their applications for clay mineralogy and petrology, soil science, and research for different branches of industry, have proceeded and developed. In the last few years the thermal investigations of clays has continued to progress and evolve. Initially the identification of minerals was the main purpose of thermal studies, but subsequently one may observe further various applications. Namely simultaneously with deeper recognition of the nature of clay minerals, thermal studies began to contribute to the detailed recognition of the structure and specific properties of clay minerals. Further attempts have

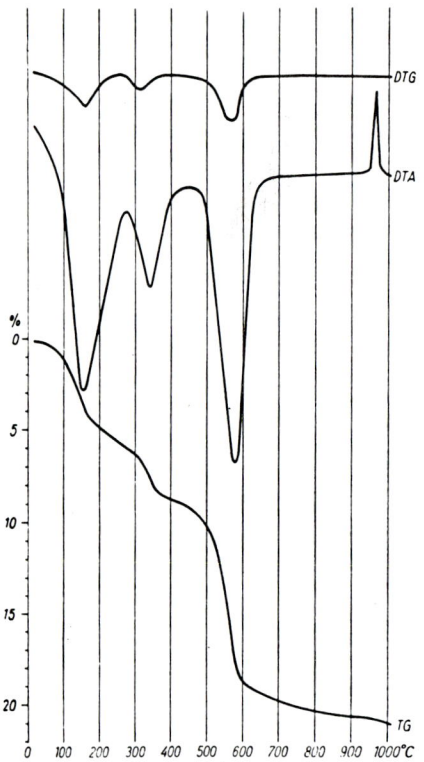

Fig. 1: Hydrated halloysite, Eureka (Utah, USA), grain size < 2 µm, showing two types of water, about 7% of gibbsite admixture.

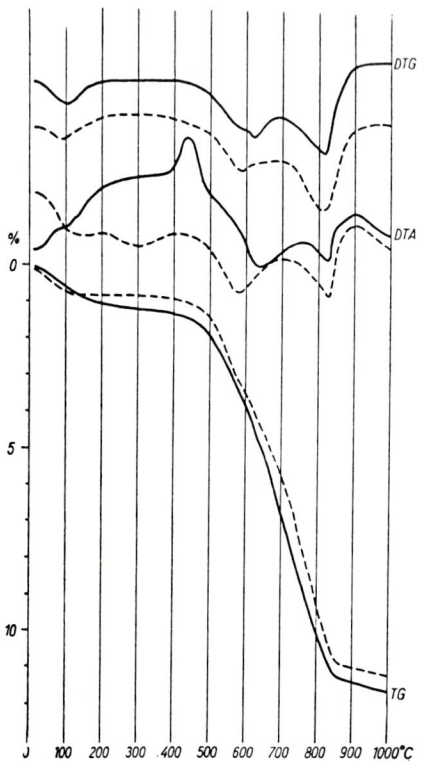

Fig. 2: Silurian graptolite shale analysed in air (continuous line) and nitrogen (broken line) atmosphere. Example for the influence of experimental conditions: DTA endothermic peaks following exothermic peak shift towards higher temperatures.

been made to develop methods and instrumentation for the quantitative determination of clay minerals in natural associations in rocks and soils and artificial mixtures (e.g. for ceramics). Simultaneously thermal methods have lost their initial significance as the rapid and simple method of the identification of component minerals in clay-containing rocks. On the other

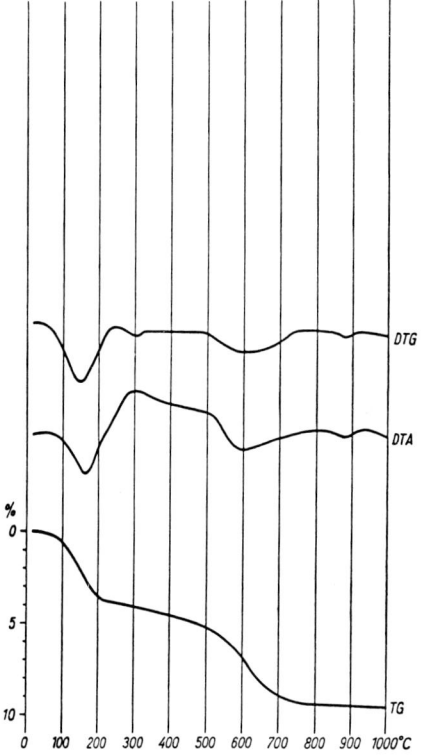

Fig. 3: Illite clay, Goose Lake (Ill., USA). Derivatograph shown for the comparison with Fig. 4.

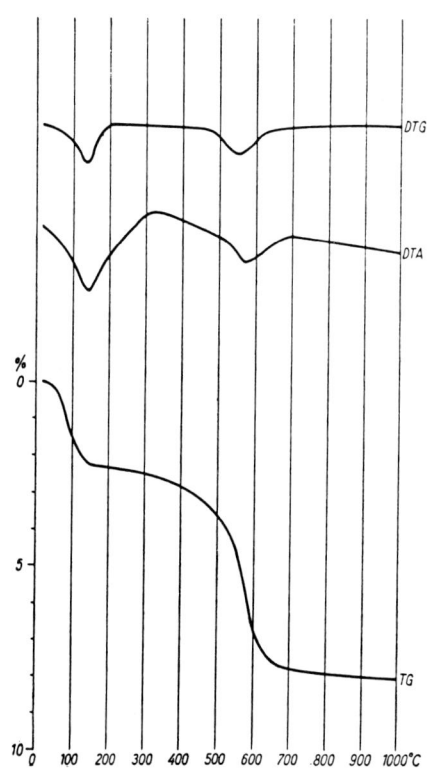

Fig. 4: A regular mixed layer I/M ratio = 1:1 (Burghersdorp, Cape Province, South Africa).

hand, although the identification of separate minerals, both natural, as found in geological occurrences or separated from rocks, has limited application, this direction of thermal analysis keeps its initial position (Fig. 1). However, the majority of identification needs deal with multiple rocks, soils and raw materials and after the initial fascination, thermal analysis results appeared to be sometimes undiagnostic and non-reproducible. This was due to dependence on variable experimental conditions, difficulties in interpretation for various reasons such as the coincidence of thermal effects originating from different clay minerals or due to variability of peak temperatures, etc. (Fig. 2-8).

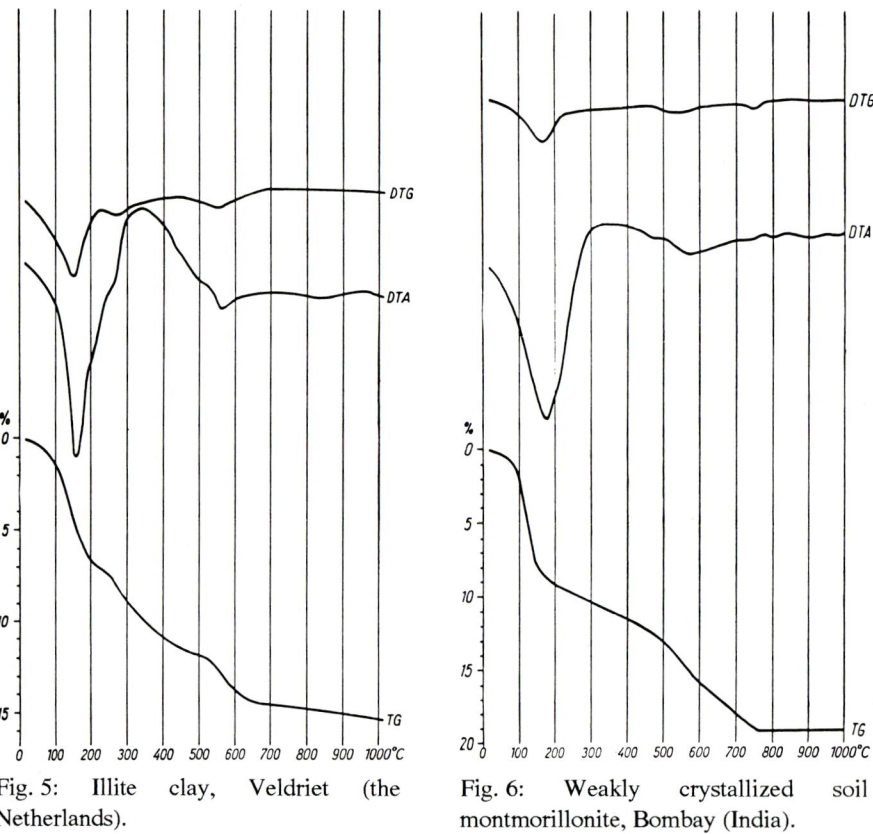

Fig. 5: Illite clay, Veldriet (the Netherlands).

Fig. 6: Weakly crystallized soil montmorillonite, Bombay (India).

Thus weakly ordered smectites, e.g. occurring in soils, show the dehydroxylation peak in the 500-600°C temperature range, which is similar to that of illite, and considerably lower than the temperature of dehydroxylation of well ordered smectites found in the majority of geological occurrences. Also the interaction between the accompanying minerals occurring in the clay rocks, e.g. between iron sulphides and carbonates when heated in nitrogen atmosphere, results in unexpected phenomena (LANGIER-KUZNIAROWA, 1969), giving a complicated run of thermal curves, sometimes making impossible to interpret them correctly. Also sometimes very remarkable influences of furnace atmosphere, e.g. CO_2 atmosphere, on thermal dissociation of carbonate minerals, frequently

Fig. 7: Silurian shale (illite and chlorite as the main minerals), containing calcite and pyrite, analysed in nitrogen atmosphere, shows the increase of weight in high temperature range due to the interaction of components.

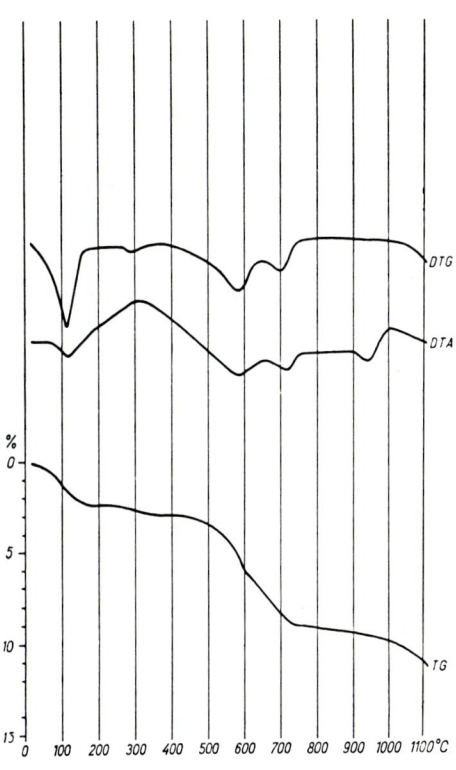

Fig. 8: K-bentonite, Oak Hall (Pennsylvania, USA), consisting of mixed-layer illite-montmorillonite, shows thermal effects as from the mixture of component minerals.

occurring as the associate minerals in clay rocks, soils and raw materials, may increase the difficulties in the right interpretation of the thermal results (compare the "PA-curves" defined by SMYKATZ-KLOSS (1974), which consider the influence of the partial pressure of CO_2 on the decomposition of carbonate minerals).

These examples show some of the reasons for the apparent regression in the application of thermal analysis for the determination of the mineralogical composition of clay-containing associations in recent times.

In spite of this tendency, at the present time one can observe an evident increase in the number of papers dealing with thermal analysis applied to clays, and used to solve many particular problems by means of various techniques, e.g. the great number of papers concerned with studies of adsorption phenomena as well as interaction of clays with various solids and reagents, especially those resulting in intercalation of clay minerals.

Main Trends in Modern Thermal Studies of Clays

Papers presented at the 9th ICTA Congress at Jerusalem in 1988 and papers published recently in thermo-analytical and clay-mineralogical journals give representative examples of modern trends and applications.

Thus SINGER & HUANG (1988) carried out DTG and isothermal heating examinations on the hydroxy Al polymer/montmorillonite complexes prepared in the presence of humic acid extracted from an Orthic Black Chernoziemic soil from Southwestern Saskatchewan, Canada. The results obtained it was concluded that some organic matter from the humic acid may have penetrated into the interlayer space of the montmorillonite, together with aluminium, and also the presence of humic acid appears to have decreased the AlOH-polymer interlayering in montmorillonite. These authors also concluded that in this organo-clay complex there are specific sites in which the organic matter is protected from thermal reactions, probably inside the interlayers. Further the results of tracing carbon elimination from humic acid in the AlOH-/montmorillonite/HA complexes have been interpreted as evidence for the interlayering of a fraction of the humic acid in montmorillonite. Two final conclusions have been drawn: that the presence of humic acid decreases AlOH-polymer interlayering in montmorillonite, and that some organic material from the humic acid may have penetrated into the interlayer space of montmorillonite together with the Al-polymer.

SHUALI and co-authors (1988) studied the mechanism of adsorption of D_2O by sepiolite and palygorskite, and structural implications resulting from thermal studies carried out by means of DTA, TG and EGA methods. They showed, that the thermal dehydroxylation of sepiolite differs from that of palygorskite with dehydroxylation taking place in two and one stages respectively. The second endothermic dehydroxylation peak of sepiolite is followed by a sharp endothermic peak. The final conclusion was that: the second stage of the dehydroxylation of sepiolite is not associated with the decomposition of residual TOT units but results from the decomposition of secondary units, formed during the first stage of dehydroxylation of this mineral. The work of YARIV, KAHR & RUB (1988) dealt with the thermal examination (by means of DTA, TG and DTG) of the mode of adsorption of rhodamine 6G by smectite minerals. Recently YARIV, MÜLLER-VONMOOS, KAHR & RUB (1989) extended these studies by using the same smectite minerals treated with acridine orange in order to identify the mechanism of thermal reactions of smectites from the behaviour of the smectites treated with the dyes under consideration. The same authors (YARIV et al., 1989) tested by thermal analysis (DTA, TG, DTG, MS and the X-ray method) the effects of adsorption of crystal violet on montmorillonite. It was also studied by HEPLER, YARIV & DOBROGOWSKA (1987) using the calorimeter method.

Also YARIV (1985) studied the adsorption of organic molecules on clay minerals to identify the types of association between water molecules, exchangeable metallic cations and organic molecules or ions formed on clay surfaces. YARIV & HELLER-KALLAI (1984) analysed by means of DTA the interactions between stearic acid and allophane, sepiolite, palygorskite, pyrophyllite or talc, and SIDHESWARAN et al. (1987); GABOR et al. (1989); HORTE et al. (1987 and 1988) studied the thermal behaviour of intercalated kaolinite.

The studies of intercalation of clays were also carried out employing the calorimetric method (SCHMALSTIEG et al., 1988; BLUMENTHAL et al., 1988). BLUMENTHAL et al. (1988) described an original method for the quantitative determination of kaolinite in natural or artificial mixtures, prepared for technological purposes, by means of DCA method using hydrazine saturated clay. Also several other works belong to the sphere of intercalation chemistry (BAHRANOWSKI, 1988; BANDOSZ et al., 1988; FIJAL, 1988). Recently minerals of the sepiolite-palygorskite group were examined after saturation with organic amines (SHUALI et al., 1989), while important agriculture and environmental protection studies were carried out on pesticides-montmorillonite complexes, by MAZA RODRIGUEZ et al. (1988) and by SANCHEZ-MARTIN & SANCHEZ-CAMAZANO (1989). MACKENZIE & RAHMAN (1987), MACKENZIE et al. (1988) and HELLER-KALLAI & MACKENZIE (1989) investigated the interaction of kaolinite with calcite on heating.

There are other examples dealing with the kinetics of dehydration and dehydroxylation, the determination of the enthalpy of dehydration and other thermodynamic values (WAZLAWSKA, 1986; KOSTER VAN GROOS & GUGGENHEIM, 1986; GIRGIS et al., 1987; GIRGIS & FELIX, 1987; GIRGIS et al., 1987; PETZOLD & RUDOLF, 1988; MINATO & NAMBA, 1985).

Independent of the above thermal investigations of clays, are a number of natural science publications demonstrating the classical approach to problems dealing with clay mineralogy, but very modern in their theoretical basis and experimental level. Thus the work of KHORAMI & LEMIEUX (1989) compared attapulgites from different sources using thermogravimetry and IR, and KIYOHIRO & OTSUKA (1989) have investigated the dehydration mechanism of bound water in sepiolite, by means of high-temperature X-ray diffractometry and TG under controlled water vapour pressure, including kinetic analysis of the resultant TG curves.

Also thermal and Mössbauer spectroscopy studies on glauconite, carried out by MACKENZIE et al. (1988) belong to this group of thermal analysis application, together with studies by MINATO (1988) on dehydration stages determined by TG and DTG and dehydration energies measured by the DSC method of 10 Å-halloysites of various geological origins and with different particle morphologies. Here should also be included the reassessment, on the basis of thermal analysis, the samples described previously in literature as volkonskoite (FOORD et al., 1987) plus the characterization of an unusual vanadium chlorite (WHITNEY & NORTHROP, 1986), the characterization of well ordered kaolinites (STÖRR & MURRAY, 1986; LOMBARDI et al., 1987), of corrensite (BRIGATTI & POPPI, 1985), chrysotile (CRAW et al.,1987), expanding clay minerals (montmorillonite, beidellite, vermiculite) characterized by NAGASAWA & OHKOCHI (1988) and by X-ray studies of dehydration and rehydration of homoionic samples. Also the dilatometric technique was used for the identification and characterization of clays (KHAJURIA et al., 1986). In some cases the appearance of some new equipment has provided an impetus for further testing in the study of clays (REISZ & INCZEDY, 1986; PAULIK F. et al., 1989; LANGIER-KUZNIAROWA, 1990).

Another group of work deals with the clay-water system in different aspects of this problem, e.g. with the influence of freezing-thawing cycles on the thermal behaviour of a clay (STEPKOWSKA, 1988) or with the phenomena occurring in clay suspensions during storage, drying and/or heating (STEPKOWSKA et al., 1988).

Another topic, namely the thermal characterization of a modified, TiO_2-cross-linked montmorillonite obtained by Ti-introduction into montmorillonite, is noteworthy (STERTE, 1986).

References

BAHRANOWSKI K. (1988) - Polish Conf. on Clays, Warszawa, Symp.-Abs. 4a.

BANDOSZ T., JAGIELLO T. & ZYLA M. (1988) - Polish Conf. on Clays, Warszawa, Symp.-Abs. 5a.

BLUMENTHAL G., OLIEW G. & KRANZ G. (1988) - 7. Tonmin.-Tagg. Greifswald, Abs. S.5

BRIGATTI M. F. & POPPI L. (1985) - Clays and Clay Min., 33, 128

CRAW D., LANDIS C. A. & KESLEY P. I. (1987) - Clays and Clay Min., 35, 43

FIJAL J. (1988) - Polish Conf. on Clays, Warszawa, Symp.-Abs. 27a.

FOORD E. E., STARKEY H. C., TOGGART J. E. JR. & SHAWE D. R. (1987) - Clays and Clay Min., 35, 139

GABOR M., PÖPPL L., IZVEKOV V. & BEYER H. (1989) - Thermochim. Acta, 148, 431

GIRGIS B. S. & FELIX N. S. (1987) - J. Thermal Anal., 32, 1867

GIRGIS B. S., EL-BARAWY K. A. & FELIX N. S. (1987) - Thermochim. Acta, 111, 9

GIRGIS B. S., FELIX N. S. & EL-BARAWY K. A. (1987) - Thermochim. Acta, 112, 265

HELLER-KALLAI L. & MACKENZIE R. C. (1989) - Thermochim. Acta, 148, 439

HEPLER L. H., YARIV S. & DOBROGOWSKA C. (1987) - Thermochim. Acta, 121, 373

HORTE C.-H., BECKER CH., KRANZ G., SCHILLER E. & WIEGMANN J. (1987) - 4th ESTAC-Symp. Jena, Workbook F-4.

HORTE C.-H., BECKER CH., KRANZ G., SCHILLER E. & WIEGMANN J. (1988) - J. Thermal Anal., 33, 401

KHAJURIA H. R., MEHTA S. K. & GUPTA A. A. (1986) - J. Thermal Anal., 31, 15

KHORAMI J. & LEMIEUX A. (1989) - Thermochim. Acta, 138, 97

KIYOHIRO T. & OTSUKA R. (1989) - Thermochim. Acta, 147, 127

KOSTER VAN GROOS A. F. & GUGGENHEIM S. (1986) - Clays and Clay Min., 34, 281

LANGIER-KUZNIAROWA A. (1967) - Termogramy mineralow ilastych.- Wyd. Geol., Warszawa, p. 3-16

LANGIER-KUZNIAROWA A. (1969) - J. Thermal Anal., 1, 47

LANGIER-KUZNIAROWA A., INCZEDY J., KRISTOF J., PAULIK F., PAULIK J. & ARNOLD M. (1990) - J. Thermal Anal., 36, 67

LOMBARDI G., RUSSELL J. D. & KELLER W. D. (1987) - Clays and Clay Min., 35, 321

MACKENZIE K. J. D., CARDILE C. M. & BROWN I. W. M. (1988) - Thermochim. Acta, 136, 247

MACKENZIE R. C. (1957) - The Differential Thermal Investigation of Clays.- Min. Soc. London., 456 pp.

MACKENZIE R. C. & RAHMAN A. A. (1987) - Thermochim. Acta, 121, 51

MACKENZIE R. C., RAHMAN A. A. & MOIR H. M. (1988) - Thermochim. Acta, 124, 119

MAZA RODRIGUEZ J., JIMENEZ-LOPEZ J. & BRUQUE S. (1988) - Clays and Clay Min., 36, 284

MINATO H. (1988) - Thermochim. Acta, 135, 279

MINATO H. & NAMBA H. (1985) - 8th Int. Clay Conf. Denver, Symp.-Abs. 157

NAGASAWA K. & OHKOCHI N. (1988) - Thermochim. Acta, 135, 285

PAULIK F., PAULIK J. & ERDEY L. (1958) - Z. Anal. Chemie, 160, 241

PAULIK F., PAULIK J., ARNOLD M., INCZEDY J., KRISTOF J. & LANGIER-KUZNIAROWA A. (1989) - J. Thermal Anal., 35, 1849

PETZOLD D. & RUDOLF B. (1988) - 7. Tonmin.-Tagg. Greifswald, Abs. S.25

REISZ K. & INCZEDY J. (1986) - J. Thermal Anal., 31, 611

SANCHEZ-MARTIN M. J. & SANCHEZ-CAMAZANO M. (1989) - Thermochim. Acta, 141, 317

SCHMALSTIEG A., BLUMENTHAL G. & WIEGMANN J. (1988) - 7. Tonmin.-Tagg. Greifswald, Abs. S.27

SHUALI U., YARIV S., STEINBERG M., MÜLLER-VONMOOS M., KAHR G. & RUB A. (1988) - Thermochim. Acta, 135, 291

SHUALI U., BRAM L., STEINBERG M. & YARIV S. (1989) - Thermochim. Acta, 148, 445

SIDHESWARAN P., GANGULI P. & BHAR A. N. (1987) - Thermochim. Acta, 118, 295

SINGER A. & HUANG P. M. (1988) - Thermochim. Acta, 135, 307

SMYKATZ-KLOSS W. (1974) - Differential Thermal Analysis. Application and Results in Mineralogy.- Springer-Verlag, Berlin, 188 pp.

STEPKOWSKA E. T. (1988) - Thermochim. Acta, 135, 313

STEPKOWSKA E. T., PEREZ-RODRIGUEZ J. L., JUSTO A., SANCHEZ SOTO P. J. & JEFFERIS S. A. (1988) - Thermochim. Acta, 135, 319

STERTE J. (1986) - Clays and Clay Min., 34, 658

STÖRR M. & MURRAY H. H. (1986) - Clays and Clay Min., 34, 689

WAZLAWSKA J. (1986) - Mineralogica Polonica, 15, 91

WHITNEY G. & NORTHROP H. R. (1986) - Clays and Clay Min., 35, 321

YARIV S. (1985) - Thermochim. Acta, 118, 49

YARIV S. & HELLER-KALLAI L. (1984) - Chem. Geol., 45, 313

YARIV S., KAHR G. & RUB A. (1988) - Thermochim. Acta, 135, 299

YARIV S., MÜLLER-VONMOOS M., KAHR G. & RUB A. (1989) - J. Thermal Anal., 35, 1997

YARIV S., MÜLLER-VONMOOS M., KAHR G. & RUB A. (1989) - Thermochim. Acta, 148, 457

DIFFERENTIAL THERMAL ANALYSIS (DTA) OF ORGANO-CLAY COMPLEXES

S. Yariv

Department of Inorganic and Analytical Chemistry
The Hebrew University of Jerusalem
Jerusalem, Israel

Abstract

DTA is used in the study of organo-clay complexes. Curves are recorded either in oxidizing or reducing environments. For the interpretation of the different peaks DTA is usually supplemented by TG and in recent years also by EGA. Most DTA curves can be divided into three regions: (I) the region of the dehydration of the clay; (II) the region of the reactions of the organic matter and (III) the region of the dehydroxylation of the clay. An oxidizing environment results in intense exothermic peaks whereas a reducing environment results in small broad endothermic peaks. Consequently, most investigations were carried out in the presence of air. The characteristic features of the exothermic peaks and the different applications of the method are discussed.

Introduction

Although DTA has been used very widely in the study of clay minerals (see e.g. MACKENZIE, 1957; MACKENZIE, 1970; SMYKATZ-KLOSS, 1974; LANGIER-KUZNIAROWA, 1989), its application for the study of adsorption of organic materials by clay minerals was relatively small (YARIV, 1985). In this respect it has been used for the following purposes: (I) to establish whether adsorption complexes were formed or whether the clay and organic material were present merely as mechanical mixtures; (II) to study the effects of the organic material on the adsorbed water content and properties; (III) to study the thermal properties of the adsorbed organic matter; (IV) to identify and differentiate between different clay minerals and (V) to establish the type of association between organic species and exchangeable metallic cations, water molecules or silicate layers.

DTA curves of organic substances recorded in oxidizing and inert environments show diagnostic exothermic and endothermic peaks which are principally associated with combustion, decomposition, dehydration, sublimation, vaporization, fusion and solid-state transitions (MITCHELL & BIRNIE, 1970). Some of these peaks disappear, the temperatures of other peaks are shifted and their relative intensities are changed as a result of adsorption of the organic substances by clay minerals.

Thermal analysis of organo-clay complexes is carried out either in an oxidizing environment (in air or under a flow of oxygen) or under a flow of an inert gas (such as nitrogen or argon). In air or under oxygen, the adsorbed organic material is oxidized, giving rise to intensive exothermic peaks. Under an inert atmosphere, weak endothermic peaks are obtained which are attributed to desorption and pyrolysis of the organic material. Since these endothermic peaks are very weak, most DTA studies of organo-clay complexes were carried out in air or oxygen atmospheres.

In most studies DTA was combined with thermogravimetry and hence, one could differentiate between peaks associated with weight loss and those associated with phase transitions. Until recently the identification of the reactions which were associated with weight loss and the attribution of most of these reactions to the various peaks, were merely speculative and the DTA curves were used only as fingerprints indicating whether there was only one or more species present and to differentiate between various types of association adsorbed on the clay mineral, such as ammonium ions, hydrated or non-hydrated, free amines or amines coordinating transition metallic cations, etc.. The differentiation between the various clay organic associations is possible because each variety gives rise to the appearence of characteristic exothermic peaks.

Recently a combined DTA-EGA (evolved gas analysis) was applied to the study of organo-clay complexes (YARIV et al., 1988; YARIV et al., 1989a; YARIV et al., 1989b; YARIV et al., 1989c; SHUALI et al., 1990; SHUALI et al, in press). In this combined technique the evolved gases (combustion or thermal decomposition products) are analyzed (e.g. by mass-spectrometry or IR spectroscopy, gastitrimetry and water detector) simultaneously with the DTA and thus an unequivocal interpretation of the peaks can be obtained (MÜLLER-VONMOOS et al., 1977; PAULIK F. et al., 1989). Information on the oxidation of carbon or nitrogen is directly obtained from CO_2 and NO_2 evolution curves respectively. The information on the oxidation of hydrogen from the organic source is more complicated, as water is evolved, in part due to the oxidation of organic matter and in part due to the dehydration of the clay. Conclusions on the combination of hydrogen atoms originating from the organic compound with oxygen can be obtained by subtracting the H_2O evolution curve of the organo-clay complex recorded under N_2 from that recorded in air. The difference between the two curves may give the required information on the oxidation of the hydrogen originating from organic matter. The H_2O evolution curve obtained in air is called the "total water evolution curve" whereas that obtained under N_2 is called the "inorganic water evolution curve". The calculated curve is called the "organic water evolution curve".

Most DTA studies of organo-clays were carried out with montmorillonites and the conclusions which are summarized in the present review are mainly valid for this mineral. This must be taken into consideration when other minerals are treated.

The Shape of DTA Curves of Organo-Clay Complexes

DTA curves of most organo-clay complexes, recorded either in an oxidizing environment or under an inert atmosphere, can be divided into three regions: (1) the region of dehydration of the clay, (2) the region of the thermal reactions of the organic material and (3) the region of the dehydroxylation and recrystallization of the meta-clay. The shape of the first region does not depend on whether the DTA is recorded in an oxidizing or an inert atmosphere. The thermal dehydration of the clay usually occurs in the same temperature range in which clay samples, not treated with organic material (untreated clays), lose their adsorbed water, that is, below 200°C-250°C. The dehydration is shown by an endothermic peak. Since some of the water is replaced by the organic matter, the size of this peak, recorded in the DTA curve of the organo-clay complex, is smaller than that recorded in the DTA curve of the untreated clay. The presence of organic molecules on the clay surface makes it hydrophobic and, consequently, the peak maximum in the curve of the organo-clay appears at a lower temperature compared to the untreated clay.

If the adsorbed organic matter has a high vapor pressure or a low boiling point, some of it is also evolved in the temperature range of this region, giving rise to endothermic peaks. For example, Mg-montmorillonite saturated with methanol, n-butanol, ethylene-glycol and glycerole give endothermic DTA peaks at 20°C, 92°C, 200°C and 238°C, respectively, in addition to the water evolution peak at 70-90°C (ELTANTAWY, 1974).

From the evolution temperatures one can get information on whether or not the organic matter is adsorbed by the clay and the strength of interaction between the organic matter and the mineral. For example, evolution of adsorbed glycols gave endothermic peaks at temperatures which were higher than the boiling points of the non-adsorbed glycols (McNEAL, 1964). Ca-montmorillonite, saturated with excess ethylene glycol, showed an endothermic peak at 150°C due to the vaporization of some free glycol condensed in the pores of the mineral, in addition to the peak at 200°C which was due to the vaporization of glycol from the interlayer space (ELTANTAWY, 1974). BODENHEIMER et al. (1966a) compared DTA curves of Na- and Cu-montmorillonites saturated with similar amounts of several glycols. In the presence of copper, the low temperature endothermic reactions were much less pronounced, presumably owing to the complexing of some of the glycols as ligands coordinating the metal cations.

The evolution peak of the adsorbed organic matter sometimes overlaps that of the water evolution and a single peak is traced in the DTA curve. For these cases the evolution of the organic matter can be detected by applying EGA technique. BODENHEIMER et al. (1963a, 1963b, 1963c, 1966a and 1966b) and YARIV et al. (1971), while recording DTA of montmorillonites saturated with various amines, showed the presence of a large endothermic peak overlapping the dehydration peak. This peak was attributed to the loss of non-adsorbed amine at relatively low temperatures, at or slightly above the boiling point of the amine. Adsorbed amines are evolved at temperatures higher than their boiling points. Liberation of amines from adsorbed ammonium ions occurs at still higher temperatures. The latter peak could be detected only when the DTA was carried out under nitrogen.

The shape of the second region depends on whether the DTA is recorded under an oxidizing or inert atmosphere. In oxidizing environments this region represents the oxidation of the organic substances and the evolution of H_2O and CO_2. The DTA shows several exothermic peaks. The exothermic peaks are not observed when the DTA is performed under an inert atmosphere. Instead, broad endothermic peaks are detected. Mass-spectrometer analysis of evolved gases (EGA) showed that in the case of sepiolite and palygorskite treated with butylamine, the following gases were detected evolving in the temperature range of the second region, when the DTA was carried out under nitrogen: butylamine (by desorption), NH_3 and CH_4 (by degradation), H_2 (due to condensation of C to charcoal), traces of CO_2 (by thermal hydrolysis) and very small amounts of propane, propanol, ethanol, acetic acid and butene (SHUALI et al., 1990). However, the resolution of the DTA endothermic peaks was very poor and it was impossible to relate the endothermic peaks to certain reactions. ELTANTAWY (1974) attributed an endothermic peak at 330°C in the DTA curves of Ca- or Mg-montmorillonite treated with ethylene glycol, recorded under nitrogen, to a pyrolysis and decomposition of ethylene glycol which was not liberated in the temperature range of the first region but remained associated with the mineral, (probably by coordination to the exchangeable Ca^{2+} or Mg^{2+}). The decomposition was accompanied by the deposition of carbon film on the mineral surface. If N_2 gas was replaced by O_2 after the 330°C endotherm, a sharp exotherm appeared immediately due to the oxidation of the deposited carbon, with a peak maximum at different temperatures dependent on the oxygen flow rate and hence the availability of O_2 for the reaction.

The third region of the DTA curves of most organo-clay complexes (above 550-650°C) shows the endothermic peak of dehydroxylation. When the dehydroxylation is followed by recrystallization, as it is in some trioctahedral clay minerals, the endothermic peak is followed by an exothermic peak. In some cases the oxidation and pyrolysis of the organic matter are not completed in the temperature range of the second region and are traced in the third region by giving rise to new peaks which overlap the dehydroxylation peaks of the clay minerals. In these cases the dehydroxylation can be traced from H_2O evolution curves and the combustion of the organic matter is best studied from CO_2 evolution curves. EGA studies showed that the dehydroxylation temperature of the organo-clay is sometimes lower than that of the untreated clay (YARIV et al., 1988; YARIV et al., 1989a; YARIV et al., 1989b; YARIV et al., 1989c). The reason for this shift is not known. With several minerals the dehydroxylation occurs below 600°C. In these cases there is an overlapping between the second and third region of the DTA curves.

Water which is evolved during the dehydroxylation of the clay mineral, may react to a very small extent, with the organic matter (CHI CHOU & MCATEE, 1969). Some authors call this reaction "oxidation of carbon", but it is more correct to define it as "thermal hydrolysis" (SHUALI et al., 1990). This reaction results in the evolution of hydro-carbon, CO and CO_2, and is detected only when the thermal analysis is carried out in an inert atmosphere.

The dehydroxylation of sepiolite takes place in two stages, giving rise to two endothermic peaks at about 525°C and 830°C. The second dehydroxylation stage is followed by an exothermic recrystallization of meta-sepiolite to clinoenstatite. During the first stage of dehydroxylation the ordered mineral undergoes a transformation into a non-crystalline meta-sepiolite phase. This transformation is not merely the result of the thermal diffusion of protons and HOH molecules from inside the bulk of the silicate phase towards the outside, but is associated with the breaking of different Si-O, Mg-O and Al-O bonds in the clayey silicate TOT units. In the new structural units of the meta-sepiolite some of these broken bonds are temporarily located at the edges of the silicate species and molecules such as CO_2 and NH_3, which have been obtained from oxidation and pyrolysis, respectively of the organic molecules, and are chemically adsorbed at these active sites. The silicate species grows while the adsorbed molecules are trapped inside the bulk amorphous phase. The second dehydroxylation stage is followed by an exothermic recrystallization of the meta-sepiolite to clinoenstatite. At this stage Si, Mg, Al and O atoms diffuse from the poorly ordered meta-sepiolite species to the well ordered units of clinoenstatite. At this stage the adsorbed CO_2 and NH_3 molecules are able to escape (SHUALI et al., 1988).

The Exothermic Peaks

In our laboratory DTA curves of several organo-clay complexes were recorded in inert and oxidizing environments. Since the oxidation reactions are highly exothermic the very intense exothermic peaks dominate the resulting curves. Reliable information on organo-clay associations was obtained mainly from DTA recorded in oxidizing atmospheres. On the other hand, only limited information could be obtained from DTA recorded under inert atmospheres, unless it was carried out simultaneously with EGA.

DTA shows that the oxidation of most organo-clay complexes proceeds in two stages, in the range 200-500°C and 400-750°C, respectively. The peak temperatures of the exotherms depend on the type of clay mineral on one hand, and on the organic compound on the other.

Two basically different theories have been proposed to account for the low and high temperature exothermic peaks on DTA curves of organo-clay complexes. The first theory was proposed by BRADLEY & GRIM (1948) and ALLAWAY (1949). They assumed that the peaks were entirely due to the oxidation of total organic materials, occurring in successive stages. ALLAWAY (1949) inferred complete oxidation of hydrogen at low temperatures, leaving a graphite layer, which is burnt off only on dehydroxylation of the clay. BRADLEY & GRIM (1948), however, postulated the initial formation of "petroleum coke" composed largely of carbon with some hydrogen which is subsequently oxidized. The second theory, proposed by RAMACHANDRAN et al. (1961a and 1961b), RAMACHANDRAN & KACKER (1964) and KACKER & RAMACHANDRAN (1964) attributes the low and high temperature exothermic peaks to oxidation of organic molecules attached to broken bonds and inside the interlayer space, respectively, the latter being more rigidly bound.

Some experimental data show that the latter theory is not always valid: (a) In organo-illite complexes the adsorbed organic molecules are located on the external surface of this mineral but DTA curves of illite treated with cationic dyes showed two exothermic peaks, at about 300°C and 500°C (RAMACHANDRAN et al., 1961a and 1961b; RAMACHANDRAN & KACKER, 1964; KACKER & RAMACHANDRAN, 1964). (b) There are many sorbed organic molecules which prove to be located in the interlayer space of the montmorillonite and give only a single low temperature exothermic peak. For example, DTA curves of copper montmorillonite saturated with etylenediamine or 1.3 propylenediamine, with amine : copper molar ratio of 1 do not show high temperature exothermic peaks (BODENHEIMER et al., 1963a and 1966a), although it has been established by X-ray measurements that these complexes contain amine in the interlayer space (BODENHEIMER et al., 1962; HELLER-KALLAI & YARIV, 1981). Also, Fe-montmorillonite saturated with pyrocatechol, gives a single exothermic peak at 330°C (YARIV et al., 1964). (c) BODENHEIMER et al. (1966a) recorded DTA curves of copper montmorillonite saturated with ethylenediamine (En) with amine : Cu molar ratio of 2, diluted with alumina in different proportions.

The DTA were carried out in the presence of air, combustion commenced at the same temperature, whatever the dilution, but the second exothermic peak shifted to higher temperatures on increasing the concentration of the organo-clay. The peak at 500°C, which first appeared as a slight hump at the 1:6 clay to alumina ratio, increased considerably at the expense of the lower temperature peaks when this ratio was raised to 1:2. (d) When more oxygen was circulated through the heating cell during the thermal treatment, the first exothermic peak increased whereas the second became very small. Samples were removed from the heating cells after they had been heated to 500°C. Thus when the thermal treatment was carried out with pure oxygen circulation through the heating cell the samples became light grey in colour. On the other hand, when the thermal treatment was carried out in air the sample became black, indicating incomplete combustion.

ALLAWAY (1949) is of the opinion that the position of the high temperature exothermic peak in the DTA curves of organo-smectite depends on the dehydroxylation temperature of the clay. This peak corresponds to the final loss of the carbonaceous material, and according to ALLAWAY, the opening up of the clay structure on dehydroxylation enables the organic residue to be oxidized completely. A survey of the literature (Table 1) does not confirm this contention. Moreover, there are many sorbed organic molecules and cations giving DTA curves with the last exothermic peaks at temperatures below that of the dehydroxylation reaction. For example, diethylenetriammonium-, triethylenetetraammonium- and tetraethylene-pentammonium-montmorillonite (Wyoming bentonite), having the final exothermic peaks at 610°C, 615°C and 585°C (BODENHEIMER et al., 1963b and 1963c). However, the oxidation reaction is incomplete at the temperature of the exothermic peak, and continues at the higher temperatures. Consequently, there is a baseline shift and the endothermic peak of dehydroxylation at 680°C is very weak or not detected.

To sum up the above discussion, the following mechanism is suggested for the occurrence of the two distinct stages during the exothermic oxidation reaction. Combustion commences at a temperature which is independent of the amount of material present, but is dependent on the activation energy of the combustion reaction. The clay mineral serves as a positive or negative catalyst of the combustion reaction and the peak temperatures should be affected by the type of bonding between the sorbate and the sorbent. If the total amount of the organic material in the DTA cell is small, oxidation will be completed at a relatively low temperature. If this amount is high, the available oxygen in the system is insufficient for a complete combustion. Oxidation of hydrogen to water and of carbon in part to CO_2 and in part to "petroleum coke" occurs at the first stage of the exothermic reaction. Oxidation of the "petroleum coke" is completed only at high temperature. The combustion temperature of the "petroleum coke" depends on the degree of cross-linking of this complex material and on the size of the polymeric species from which it is composed. These are affected by (a) the composition, size and shape of the parent organic molecule and (b) the type of clay mineral, which acts as a catalyst. It is not clear whether air penetrates into the interlayers and the "petroleum coke" is oxidized inside this space or whether the coke decomposes into small fragments which are liberated from the clay phase prior to combustion.

Recently a combined DTA-EGA was used for the study of butylamine and pyridine complexes of sepiolite and palygorskite (SHUALI et al., 1990 and in press). The combustion products were analyzed by mass-spectrometry simultaneously with the DTA and EGA curves were thus obtained which were useful for the interpretation of the exothermic peaks. The results can be summarized as follows. The profile of the DTA curves seems to be determined by the oxidation of both C and H elements. For clay treated with aliphatic amines, due to the fact that the organic molecule has a high H to C atomic ratio, the DTA curve has a profile similar to that of the "organic water evolution curve" (hydrogen oxidation). However, the temperatures of the principal exothermic peaks of the DTA curves are determined by the rate of C oxidation, whereas temperatures of exothermic shoulders are determined by the rate of interaction between H and O. For

Table 1: DTA exothermic peaks (in °C) of different organo-clay complexes and dehydroxylation endothermic peak (in °C) of the clay mineral. Data collected from the literature.

Clay mineral	Source	Adsorbed organic compound	Exothermic peak temperatures [°C]	Dehydroxylation peak temperature [°C]
Non-expanding				
Kaolinite	Amortex, India	MB	250, 420	590
"	"	MG	---, 430	590
Allophane	Barcelona, Spain	P	260, 440	---
Illite	Fithian, USA	MB	260, 465	560
"	"	MG	345, 465	560
"	"	CV	343, 453	560
"	"	P	218, 403	560
Expanding				
Montmorillonite	Wyoming, USA	MB	250, 660	710
"	"	MG	360, 670	710
"	"	P	300, 680	710
"	"	R6G	415, 515	680
"	"	AO	350, 480, 670	680
"	"	CV	365, 480, 575, 680	680
"	"	CHA	312, 660	680
"	"	TEA	390, 670	700
"	Miss., USA	CHA	305, 580	663
"	"	P	198, 255, 437, 539	663
"	Almeria, Spain	P	317, 592	517, 650
"	Perthshire, Scot.	P	321, 670	560
"	Bedfordshire, Eng.	P	243, 404, 569	532
"	"	TEA	313, 400, 600	532
"	B.C., Canada	TEA	380, 690	Unknown
"	Antigua	TEA	324, 587	Unknown
Nontronite	Gujarat, India	MB	250, 595	580
"	"	MG	343, 442, 585	580
"	"	CV	343, 455, 600	580
"	"	P	280, 525	580
"	Unknown	TEA	300, 515	Unknown
Saponite	Glasgow, Scot.	TEA	345, 570	904
"	AlltRibhein, Scot.	CHA	355, 385, 645	858
Beidelite	Burmah Oil Co., Ind.	TEA	324, 580	550
Hectorite	Calif., USA	TEA	390, 640	830
"	Syn. Baroid Div.	CHA	335, 615	850
Laponite	Laporte, Eng.	R6G	405, 510	730
"	"	AO	325, 570	730
"	"	CV	290-390, 515	730

```
Abreviations used in Table 1:  P    = piperidine
                               MB   = methylene blue
                               MG   = malachite green
                               CV   = crystal violet
                               AO   = acridine orange
                               CHA  = cyclohexylamine
                               TEA  = triethylamine
```

clays treated with aromatic amines, due to the fact that the organic molecule has a low H to C atomic ratio, the DTA curve has a profile similar to that of the CO_2 evolution curve (carbon oxidation), and the peak maxima are, in principal, those of the CO_2 evolution curve. Peak maxima of the "organic H_2O evolution curve" appear only as shoulders in the DTA curves.

The gas evolution curves prove the assumption of BRADLEY & GRIM (1948) who claimed that the hydrogen reaches a high rate of oxidation before the carbon. More H_2O than CO_2 is evolved in the first stage of the exothermic reactions, indicating that only part of the equivalent carbon is oxidized to CO_2. The rest of the carbon probably forms the "petroleum coke". The high temperature peaks in CO_2 evolution curves present the oxidation of the "petroleum coke". The evolution of "organic H_2O" continues almost up to the last stages of CO_2 evolution proving that the initially formed "petroleum coke" is indeed composed largely of carbon with some residual hydrogen as suggested by BRADLEY & GRIM (1948).

Carbonization of organic matter in the interlayer space of montmorillonite was recently investigated by graphitizing polymers between the lamellae of this clay (SONOBE et al., 1990). The coke obtained was a film shape and highly stacked structure with small interplanar spacing of 0.337 nm. The formation of such a unique coke is attributed to the peculiar carbonization method where the two dimensional space between the lamellae serves as a unique field for carbonization.

The last stages of the oxidation and the temperature of the last exothermic peak depend on (a) the mineral, (b) the organic compound and (c) the type of bonds which exist between the "petroleum coke" and the clay mineral.

(a) Non-expanding clay minerals adsorb only small amounts of organic matter. "Petroleum coke" is not formed, or, is formed with a small degree of cross-linking, and thus the oxidation is complete below 500°C. The expanding clay minerals adsorb big amounts and "petroleum coke" is formed during the first stages of the DTA run with a high degree of cross-linking, and thus oxidation occurs at higher temperatures.

(b) Very little systematic work has been done to establish the relation between the high temperature peak and the nature of organic compounds. In the study of aliphatic amines adsorbed by Wyoming bentonite, it is possible to see that an increase in the ratio of C to N leads to a higher peak temperature (YARIV, 1985). This peak appeared during or after the dehydroxylation (above 650°C) only with C/N ≥ 4.

(c) At this instance we shall point to the effect of π interaction between aromatic organic matter and the oxygen plane of the silicate layer. A combined DTA-EGA study of montmorillonite and laponite complexes of the cationic dyes, crystal-violet acridine orange, showed that this type of interaction increases the thermal stability of the organic molecule against oxidation (YARIV et al., 1989a, 1989b and 1989c). A spectrophotometric study revealed that in montmorillonite, but not in laponite, there are π interactions between the aromatic rings of the cationic dyes and the oxygen plane (YARIV et al., 1990). The last exothermic peak in the DTA curves of montmorillonite-dye complex occurs at higher temperatures than in laponite.

A combined DTA-EGA study of montmorillonite and laponite complexes of the cationic dye, rhodamine 6G, was also carried out (YARIV et al., 1988). π interactions between the oxygen plane of montmorillonite and rhodamine 6G cannot occur due to steric hindrance; in this molecule the phenyl ring is sterically constrained to be roughly perpendicular to the planar xanthene group (GRAUER et al., 1984). The DTA curves of rhodamine 6G treated laponite and montmorillonite are similar (Table 1) and most of the organic matter is oxidized before the dehydroxylation of these clays. Only a very small peak was observed in the CO_2 evolution curves of rhodamine 6G-montmorillonite at 670°C and the endothermic DTA peak of dehydroxylation became very small. The presence of this small CO_2 evolution peak results from the formation of small amounts of "petroleum coke" in the interlayer space of montmorillonite during the first exothermic reaction.

Based on these observations it was postulated that the temperature in which the last stage of oxidation occurs depends on whether or not π bonds are formed between the clay mineral surface and the "petroleum coke" which has been formed during the first exothermic reaction.

Table 1 shows that the last exothermic peak is associated with the dehydroxylation peak only with beidellite and nontronite in addition to several montmorillonites, but not with saponite, hectorite or laponite. According to YARIV (1988) π interactions between aromatic rings and the clayey oxygen plane can be formed only if the aromatic ring is positively charged and there is some substitution of Si by Al in the tetrahedral sheet. This can explain the behaviour of nontronite, beidelite and many of the montmorillonites (with tetrahedral substitution), on one hand and that of laponite, and hectorite and some montmorillonites (with no tetrahedral substitutions) on the other. It does not explain the behaviour of saponite, which also has much tetrahedral substitution and an endothermic peak of dehydroxylation at 858°C. The cause of the thermal behaviour of the saponite requires further study.

Application of DTA in the Study of Organo-Clay Complexes

Since several exothermic and endothermic reactions occur simultaneously and since some of these reactions depend on the total amount of the organic matter, the absolute concentration of the organic matter in the DTA cell should affect the DTA curves and in some cases wrong interpretations can be made. To avoid this error, it is suggested to dilute the organic-clays with inert material and to record curves in several clay concentrations in order to be able to differentiate between the effects of absolute concentration of the organic material in the heating cell and the intrinsic thermal properties of the organo-clay complex. Calcined kaolinite or calcined alumina are used in our laboratory as diluents. The concentration of the clay sample profoundly influences the nature of the DTA curves. At low concentrations the exothermic peak corresponding to the second oxidation stage is less intense than that corresponding to the first oxidation stage. It increases with increasing concentration until it becomes the dominant exothermic peak. Conversely, at high clay concentration, some lower temperature exothermic effects are obscured by adjacent peaks. Some peaks cannot be detected unless the concentration of the organo-clay complex is very low. Endothermic peaks are shifted to higher temperatures, when the clay concentrations are increased. On the other hand, the positions of the maxima of exothermic peaks are not affected by changes in concentration, unless there are endothermic peaks in their vicinity (BODENHEIMER et al., 1966a).

1. *The Study of Mixtures of Clay Minerals and Solid Organic Matter*

DTA is used to establish whether there are interactions between clay minerals and solid organic compounds. In this technique the DTA curve of a mixture containing a clay mineral and organic matter is compared to the DTA curves of the clay mineral and the organic matter, both heated individually. If the DTA curve of the clay-organic matter mixtures differ from the sum of those of the clay mineral and the neat organic matter heated individually, it is a proof that chemisorption of the organic matter, on the clay phase occurred during the preparation of the mixture or in the course of heating. As an example for this application, the study of the interaction between stearic acid and the clay minerals, sepiolite, palygorskite, talc, pyrophyllite and allophane will be mentioned (YARIV & HELLER-KALLAI, 1984; HELLER-KALLAI et al., 1986; YARIV et al., 1988).

The study of the adsorption of fatty acids by clay minerals raises some technical difficulties due to the fact that both components, the adsorbent and the adsorbate are solid. Most of the clay minerals have relatively small surface areas and the degree of adsorption is low. There is, as well, no way to separate between the adsorbed and the excess non-adsorbed acid because the former is washed out together with the latter by repeated washings with organic solvents. For these mixtures, DTA can serve as a useful tool to establish whether adsorption complexes are formed or whether the clay and organic material are present merely as mechanical mixtures. In this technique DTA curves of the pure acid, on one hand, and of the clay mineral-stearic acid mixture, on the other, are recorded either in nitrogen or oxygen atmosphere. The curves of the mixtures show all the peaks that occur in the curve of the pure acid, but, additional peaks appear indicating that another phase is present. It is reasonable to infer that the additional peaks which are not due to the clay minerals themselves, but arise from adsorbed stearic acid. When DTA is carried out together with TG, semiquantitative information on the adsorption can be obtained.

This technique is applicable for the study of the interaction between clays and herbicides, pesticides or other pollutants. For example, HERMOSIN et al. (1985) showed that the pesticide chlordimeform is adsorbed by montmorillonite. From the thermal analysis curves it can be seen that the thermal stability of this compound is increased due to its adsorption by montmorillonite.

A similar technique was used to study the interaction between benzoic acid and calcined kaolinite, alumina or diatomaceous earth (celite, amorphous silica) (YARIV et al., 1967). DTA runs using calcined kaolinite or alumina showed exothermic peaks with maxima between 420°C and 530°C, indicating retention of organic matter by the inorganic diluent. These exothermic peaks were enhanced when the mixtures of the acid and the solid substances were vigorously ground. In contrast to alumina and calcined kaolinite, diatomaceous earth retained no benzoic acid. In view of the DTA curves it was concluded that benzoic acid is adsorbed by alumina and calcined kaolinite, but not by diatomaceous earth. Adsorption of the acid increases with the time of grinding.

2. *The Effect of Organic Material on the Adsorbed Water Content and Properties*

The evolution of adsorbed water is displaced in the first region of the DTA curves of the organo-clay complexes. Water is determined from TG curves and is equal to the weight-loss occurring during the temperature range of the first region. Since small amounts of organic matter are also desorbed at this stage, accurate determination of water can be done only if the TG is supplemented by EGA.

Already in 1949, JORDAN demonstrated that DTA curves of amine treated montmorillonite can give useful information on the effect of the adsorbed organic molecule on the hydration of the clay. These curves showed a progressive decrease in the height of the low-temperature endothermic peak with an increase in the size of the aliphatic chain. This endothermic peak characterizes the thermal loss of interlayer water and JORDAN (1949) concluded that the adsorption of amines was associated with increasing hydrophobic character of the clay surface and the loss of adsorbed water. Water content of montmorillonite treated with amines decreased with the size of the aliphatic chain.

In the case of organo-sepiolite and palygorskite complexes, only zeolitic and interparticle water is evolved in the first region of the curve (MARTIN-VIVALDI & FENOLL HACH-ALI, 1970). Water, bound to Mg inside the channels, is evolved in the second region only, together with thermal reactions of the organic matter. From the size of the characteristic endothermic peaks recorded under inert atmosphere, supplemented by weight-loss data (TG-curves) and H_2O evolution curves, it was concluded that the amounts of zeolitic and bound water became very small due to the penetration of the amine into the pores. Sepiolite, with wider pores, adsorbed more organic material than palygorskite and thus lost more water due to this adsorption (SHUALI et al., 1990 and in press).

3. Thermal Properties of Adsorbed Organic Matter

The thermal properties of organic matter can be changed as a result of adsorption. DTA can be applied to study these changes and especially when it is supplemented by EGA. In this study the thermal analysis curves of the neat organic matter are compared to those of the organo-clay complexes. For example, DTA study showed that in the presence of allophane, talc and pyrophyllite oxidation of stearic acid started at lower temperatures than with the pure acid, but the most intense exotherm occurred at a higher temperature (YARIV & HELLER-KALLAI, 1984; HELLER-KALLAI et al., 1986; YARIV et al., 1988). It appears that these clay minerals play a dual role in these reactions: (a) they act as catalyst, reducing the temperature at which oxidation commences and (b) they delay oxidation of part of the organic matter to higher temperatures.

Another example is the study of THIELMANN & MCATEE (1975) who proposed the use of metal-tris (ethylenediamine) montmorillonites as gas chromatographic packing materials. For this purpose, heating effects were studied by DTA to determine thermal stability of the different complexes and the changes in retention times were compared with various structural changes occurring during the heating of the diamine clay complex.

Simultaneous DTA-EGA was recently used to study the thermal decomposition of butylamine and pyridine adsorbed on sepiolite and palygorskite in an inert atmosphere (SHUALI et al., 1990 and in press). This combined technique is applicable for the study of the catalytic properties of clay minerals and their contribution to different petro-chemical reactions.

4. Identification of Clay Minerals

ALLAWAY (1949) proposed to saturate smectites with piperidine to DTA examination. He suggested that the last exothermic peaks at 700°C, 600°C and 400-500°C are characteristic for montmorillonite, beidellite and nontronite, respectively, and can be used to distinguish between these minerals or between smectites and other clay minerals that give the last exothermic peak below 500°C. Several investigators (CARTHEW, 1955; OADES & TOWNSEND, 1963; MACKENZIE, 1970) showed that the last exothermic peak in different montmorillonites saturated with piperidine appears at different temperatures ranging between 530°C and 700°C. This is not surprising if one assumes that some of the montmorillonites have tetrahedral substitution which is essential for π interactions between the organic matter and the oxygen plane, whereas the other montmorillonites have no tetrahedral substitution.

KACKER & RAMACHANDRAN (1964) proposed to saturate the clay sample with cationic dyes, such as methylene blue or malachite green, prior to DTA examination. They suggested that the last exothermic peaks at 670°C, 590°C, 465°C and 425°C are characteristic for montmorillonite, nontronite and kaolinite and can be used to distinguish between these minerals when they are present in mixtures.

5. The Study of Types of Associations Between Adsorbed Organic Species and Functional Sites (Water Molecules and Exchangeable Metallic Cations) on Clay Surface

Exothermic peaks of DTA curves recorded in air are applicable for the study of the adsorption of organic compounds by clay minerals. The interpretation of different DTA peaks is complicated and must be supplemented by information from EGA curves. However, only recently combined EGA-DTA studies of organo-clay complexes have been performed and the available information is limited (YARIV et al., 1988, 1989a, 1989b, 1989c; SHUALI et al., 1990 and in press). Thus the DTA curves were used merely as fingerprints for the identification of different associations which are obtained on the clay surface. For example, LIBOR et al. (1970 and 1971) compared DTA curves of urea, H- and Na-montmorillonites and montmorillonites treated with urea. From differences between these DTA curves they concluded that the two different exchangeable cations have differently bonded the urea to the montmorillonite surfaces.

In our laboratory characteristic DTA curves were recorded for montmorillonites saturated with ionic and molecular aliphatic and aromatic amines. Aromatic amines are characterized by their tendency to develop an exothermic peak at about 700°C. Unlike aliphatic complexes, small exothermic peaks at this temperature persisted with the aromatic complexes even when the total organic matter in the heating cell was very small (BODENHEIMER & HELLER, 1968; YARIV et al., 1968; SOFER et al., 1969). DTA curves were used as fingerprints to identify coordination of mono-, di- and poly-amines to exchangeable metallic cations, such as Cu, Ni, Zn, Cd and Hg (BODENHEIMER et al., 1963a, 1963b, 1963c and 1966b), glycols to Cu (BODENHEIMER et al., 1966a), pyrocatechol to Fe (YARIV et al., 1964) and to determine coordination numbers. The subject was recently reviewed (YARIV, 1985).

Several authors applied DTA or DSC to measure decomposition enthalpies of interlayer complexes of exchangeable cations with organic molecules. For example, BRUQUE et al. (1982) calculated decomposition enthalpies of amino lanthanide montmorillonites from the areas of endothermic peaks at 25-220°C. For each amine, the area of the endothermic peak increased as the ionic radius of the exchangeable cation decreased. From the results the authors concluded that the bond between the amine and the lanthanide cation can be attributed to the ion-dipole type.

Conclusion

The present communication demonstrates that DTA curves are applicable for the study of the adsorption of organic compounds by clay minerals. The interpretation of the DTA curves is complicated and must be supplemented by information from more sophisticated techniques. The combined DTA-EGA technique, which was applied recently to the study of a few organo-clay samples, seems to be useful for a better understanding of the DTA peaks.

References

ALLAWAY W. H. (1949) - Proc. Soil Sci. Soc. Amer., 13, 183

BODENHEIMER W. & HELLER L. (1968) - Isr. J. Chem., 6, 307

BODENHEIMER W., HELLER L., KIRSON B. & YARIV S. (1962) - Clay Miner. Bull., 5, 145

BODENHEIMER W., HELLER L., KIRSON B. & YARIV S. (1963b) - Isr. J. Chem., 1, 391

BODENHEIMER W., HELLER L., KIRSON B. & YARIV S. (1963c) - Proc. Intern. Clay Conf. Stockholm, 2, 351

BODENHEIMER W., HELLER L., KIRSON B. & YARIV S. (1966b) - Proc. Intern. Clay Conf. Jerusalem, 1, 251

BODENHEIMER W., HELLER L. & YARIV S. (1966a) - Clay Minerals, 6, 167

BODENHEIMER W., KIRSON B. & YARIV S. (1963a) - Isr. J. Chem., 1, 78

BRADLEY W. F. & GRIM R. E. (1948) - J. Phys. Colloid Chem., 52, 1404

BRUQUE S., MORENO-REAL L., MOZAS T. & RODRIGUEZ-GARCIA A. (1982) - Clay Minerals, 17, 208

CARTHEW A. R. (1955) - Soil Sci., 80, 337

CHI CHOU C. & MCATEE J. L. JR. (1969) - Clays Clay Minerals, 17, 339

ELTANTAWY I. M. (1974) - Bull. Groupe Franc. Argiles, 26, 211

GRAUER Z., AVNIR D. & YARIV S. (1984) - Can. J. Chem., 62, 1889

HELLER-KALLAI L. & YARIV S. (1981) - J. Colloid Interface Sci., 79, 479

HELLER-KALLAI L., YARIV S. & FRIEDMAN I. (1986) - J. Thermal Anal., 31, 95

HERMOSIN M. C., CORNEJO J. & PEREZ-RODRIGUEZ J. L. (1985) - Clay Minerals, 20, 153

JORDAN J. W. (1949) - Mineralog. Mag., 28, 598

KACKER K. P. & RAMACHANDRAN V. S. (1964) - 9th Intern. Ceram. Congress, Brussels Transp., 483

LANGIER-KUZNIAROWA A. (1989) - Thermochim. Acta, 148, 413

LIBOR O., GRABER L. & DONATH E. P. (1970) - Agrokemia es Talajtan, 19, 293

LIBOR O., GRABER L. & DONATH E. P. (1971) - Acta Mineralogica-Petrographica (Szeged), 20, 97

MACKENZIE R. C. (ed.) (1957) - The Differential Thermal Investigation of Clays.- Mineralogical Society, London.

MACKENZIE R. C. (1970) - in Differential Thermal Analysis.- (MACKENZIE R. C. (ed.)), Academic Press, London, Vol. 1, 498

MARTIN-VIVALDI J. L. & FENOLL HACH-ALI P. (1970) - in Differential Thermal Analysis.- (MACKENZIE R. C. (ed.)), Academic Press, London, Vol. 1, 553

MCNEAL B. L. (1964) - Soil Sci., 97, 96

MITCHELL B. D. & BIRNIE A. C. (1970) - in Differential Thermal Analysis.- (MACKENZIE R. C. (ed.)), Academic Press, London, Vol. 1, 611

MÜLLER-VONMOOS M., KAHR G. & RUB A. (1977) - Thermochim. Acta, 20, 387

OADES J. M. & TOWNSEND W. M. (1963) - Clay Miner. Bull., 5, 177

PAULIK F., PAULIK J., ARNOLD M., INCZEDY J., KRISTOF J. & LANGIER-KUZNIAROWA A. (1989) - J. Thermal Anal., 35, 1849

RAMACHANDRAN V. S., GORG S. P. & KACKER K. P. (1961a) - Chem. Ind., 790

RAMACHANDRAN V. S. & KACKER K. P. (1964) - J. Appl. Chem., 14, 455

RAMACHANDRAN V. S., KACKER K. P. & PATWARDHAN N. K. (1961b) - Nature.- London, 191, 696

SHUALI U., YARIV S., STEINBERG M., MÜLLER-VONMOOS M., KAHR G. & RUB A. (1988) - Thermochim. Acta, 135, 291

SHUALI U., STEINBERG M., YARIV S., MÜLLER-VONMOOS M., KAHR G. & RUB A. (1990) - Clay Minerals, 25, 107

SHUALI U., STEINBERG M., YARIV S., MÜLLER-VONMOOS M., KAHR G. & RUB A. (in press) - Clay Minerals.

SMYKATZ-KLOSS W. (1974) - Differential Thermal Analysis. Application and Results in Mineralogy.- Springer Verlag, Berlin-Heidelberg-New York.

SOFER Z., HELLER L. & YARIV S. (1969) - Isr. J. Chem., 7, 697

SONOBE N., KYOTANI T. & TOMITA A. (1990) - Carbon, 28, 483

THIELMANN V. J. & MCATEE J. L. JR. (1975) - Clays Clay Minerals, 23, 173

YARIV S. (1985) - Thermochim. Acta, 88, 49

YARIV S. (1988) - Intern. J. Tropic. Agric., 6, 1

YARIV S., BIRNIE A. C., FARMER V. C. & MITCHELL B. D. (1967) - Chem. Ind., 1744

YARIV S., BODENHEIMER W. & HELLER L. (1964) - Isr. J. Chem., 2, 201

YARIV S. & HELLER-KALLAI L. (1984) - Chem. Geol., 45, 313

YARIV S., HELLER-KALLAI L. & DEUTSCH Y. (1988) - Chem. Geol., 68, 199

YARIV S., HELLER L., DEUTSCH Y. & BODENHEIMER W. (1971) - "Thermal Analysis".- Proc. 3rd ICTA, Davos, Birkhäuser Verlag, Basel, 3, 663

YARIV S., HELLER L., SOFER Z. & BODENHEIMER W. (1968) - Isr. J. Chem., 6, 741

YARIV S., KAHR G. & RUB A. (1988) - Thermochim. Acta, 135, 299

YARIV S., MÜLLER-VONMOOS M., KAHR G. & RUB A. (1989a) - Thermochim. Acta, 148, 457

YARIV S., MÜLLER-VONMOOS M., KAHR G. & RUB A. (1989b) - J. Thermal Anal., 35, 1941

YARIV S., MÜLLER-VONMOOS M., KAHR G. & RUB A. (1989c) - J. Thermal Anal., 35, 1997

YARIV S., NASSER A. & BAR-ON P. (1990) - J. Chem. Soc. Faraday Trans., 86, 1593

THERMAL ANALYSIS IN ENVIRONMENTAL STUDIES

W. Smykatz-Kloss, A. Heil, L. Kaeding & E. Roller

Mineralogisches Institut der Universität
Karlsruhe, Germany

Dedicated to the 60th birthday of Slade St. J. Warne

Abstract

Thermal analysis offers suitable methods for environmental studies. Impurities of toxic elements in minerals, rocks, and raw materials can be determined by DTA, TG, CSA-CWA or EGA techniques (S in clays and coals, Cr or Cd in oxides and sulphides etc.). The DTA of the illite or smectite crystallinity shows to be a good controlling measure for barrier clays around waste disposals. The DTA/TG investigation of organo-clay mineral complexes contributes to the estimation of erosion tendencies in agricultural environments. This is discussed for several pesticide-bentonite complexes.

Introduction

Increased efforts are made by public authorities, engineers and scientists to prevent the environment from growing pollution. Industrial and traffic emissions, garbage and sewages contaminate air, soils and waters. Among the numerous and different controlling measures against environmental pollution the *thermal methods* have their particularly important place.

Thus, special investigations are made by the chemical and building industry to decrease the amount of dangerous impurities in raw materials for the cement or power plants. Mainly (iron) sulphides are dangerous admixtures (even when abundant in traces) because of producing sulphur oxides after combustion. Traces of sulphides contained in coals or clays sum up to thousands of tons of sulphuric acid which damage nature and cities heavily. Reports on "acid rain" and its bad contribution in destroying houses and cathedrals, in deteriorating life conditions for forests and animals are to be found in our newspapers daily.

Thermal Determination of Impurities in Minerals and Rocks

By means of simple (SMYKATZ-KLOSS, 1974 and 1984) or - in presence of organic matter - variable atmosphere differential thermal analysis (WARNE & FRENCH, 1984; AYLMER & ROWE, 1984; WARNE, this volume) very low amounts of pyrite and other sulphides can be determined in rocks, soils, raw materials, e.g. contents of 0.1 mass-% or even less. Traces of sulphur or sulphides in raw materials for industry and for energy production can be thermally identified very quickly by means of carbon/sulphur (CSA) and carbon/water analysis (CWA) equipment (for details see STARCK, this volume). These identification of traces of S or organic matter by means of thermal methods is possible because of the very large ΔH of the occurring oxidation effects ($S \rightarrow SO_3$; $C \rightarrow CO_2$).

Most minerals contain crystal chemical impurities. The host substances may be harmless when used for industrial or technical purposes. But certain impurities (e.g. trace metals incorporated in the structure of the host minerals) could damage life and nature when they are released from the host minerals (e.g. after heating or burning). This may be due for Se, Te, Cd, Cr or radioactive tracers occurring as impurities in sulphides, oxides or silicates. These and other toxic elements can - of course - be precisely determined by modern sophisticated methods like ICP-MS or NAA. But they may be determined by means of thermal methods as well, - less expensive, quicker and less complicated (but not in all cases). Crystal chemical impurities influence the thermal properties of minerals, e.g. the temperatures of structural or magnetic transformations. For example, the

Curie temperature of magnetites is lowered by incorporated Cr, Ti, Mn. By means of standardized DTA methods and in a reducing furnace atmosphere these substituents can be determined (SMYKATZ-KLOSS, 1974 and 1984). LAPHAM (1958) determined small amounts of chromium in chlorites and other clay minerals by means of DTA. Traces of metal substituents in sulphides lower the temperatures of high-low inversion effects. Thus, the DTA determination of inversion temperatures of galena, sphalerite, pyrite and related minerals (inert furnace atmosphere, ceramic crucibles or glass ampullas) may give some information about the presence of toxic metals in these minerals (SMYKATZ-KLOSS, 1984; NIESZERY-HAUSMANN & SMYKATZ-KLOSS, in preparation).

Very valuable methods to determine type and amount of gaseous reaction products occurring in heating processes are *evolved gas analyses* (EGA), combined methods *DTA/mass spectrometry*. These methods can be applied to the detection of trace amounts of S or C in geological materials as well (MÜLLER-VONMOOS et al., 1977; YARIV et al., 1989).

The Crystallinity of Clay Minerals as a Controlling Measure for the Stability of Geological Barriers

The "crystallinity" of clay minerals expresses their structural stability, e.g. the degree of (dis-) order. For *illites* KUBLER (1964 and 1966) found an interrelation between the degree of disorder and the geological evolution of the illite-bearing rocks. With increasing pressure/temperature conditions this "illite crystallinity" is increasing as well (KUBLER, 1966; CHAMLEY, 1967; DUNOYER DE SEGONZAC et al., 1968; FREY, 1969; WEBER, 1972; SMYKATZ-KLOSS & ALTHAUS, 1975). Generally, the illite crystallinity has been measured by means of X-ray diffractometry (KUBLER, 1966; WEBER, 1972), rarely by IR spectroscopy (FLEHMIG, 1973). But the full width at half maximum (FWHM, "Halbwertsbreite") which was measured for the (00l) interferences is dependent on grain size and other factors. And the IR spectroscopic measurements have in some cases shown not being exact enough. To overcome these difficulties a *standardized differential thermal analysis* method was outlined which enables the exact determination of the degree of disorder for kaolinites (SMYKATZ-KLOSS, 1975), illites and smectites (KRÜGER & SMYKATZ-KLOSS, 1984 and 1985). The method seems to be quite valuable in environmental science, e.g. for the estimation of the stability of geological barriers. "Clays" are widely regarded as impermeable to seepage waters from waste disposals. This is why clays recently are recommended and used for basic layers and walls under and around (hazardous) waste disposals.

Experimental investigations on the interrelation between seepage waters from waste disposals and three-layer clay minerals (illites, smectites) showed that structural stabilities and physical properties of these clay minerals were affected by low-pH waters and solutions contaminated with special organic molecules as well (KRÜGER & SMYKATZ-KLOSS, 1984 and 1985; SMYKATZ-KLOSS & KAEDING, 1988). Fig. 1 shows the decrease in illite crystallinity with decreasing pH-value of the contact solutions. pH-values around 4 are quite normal for the first months of a household waste disposal. Temperature and shape of the endothermic dehydroxylation effect in DTA curves (Fig. 1) mirror the degree of disorder of the barrier illite. The lower the peak temperature, the weaker the (structural) stability of the barrier clay. This is due for other three-layer silicates, too, the swelling clay mineral species included.

Organo-Clay Mineral Complexes and Their Possible Influence on Erosion

Organic substances, brought into contact with fine-grained mineral particles exhibiting a comparably large surface area, tend to be adsorbed on these particles even above the boiling point of the organic molecules. In contact with swelling clay minerals (smectites, vermiculites, certain interstratifications) there may result a stronger fixation, especially when the

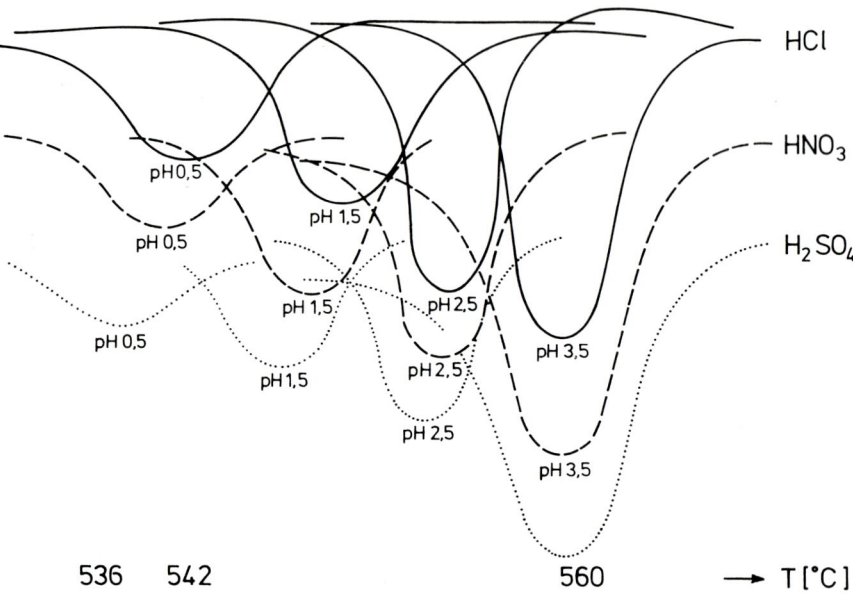

Fig. 1: Dehydroxylation effect of illites in dependance on the pH of contact solutions.

organic molecules exhibit a polar character (e.g. amines, MORILLO et al., 1991). In these cases the organic molecules penetrate partly the interlayer space, replacing the abundant water and changing the physical properties of the layer silicate. The most striking change may occur in the *volume*: both large swelling and dramatic shrinking effects have been observed for organo-clay mineral complexes (e.g. HASENPATT, 1988). The *density* and the *rheologic behaviour* may be changed as well. Generally, swelling clay minerals may be affected strongly in their (structural) stability by interrelations with certain organic solutions. Clays get into contact with organic solutions

a) in the environment of (hazardous) waste disposals,

b) in agricultural environments.

In order to estimate the physical processes which may be initiated by these contacts, the properties and stabilities of the occurring organo-clay complexes should be known. *Shrinking* effects may lead to a remarkable increase in permeability of barrier clays. A lowering in *density* may lead to certain flow phenomena. And different *rheologic* properties may cause soil movement or rock sliding.

In spite of numerous investigations on organo-clay complexes (e.g. YARIV & HELLER-KALLAI, 1984; HELLER-KALLAI et al., 1986; HASENPATT, 1988; YARIV et al., 1989; SHUALI et al., 1990; MORILLO et al., 1991; YARIV, this volume) their bonding character, stability relations and transformation reactions are still not cleared up completely. When heated, most organo-clay complexes exhibit two strong exothermic effects (Fig. 2-8; compare with YARIV, this volume). The first is due to the transformation of the original organic molecules into a kind of "petroleum coke" (YARIV, this volume). This process is accompanied by a ± spontaneous weight loss (caused by dehydration and loss of some CO_2) and occurs - depending on the type of organic compound and on the character of bonding - between 200°C and 400°C. The second one occurs around 500°C; due to Yariv and coworkers it seems to be caused by the burning of the "petroleum coke". Heating in a reducing or inert atmosphere will suppress the oxidation of carbon.

The possible physical alterations of organo-clay complexes may be considered for the case of pesticide-bentonite complexes. The used bentonite is a Ca-montmorillonite from Bavaria. The sample has been dried at 80°C for two days. The pesticide-bentonites were obtained after contacting the clay with diluted pesticide solutions for some days. Preparation and DTA runs (Netzsch apparatus STA 409 EP) were made under standardized conditions.

The X-ray and thermal characterization of the six complexes are listed in Table 1. With the exception of the "Goltix® (Bayer AG)" sample all complexes show only slight differences in volume, compared with the pure, dried bentonite. But all organo-bentonites show altered structures: from nearly ordered (e.g. the untreated Ca-bentonite) to strongly disordered types, reflecting *different degrees of mixed-layering* (Table 1). There is some evidence for a good correlation between the character of structural (dis-) order and the temperature of transformation of the organic molecules (Table 1): specimens exhibiting strong disorder (e.g. isoproturone, Stomp® (American Cyanamid Company) and Gardoprim® 500 (BASF AG)) show the lowest transformation temperatures. Some of the organic species exhibit a distinct high-low inversion, the temperature of which may be of some diagnostic interest (Table 1, Figs. 2-8).

The Goltix®-bentonite shows some extraordinary properties. Its X-ray diffractogram (Fig. 9) reflects a comparably well-ordered, but quite complicated structure. Evidently, the layer sequence includes some domains of swelling and some more of shrinking. But the resulting complex seems to be a kind of regular interstratification.

The most striking effect of the organic contacts is a remarkable *lowering* in the *dehydroxylation temperature* (Table 1), from 9°C for the isoproturone specimen to about 40 (!) degrees for the Goltix® specimen. The processes of interrelations and reactions between the different organic molecules and the Ca-bentonite are not yet completely clear.

Thermal Analysis in Environmental Studies

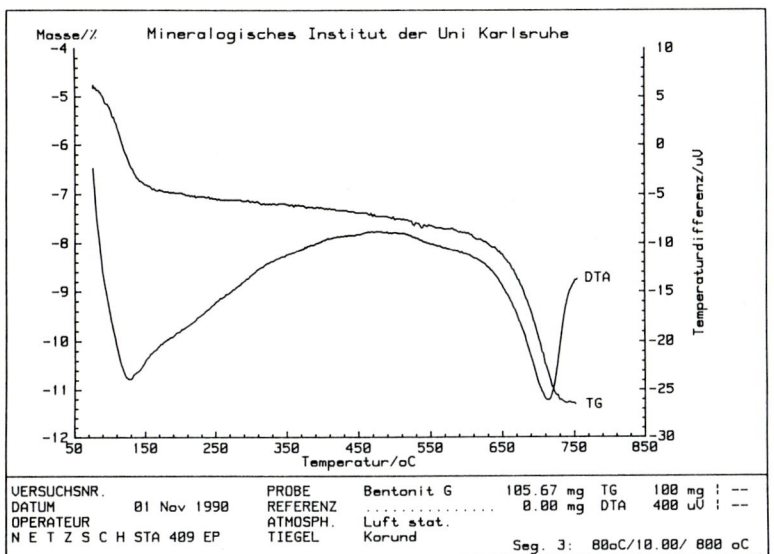

Fig. 2: Simultaneous DTA / TG curves of a Ca-bentonite (Moosburg, Bavaria).

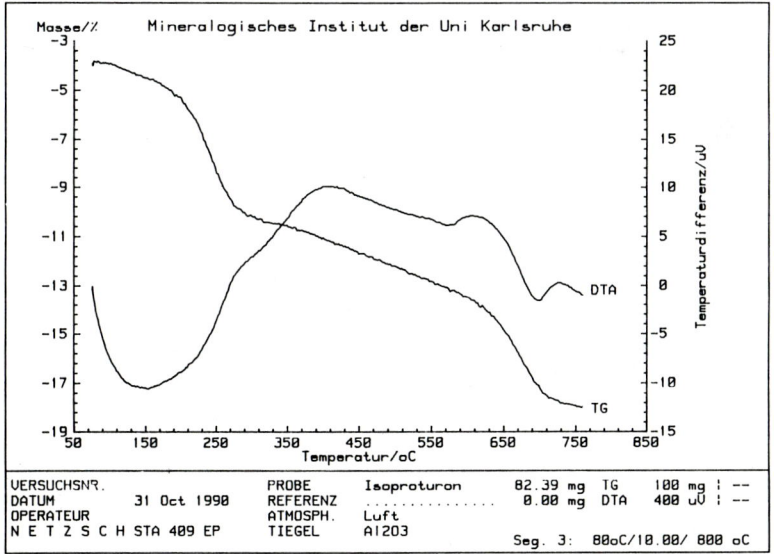

Fig. 3: DTA / TG curves of an isoproturone / bentonite complex.

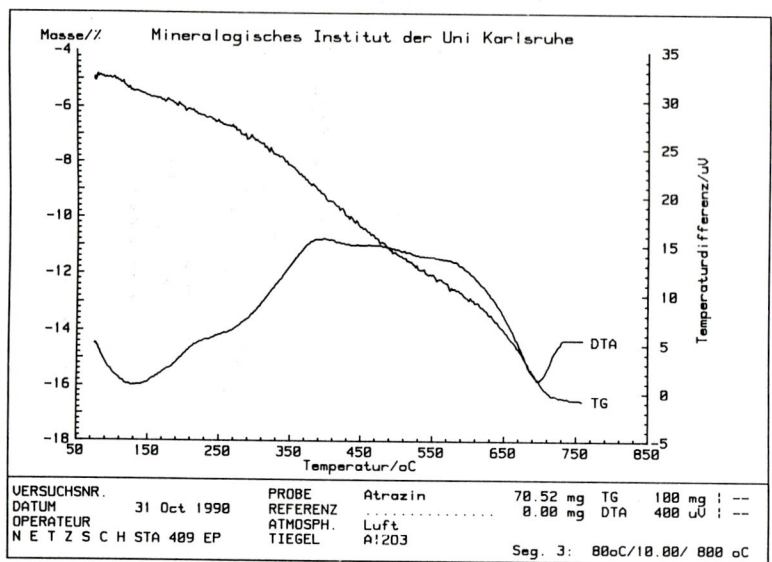

Fig. 4: DTA / TG curves of an atrazine / bentonite complex.

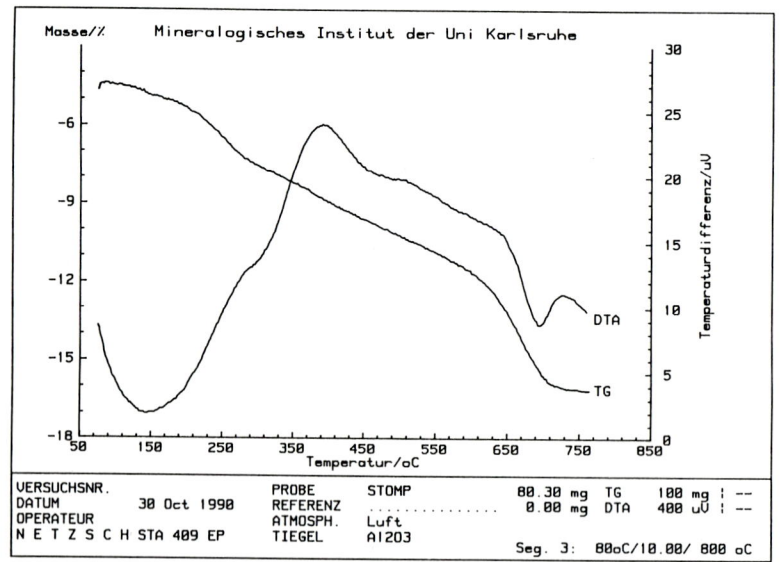

Fig. 5: DTA / TG curves of a Stomp® / bentonite complex.

Thermal Analysis in Environmental Studies

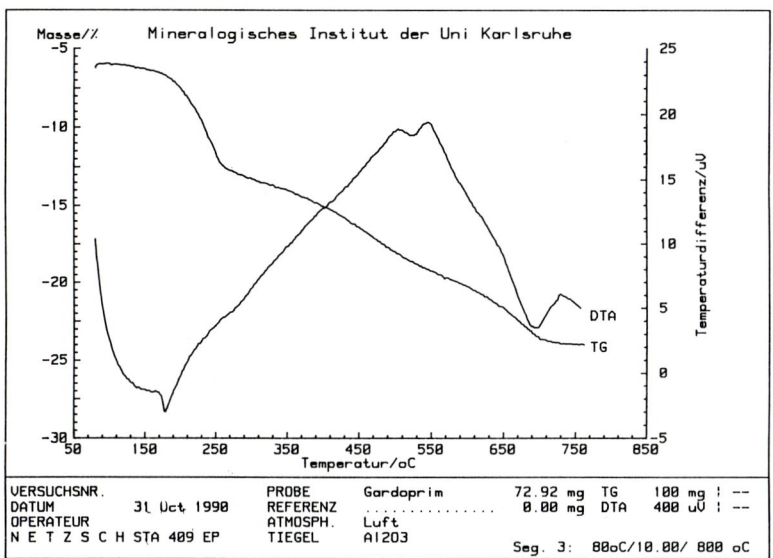

Fig. 6: DTA / TG curves of a Gardoprim® 500 / bentonite complex.

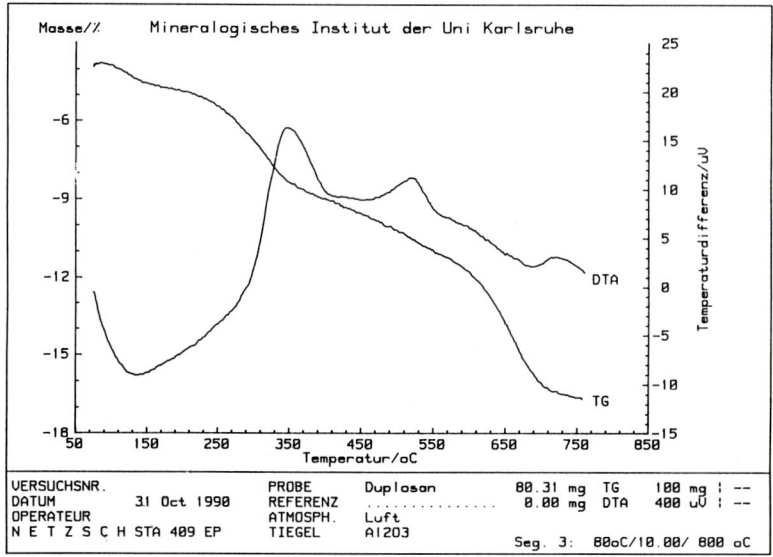

Fig. 7: DTA / TG curves of a Duplosan® KV / bentonite complex.

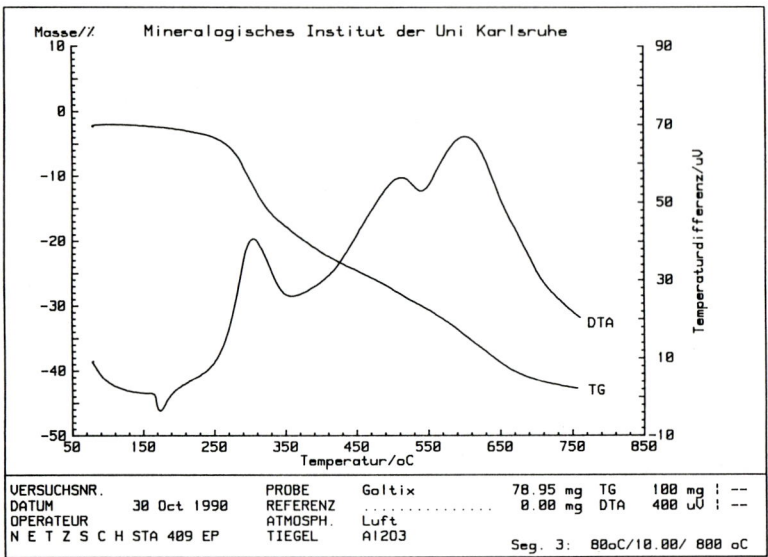

Fig. 8: DTA / TG curves of a Goltix® / bentonite complex.

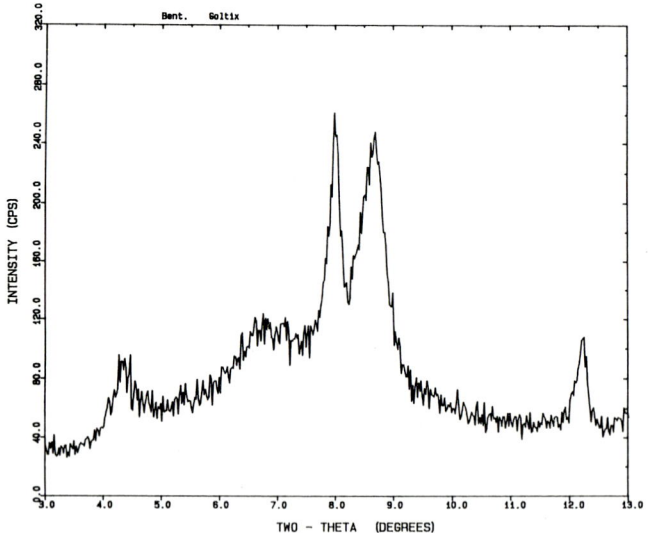

Fig. 9: X-ray diffractogram of a Goltix® / bentonite complex.

Table 1: Characterization of bentonite / pesticide complexes

Type	X-ray diffraction (001), Å	mixed layering	high-low inversion, [°C] DTA	organ. trans-formation, [°C] TG	bentonite dehydrox. [°C], DTA*
pure Ca-bentonite	12.6	----	----	----	707
Isoproturone (urea deriv.)	15.2-14.7	strong	159	230-260	698
Atrazine	13.0	weak	158	280-420	698
Stomp [1] (dinitroaniline)	13.0	strong	----	240-300	693
Gardoprim 500 [2] (triazine)	13.4	remarkable	179	220-260	692
Duplosan KV [3]	12.1	weak	----	260-360	688
Goltix [4]	10.3 13.4 20.5	partly strong shrinkage	174	270-330	<670 !

*: corrected for 80 mg sample amount

[1]: registed trademark of American Cyanamid Company
[2]: registed trademark of BASF AG
[3]: registed trademark of BASF AG
[4]: registed trademark of Bayer AG

The following *conclusions* can be drawn:

(1) The interrelations between pesticides and swelling clay minerals change the physical properties of the layer silicates, e.g. the volume, the dehydroxylation temperature and the structural (dis-) order.

(2) These alterations lead to a remarkable lowering of the structural stability.

(3) The decrease of the structural stability favours the decomposition or at least the transformation of the layer silicate into unstable interstratifications (mixed layers).

(4) Strong mixed-layering favours the transformation of the organic compound.

(5) The increased tendency of decomposition, associated with alterations in density and rheologic properties (which have been observed for some different types of organo-clay complexes, e.g. SMYKATZ-KLOSS & KAEDING, 1988) may lead to an increase in erosion processes of agricultural grounds.

(6) Thermal methods of investigation, namely DTA and TG, are very suitable for the characterization of organo-clay mineral complexes and their properties, and thus for the estimation of clay minerals and their stability in agricultural environments.

References

AYLMER D. M. & ROWE M. W. (1984) - A new method for simultaneous determination of pyrite content and proximate analysis of coal.- Thermochim. Acta, 78, 81-92

CHAMLEY H. (1967) - Possibilité d'utilisation de la cristallinité d'un minéral argileux (illite) comme témoin climatique dans les sédiments.- C. R. Acad. Sci. Paris, 265, 184-187

DUNOYER DE SEGONZAC G., FERRERO J. & KUBLER B. (1968) - Sur la cristallinité de l'illite dans la diagenése et l'anchimetamorphose.- Sedimentology, 10, 137-143

FLEHMIG W. (1973) - Kristallinität und Infrarotspektroskopie natürlicher dioktaedrischer Illite.- N. Jb. Miner., Mh. Jg. 1973, pp.351-361

HASENPATT R. (1988) - Bodenmechanische Veränderungen reiner Tone durch Adsorption chemischer Verbindungen (Batch- und Diffusionsversuche).- Mitt. Inst. Grundbau u. Bodenmechanik ETH Zürich, Nr. 134, 146 p.

HELLER-KALLAI L., YARIV S. & FRIEDMAN I. (1986) - Thermal analysis of the interaction between stearic acid and pyrophyllite ot talc. IR and DTA studies.- J. Thermal Anal., 31, 95-106

KRÜGER E.-M. & SMYKATZ-KLOSS W. (1984) - Mineralogical study of the interactions between clay minerals and acid soil waters.- Proc. 3rd Intern. Symp. on "Interactions between sediments and water", Geneva, pp.293-296

KRÜGER E.-M. & SMYKATZ-KLOSS W. (1985) - Differential thermal analysis as an indicative method for the determination of soil mineral damage.- Thermochim. Acta, 83, 107-112

KUBLER B. (1964) - Les argiles, indicateurs de métamorphisme.- Rev. Inst. Fr. Pétrole, 19, 1093-1112

KUBLER B. (1966) - La cristallinité de l'illite et les zones tout à fait supérieures du metamorphisme.- In: Coll. Et. Tectoniques, A la Baconnière, p. 105-122

LAPHAM D. L. (1958) - Structural and chemical variation in chromium chlorite.- Am. Mineralogist, 43, 921-956

MORILLO E., PÉREZ-RODRIGUEZ J. L. & MAQUEDA C. (1991) - Mechanisms of interaction between montmorillonite and 3-aminotriazole.- Clay Minerals, 26, 269-279

MÜLLER-VONMOOS M., KAHR G. & RUB A. (1977) - DTA-TG-MS in the investigation of clays. Quantitative determination of water, carbon monoxide and carbon dioxide by evolved gas analysis with mass spectrometer.- Thermochim. Acta, 20, 387-393

NIESZERY-HAUSMANN K. & SMYKATZ-KLOSS W. (in preparation) - DTA investigations on polymorphic copper and silver sulphides.- To be subm. to J. Thermal Anal.

SHUALI U., STEINBERG M., YARIV S., MÜLLER-VONMOOS M., KAHR G. & RUB A. (1990) - Thermal analysis of sepiolite and palygorskite treated with butylamine.- Clay Minerals, 25, 107-119

SMYKATZ-KLOSS W. (1974) - Differential Thermal Analysis. Application and Results in Mineralogy.- Minerals and Rocks, vol. 11, K. Springer Verl., Berlin-Heidelberg-New York, 185 p.

SMYKATZ-KLOSS W. (1975) - The DTA determination of the degree of (dis-) order of kaolinites: Method and application to some kaolin deposits of Germany.- In: BAILEY S. W. (ed.): Proc. Intern. Clay Conf. Mexico-City, pp. 429-438, Appl. Publ., Wilmette Ill./USA.

SMYKATZ-KLOSS W. (1984) - Determination of impurities in minerals by means of standardized differential thermal analysis.- In: BLAINE R. L. & SCHOFF C. K. (eds.): Purity Determinations by Thermal Methods. ASTM STP 838, Amer. Soc. for Testing and Materials, pp. 121-137

SMYKATZ-KLOSS W. & ALTHAUS E. (1975) - Experimental investigation of the temperature dependence of the "crystallinity" of illites and glauconites.- Bull. Groupe franç. Argiles XXVI, 319-325

SMYKATZ-KLOSS W. & KAEDING L. (1988) - Interactions between clay minerals and organic compounds around hazardous waste landfills.- In: ANDERSON L. & MÖLLER J. (eds.): Proc. 5th Intern. Solid Waste Conf., Copenhagen 1988, vol. 2, pp. 281-286, Ácad. Press London.

STARCK S. (1991) - Application of thermal methods in raw material control and during the production process.- This volume, pp. 234-243

WARNE S. ST. J. (1991) - Variable atmosphere thermal analysis - methods, gas atmospheres and applications to geoscience materials.- This volume, pp. 62-83

WARNE S. ST. J. & FRENCH D. H. (1984) - Siderite, pyrite and magnesite identification in oil shale by variable atmosphere DTA.- Thermochim. Acta, 79, 131-137

WEBER K. (1972) - Kristallinität des Illits in Tonschiefern und andere Kriterien schwacher Metamorphose im nordöstlichen Rheinischen Schiefergebirge.- N. Jb. Geol. Paläont., 141, 333-363

YARIV S. (1991) - Differential thermal analysis (DTA) of organo-clay complexes.- This volume, pp. 328-351

YARIV S. & HELLER-KALLAI L. (1984) - Thermal treatment of sepiolite and palygorskite stearic acid associations.- Chem. Geol., 45, 313-327

YARIV S., MÜLLER-VONMOOS M., KAHR G. & RUB A. (1989) - Thermal analytical study of the adsorption of acridine orange by smectite minerals.- J. Thermal Anal., 35, 1997-2008

Subject Index

Acid rain 353
activation energies, coals kinetic analysis 41
activation energy, CRTA determination 145 ff.
-, kinetic analysis of melt crystallization 173 ff.
additives, clay properties improvement 247 f.
adsorption complexes, clay minerals / organic molecules 328 ff.
adsorption, organics on clay minerals 355 f.
adsorption water 84 f.
Al-polymer, montmorillonite 320 f.
Al-spinel, kaolinite heating 208 f.
albite nucleation, melt 178
aliphatic amines, montmorillonite 347
allophane 254
-, adsorbed organics 338
-, heating-rate traces 11
alloys, solidification 8
aluminite, water binding 86
alunite, dehydroxylation 96
amines, Cu-montmorillonite 335
-, montmorillonite 332
-, smectites 356
amorphisation, thermal 130 f.
amorphous solids, crystallization 122 f.
anathorite formation, montmorillonite 301
ankerite, decomposition enthalpy 33 ff.
-, DTA 69 f.
annealing 153
anorthite, nucleation, melt 178
anthracite, pyrolysis 42
antigorite, dehydroxylation 96
aphthitalite 156 f.

aquo complexes, size control in mineral precipitation 92
aragonite, TA curves 93
arcanite 156 f.
archaeometry, TA 225
aromatic amines, montmorillonite 347
Arrhenius definition, activation energy 145, 173
ash, coal oxidation 75, 78
asymmetry, decomposition peak 98 f.
atmosphere, influence on peak shape 23
-, organo-clays 329 ff.
atmosphere control, DTA 11, 14
atrazine-bentonite, DTA / TG 360 f.
Au, melting point 11
autocatalysis, coal pyrolysis 41

Baddeleyite 201 f.
-, martensitic transformation 121
barriers, geological, stability control 355 f.
baryte, decrepitation 91
base line delineation, DSC enthalpy measurement 22
base line drift, DTA 11
base line extrapolation 22
bauxite assessment, DTA 12
bed ash, multiple-gas TG 76
beidellite 289
-, dehydroxylation 96
-, DTD curve 230
-, organo-complexes, exothermic peak temperatures 338
bentonite 357
-, Black Hills 294 f.
-, use and TA application 226
bentonite-pesticide complexes, thermal characterization 363

Subject Index

binary oxide systems, phase studies 153 ff.
birnessite, DSC 30
bitumen, oil shale 44
blowpipe 5
boehmite, dehydroxylation 96, 99
-, Q-TG curve 97
boiling points, water / S / Se 11
bonding strength, thermal dissociation 95
borates, decomposition 125 f., 129
breadth, decomposition peaks 98 f.
brick formation 206 f.
brown coal, TG / multiple gas sequence 76
brucite, dehydroxylation 96
building stone decay 225
bulk diffusion, melt crystallization 180
buserite, DSC 29 f.
butylamine, sepiolite 332, 337

C, trace detection 354
Ca-bentonite, DTA / TG 359 f.
Ca_2SiO_4-Ca_2GeO_4, phase diagram study 161 ff.
calcite 76
-, coals, DSC 33
-, decomposition 65, 69
-, dissociation 194
-, illite 283
-, ilmenite sintering 205 f.
-, impurity in hectorite 291
-, ion mobility 120
-, thermobalance assessment 11
calibration, DSC 19 f.
-, DTA 292
calorific enhancement, siderite in coals 33
calorific values, coals 43
capillary condensation 86
carbonates, illitic clays 276, 279 ff.

carbon species, differentiation 235
carnegieite 12
catalyst, organo-clay combustion 337, 345
celite, benzoic acid interaction 343
Celsius degrees, relation to degrees Wedgwood 7
ceramic production, materials and problems 207 f., 226 ff.
ceramic stabilizers, concrete 210 f.
cerussite, dissociation steps 192
chalcanthite, TA curves 94
chalcocite, oxidation 195
chalcopyrite, roasting 195
-, weathering 103 f.
charcoal, TG / multiple gas sequence 76
Cheto type, montmorillonite 293 f.
china clay 6, 208
chlorite, Cr determination by DTA 354
-, use 227, 232
-, DTA classification 12
chlorite dehydroxylation, volume increase 126
chromite, ceramic stabilizer 211
chrysotile, dehydroxylation 96
clay, ceramic processing 207
clay minerals, dehydroxylation 10
-, interaction with seepage waters 355
clay raw materials, TA methods application 226
clays, thermal investigations 314 ff.
clay-water system, TA 323
cleaning process, bentonite, TA control 302
clinoenstatite formation, hectorite 310
-, sepiolite 334
clinoptilolite, dehydration parameters 99
CO_2, purge gas atmosphere, illites 281 f.
CO_2 atmosphere, influence on carbonate decomposition 319 f.
CO_2 partial pressure 65

Subject Index

CO_2 partial pressure, influence on carbonate decomposition 280 f.
coal, illites 271
coal combustion 64 ff.
coal minerals, DSC 32 f.
coal proximate analysis 64, 75
coal pyrolysis 39 ff.
coal rank indicator, DSC 39
cobalt arsenates 73
coke, TG / multiple gas sequence 76
colemanite, decomposition 125, 129
combustion 65 ff.
-, organic molecules 330 ff.
-, solid fuels 42 f.
combustion efficiency 67
composites, clay / fly ash, water sorption and drying 256 ff.
concrete, refractory 210
condensation 122
consolidation pressure, bentonite 250
contraction, kaolinite 126
controlled transformation rate TA 134 ff.
cooling rate data, metal solidification 8
-, melt crystallization 175
cordierite formation, montmorillonite 301
corrensite 323
corundum 196 f., 201 f.
-, fluorination 203
-, topaz formation 217 f.
corundum production, kaolinite sintering 201
covellite, oxidation 195
Cr, chlorites, DTA determination 354
CR-EGA, principle 143
CR-EGD, kinetical study 141 f.
cristobalite, crystallization 130 f.
cristobalite formation, montmorillonite 300
cross-linking, organo-clays 337

CRTA, kinetics 135 ff.
-, principles 137
cryosuction 253
cryptohalite 196
crystal growth rate, melt crystallization 176, 179 f.
crystallinity, clay minerals, DTA determination 355
-, -, improvement by fly ash 263
crystal water, dehydration 92
Cu-montmorillonite, amines 335
Curie point, detection in TM 74
Curie temperature, magnetites, lowering by impurities 354

Dawsonite, oil shale 44
decomposition centre 124
decomposition enthalpies, carbonates 33 ff.
-, interlayer complexes 348
decomposition reactions, carbonates, partial pressures 65
defect structure, influence on thermal reactions 120
degree of crystallinity 118
dehydration, gypsum 28
-, montmorillonite 295 ff.
-, sulphates 103 ff.
dehydration enthalpies, clays 29
dehydration kinetics 322
dehydroxylation, illite 273
-, illites 355 f.
-, kaolinite 120
-, montmorillonite 299
-, organo-clays 333 f.
-, Q-TG curves 97
-, sepiolite and palygorskite 321, 333
-, sulphates 103 ff.
-, tri-octahedral smectites 304
-, volume increase 126

Subject Index

dehydroxylation temperature, organo-bentonites 358 f.
denitration, uranyle nitrate, kinetic mechanism 150
density, organo-clay complexes 357
depolymerization, Si-O-network 130
derivative DTA, detection 11
derivatograph, clays 315
detection limit, carbonates 66
ΔH values, influencing factor in DSC 22
diamond, TG / multiple gas sequence 76
diatomaceous earth, benzoic acid interaction 343 f.
dicalcium ferrite, Fe-dolomite decomposition 34 f.
dickite, dehydroxylation 96, 126
diffusion, activation energy 174, 179
-, crystal structure 120
-, isokinetical curve 149
diffusion rate 169
dilution, organo-clays, DTA 342
diopside formation, hectorite 310
disc model, topochemical decomposition 124
disorder, illites 354 f.
-, kaolinite, DTA determination 355
-, -, influence on spinel formation 209
-, organo-bentonites 358 f.
-, smectites 355
dissociation, skutterudite 73
dissolution 122
dolomite, coal analysis 69 f.
-, CRTA 146
-, decomposition enthalpy 33 ff.
-, overlapping DTA peaks 14
-, substitution, influence on decomposition 34 f.
drying rate, bentonite suspensions 247 ff.
-, fly ash composites 260 f.
DSC, instrumentation 17 f.
-, introduction 14, 16
-, principles 18
DTA, device 10
-, historical development 11 f.
-, mineral detection 189
-, prospecting tool 12
DTA-EGA, organo-clay complexes 330 f., 340
Duplosan®-bentonite, DTA / TG 361 f.
dust investigation 225
dynamic methods 154

Efflorescences, sulphates 103
EGA, organo-clays 332 f.
electronegativity, influence on thermal dissociation 95 f.
energy barrier, nucleation 178
engobes 232
enthalpies, halloysite dehydration 29
enthalpy change, hectorite endotherms 308
-, melt crystallization 175
-, metallurgy 10
enthalpy difference, DSC 17 f.
enthalpy measurement, DSC 19 f.
enthalpy standards, DSC 22
environmental studies, thermal analysis 352 ff.
epsomite, dehydration parameters 99
equilibrium DTA signals 155
erosion tendency, organo-clay complexes 364
ethylene-glycol montmorillonite, DTA 331
eutectic reactions 154 f.
evolved gas detection 6
-, controlled transformation rate 137 ff.
exchange of components 122
expanding clay minerals 323
expansion, smectites 290
exsolutions 159

Subject Index

F, hectorite 291
Fahrenheit temperature scale 6
fatty acids, adsorbed on clays 343
Fe determination, coals 75
Fe, substitution in dolomite, influence on decomposition enthalpy 34
firing temperature range, illitic clays 276
first-order transformation, melt crystallization by cooling 174 f.
first-order transformations 154 f.
Fischer Assay, oil shales 44
Fithian, Illinois, illite 276 f.
Fithian illite, organics adsorption 338
flue gas - N_2 75
fluorite, thermoluminescence 4
fly ash 63, 67
-, clay properties improvement 247 ff.
forsterite formation, hectorite 310
framework lattice 119
Frankenheim method 8
free enthalpy, melt crystallization 177
freezing, clays 252
furnace atmosphere, control 63
fusion heat, melt crystallization 175

Galena, roasting 191
-, weathering 103 f.
Gardoprim®-bentonite 358 f.
gas atmosphere, variable 63 ff.
gas evolution, heating 5 f.
-, organo-clay complexes 339 f.
gaseous products, internal formation 124
gas flow rate, DSC 24
gel crystallization 123
germanates, glass formers 164
Gibbs enthalpy relation, melt crystallization 177
gibbsite, dehydroxylation 96, 99
glaserite 156 f.
glass, crystallization 129
-, melt formation 176 f., 181

-, thermal behaviour 212 f.
glass crystallization 123
glass formers, melts 164
glass modifiers 128
glass transition 181 f.
glauconite, dehydroxylation 96
-, TA / Mössbauer study 323
glaze 232
glycol, montmorillonite 331
goethite, dehydroxylation parameters 99
Goltix®-bentonite 358 f., 362
grain size, influence on decomposition 120
-, influence on water detection 239 f.
granite, weathering 103 f.
graphite, ilmenite sintering 205 f.
-, intercalation 123
-, TG / multiple gas sequence 76
Green River oil shales 44
grinding, influence on kaolinite dehydroxylation 209
gypsum, coal 76
-, coals, DSC 32
-, dehydration 9, 28

Halloysite, dehydration 28 f., 316, 323
-, dehydroxylation parameters 99
-, heating-rate traces 11
heat capacity concept 7
heat flow, specific heat determination 25
heat flow measure, DSC 19
heat-flux difference, DSC 17
heat-flux DSC 17
heating-curve method, gypsum dehydration 9
heating rate, influence on illite endotherms 275
heating rate determination, Le Chatelier method 10
heating rate variation, melting point determination 167, 169

heats of combustion, coals 43
hectorite 290 f.
-, dehydroxylation 304 ff.
-, endothermic effects, total enthalpy change 308
-, industrial cleaning, TA control 306
-, ordering exotherm 310
-, organo-complexes, exothermic peak temperatures 338
-, use 291
hematite, pyrite decomposition 73
hemihydrate, gypsum dehydration 9, 28
heterogeneous nucleation, melt crystallization 176 ff.
homogeneous nucleation, melt crystallization 176 ff.
hot-stage microscopy 182 f.
humic acid / montmorillonite interaction 320
hydrated sulphates, TA characterization 102 ff.
hydrogen, diffusible, welding flux 234
hydrogenation, coals 42
hydromicas, dehydroxylation 96
hydromuscovite, DTD curve 230
hydroxides, dehydroxylation 96
hysteresis, phase diagram studies 163

Illite, adsorbed organics 338
-, characterization 271 f., 284
-, dehydroxylation 96, 99
-, derivatogram 317 f.
-, DTD curve 230
illite crystallinity, determination methods 354 f.
illite-smectite mixed layer 276
ilmenite, sintering 205 f.
impurities, crystal chemical, DTA determination 353 f.
incomplete combustion, organic matter 43

Indus Valley civilization, pottery kiln 4
inert gas atmosphere 65, 68
intercalation 122 f.
-, organics in clay minerals 322
interface reactions, isokinetical curve 149
interface, crystal-melt, kinetics 179
interlayering, humic acid / montmorillonite 320 f.
interlayer space, montmorillonite, organics 335
interlayer water, clay minerals, dehydration parameters 99
-, phyllosilicates 86, 88
-, smectites 290
interlayer water loss, amine-smectites 344
internal processes 122 ff.
internal surface, water binding 86 f.
internal thermal decomposition 123 ff.
internal thermal dissociation 123 f.
internal thermal reactions 119 ff.
interval transitions 154 f.
intracrystalline decomposition 125 f.
intraframework reactions 121 f.
inversion temperature, sulphides, lowering by substitutions 354
iron ores, DTA prospection 12
IR spectroscopy, water detection 235
iso-conversion points, CRTA 145
isokinetical curves 149
isoproturone-bentonite 358 f., 363
isothermal experiments, kinetics 145
isothermal transitions 154

Jarosite, dehydroxylation 96, 99

K value, DSC calibration 19 f.
K-bentonite, derivatogram 319
kandites, decomposition 119
kaolin, ceramics processing 207 f.

Subject Index

kaolin, use 226, 232
kaolinite, adsorbed organics 338
-, coals, DSC 32
-, decomposition 6
-, dehydroxylation 14, 96, 99, 120 f., 126
-, fluorination 196 f., 201 f.
-, heating-rate traces 11
-, Kalabsha 208
-, Q-TG curve 97
kaolinite-calcite interaction 322
kaolinite-dickite distinction 126
kerogen, CR-EGA 144
-, oil shale 44 f.
kieserite transformation, activation energy 174
kinetic effect, peak broadening 164
kinetic equation, constants 98 f.
kinetics, CRTA approach 135 ff.
-, phase diagram studies 152 ff.

Laponite, organo-complexes, exothermic peak temperatures 338
Le Chatelier method, clay minerals dehydroxylation 10
lignin, TG / multiple gas sequence 76
lignite, hydrogenation 42
limestone, firing 4

Magnesioferrite, coals 33
magnesite, ceramic stabilizer 211
-, decomposition enthalpy 33 ff.
-, oil shale 72
-, pyrite decomposition 73
magnetic phases, detection 73
magnetite, Curie temperature lowering by impurities 354
malachite green, smectites 346
manganates, enthalpies of dehydration 30
manganese ores, DTA prospection 12
manganite, decomposition 73

-, dehydroxylation 96
marine clay, geotechnical properties 247 ff.
mass-change determination, isothermal 9
mechanisms, thermal decomposition of solids 148
melt, lead silicates 164
melt crystallization, kinetics 172 ff.
melting 122, 153 f.
melting, DSC calibration 21 f.
melting temperature reduction, silica 212
metacristobalite 121, 130
metakaolinite 208
metallurgy, TA history 10
meta-sepiolite 334
methylene blue, smectites 346
mica, use 226
mica dehydroxylation 126
minerals, chemical processing 189 ff.
mixed-layer, derivatogram 317
mixed-layering, organo-bentonites 358 ff.
mixtures, clay-organics 342
molybdenite, roasting 196
Monte Carlo simulation, melt crystallization 180
montmorillonite, characterization and use 289 f.
-, dehydroxylation 299
-, dehydroxylation 96, 99
-, DTD curves 228, 230
-, heating-rate traces 11
-, intercalation of humic acid 320
-, interlayer water 88
-, intraframework re-arrangement 129
-, organo-complexes, exothermic peak temperatures 338
-, thermal analysis 293 ff.
-, use 232
montmorillonite group, characterization 288

Subject Index

mordenite, zeolitic water 88
morphology, influence on decomposition 121
mullite 196 f., 202, 220
multiple purge gas sequences 75 ff.
multistage decomposition, colemanite 125
multistage dehydration, enhanced separation of steps 137 ff.
muscovite, calcined, water release plot 236 f.
-, dehydroxylation 96
-, DTD curve 230

N_2, carbonate analysis 69 ff.
Na_2SO_4, modifications 156
Na_2SO_4-K_2SO_4, phase diagram study 156 ff.
nahcolite, oil shale 44
NH_4F, kaolinite sintering 197 f.
nontronite 289
-, dehydroxylation 96
-, DTD curve 230
-, organo-complexes, exothermic peak temperatures 338
nucleation, isokinetical curve 149
nucleation-growth 131
nucleation process determination, melt crystallization 180
nucleation rate, crystallization 167
-, influencing factors 180
-, melt crystallization 175 ff.
nucleation rate determination 180 f.

Occlusion, water in aragonite 92
oil drilling fluids, bentonite 302
oil sands, pyrolysis 46 f.
oil shale, DSC 44 f.
-, DTA of carbonates 72 f.
opals, dehydration 87, 90, 99
order-disorder transitions 155

-, montmorillonite 300
ordinate displacement correction, specific heat determination 27
organic matter, combustion 67
organo-bentonites, thermal characterization 358 ff.
organo-clay complexes, DTA 328 ff.
-, exothermic peak temperatures 338
organo-illite complex 335
organophilic clays 302 f.
organo-sepiolite, water evolution 344 f.
overheating effects, phase diagram study 163
oxidation 65 f., 75 f., 122
-, incomplete, organic matter 43
-, N and C 330
-, organo-clay complexes 334 ff.
oxide systems, binary, TA 153 ff.
oxyhydroxides, dehydroxylation 96

PA curves 67
-, carbonates 320
palygorskite, D_2O adsorption 321
-, use 227, 232
-, water binding 86
pandermite, decomposition 126, 129
partial pressure, furnace atmosphere TA 64 f.
-, mineral decomposition 127
partial TG curves 142
particle size, influence on DSC peak area 23
particle size effect, specific heat of coals 48
Pb_3SiO_5-Pb_3GeO_5, phase diagram study 164 ff.
peak area measurement 22
peak intensity, DSC, heat flow measure 19
peak separation, variable gas atmosphere 71 f.

Subject Index

peritectic reactions 154 f.
perlite 225
-, water distribution curves 87, 89
perovskite formation 205
pesticide-bentonite complexes 357 ff.
pesticides-montmorillonite complexes 322
petroleum coke 357
petroleum coke formation, organo-clays 335 f.
phase boundary diffusion, melt crystallization 180
phase boundary kinetics 179
phase diagram determination, binary oxide systems 153 ff.
phase transitions 122 ff.
-, minerals 9
phlogopite, dehydroxylation 96
phyllosilicates, dehydroxylation 96
phytoclasts 67
π interactions, dyes / montmorillonite 340 f.
pine cellulose, multiple gas sequence 76
piperidine-smectites, DTA 346
pitchstone 225, 231
polar surface, internal, minerals 84
polymorphic transformations 121
polymorphism 153
-, Na_2SO_4 159
Portland cement concrete 210
portlandite, dehydroxylation 96
-, Q-TG curve 97
pottery kiln 3 f.
power-compensated DSC 17 f.
-, temperature range 18
pressure gradients, kinetics 136 f.
processes, heterogeneous, classification 122
pseudomorphism 123
purge gas atmosphere 67 ff.

pyrite, clays 228 f.
-, oil shale 44, 72
-, trace determination 353
-, weathering 103 f.
pyrite / pyrrhotite transformation, coal pyrolysis 42
pyrite removal, coals 73, 78
pyrobitumen, shale combustion 46
pyrolysis, DSC peak shape distortion 23
pyrometer, development 10
pyrophyllite, dehydroxylation 96
-, Q-TG curve 97
-, use 226
pyrrhotite, pyrite decomposition in N_2 73

Q-derivatography 119
Q-TG curves 88, 97
quadrupole gas analyzer, CR-EGA 141
quality control 225
quartz, α-β inversion, particle size influence on DSC peak area 23
-, α-β transformation 121
-, coals, DSC 32
-, inversion 215
-, topaz formation 217 f.
quartz-cristobalite transformation 130
quartz detection, porcelains 10
quasi-isothermal heating technique 87 f.

Radiation loss, DSC heating 23
rate-jump method, CRTA determination of activation energies 145 ff.
rate of reaction, influencing factors 135 f.
-, modulation, CRTA 146 f.
ratio method, specific heat determination 26
reaction front 124
reaction rate, decomposition, influencing factors 121

Subject Index

reaction separations, carbonates 69 f.
reaction suppression, iron minerals 65 f.
re-constitution, crystal structure 128
recrystallization 122
refractories, processing 206 f.
refractory bricks 207
regions, thermal, DTA of organo clays 330 f.
rehydration, illite 273
rheologic behaviour, organo-clay complexes 357
roasting, sulphides 190 ff.
-, temperature regulation 194
Rudberg apparatus 8

S, boiling point 11
-, DSC 33
-, trace detection 354
Saladin system 10
salpetre, cooling data 8
salt minerals, DTA prospection 12
sample geometry, DSC 23
sample preparation, DSC 23
saponite 289, 304
-, dehydroxylation 96
-, organo-complexes, exothermic peak temperatures 338
-, weight loss 231
sassolite, dehydroxylation 96
Se, boiling point 11
second-order transitions 155
segregation, synthesis 122
self-cooling, dehydration 135
self-diffusion, water in solid solution 91 f.
sepiolite, D_2O adsorption 321
-, dehydration 322
-, organo-complex, water evolution 344 f.
-, pyridine-complex 337
-, use 227

sepiolite dehydration, CRTA 139 f.
serpentines, dehydroxylation parameters 99
-, DTA classification 12
-, use 226
shale combustion enthalpies, DSC 45
sharpness, decomposition peak 98 f.
shrinkage, ceramics 6
-, clay / fly ash composites 258 ff.
shrinkage limit, clays 208 f.
shrinking, organo-clay complexes 357
siderite, decomposition enthalpy 33 ff.
-, illite 283
-, oil shale 72
siderite decomposition 67, 74
silica, melting temperature reduction 212
silicate melt, kinetic analysis 172 ff.
Silver Hill, Montana, illite 273
single purge gas conditions 67
sintering 120
-, kaolinite / NH_4F 196 f., 201 f.
size control, water in sulphates 92
skutterudite, decomposition 72 f.
slags, thermal investigation 225
smectites, dehydroxylation 96
-, disorder 318
-, DTD curves 230
-, organo-complexes, π interactions 341
-, rhodamine adsorption 321
-, thermal characterization 288 ff.
smithsonite, decomposition 194
SO_2, partial pressure 70
solid solutions 154 ff.
solution-transport mechanism 173
specific heat, coals 47 f.
-, oil sands 49
specific heat, thermodynamics 25 ff.
sphalerite, roasting 194
-, weathering 103 f.
spinel crystallization, illite 274 f., 284
-, montmorillonite 300

Subject Index

sponoidal effects, glass 123
SRD test, flue gas sorbents 77
standards, DSC 17, 22
static methods, phase diagram studies 153 f.
stearic acid, clay minerals 322, 343 f.
steel, water content 234
Stomp®-bentonite 358 f.
structural factors, influence on thermal reactions 119 ff.
structural reorganization, montmorillonite 301
structural thermochemistry 119 f.
sublimation 122
substitution, ankerite 70
-, carbonates 34 f.
-, montmorillonite 289 f.
-, silica phases 130
suction, bentonite 250
sulphates, hydrated 102 ff.
-, water types 104
sulphation, SRD test 77
sulphide ores, roasting 190 ff.
sulphides, combustion 353
sulphur, cooling rate data 8
surface energy, melt crystallization 176 f.
suspensions, bentonites, drying properties 249 ff.
sweeling, organo-clay complexes 356 f.
-, clay / fly ash composites 258 ff.
-, layer silicates 126 f.
synthesis, topaz 215

Talc, dehydroxylation 96
-, use 227, 232
temperature control 6 f.
-, history 8 f.
temperature gradients, kinetics 136 f.
temperature-jump method, kinetics 146

temperature range, power-compensated DSC 18
TG / gas sequence 75
thenardite 156
Theophrastus 4
thermal analysis, definition 3, 6
-, prehistory 3
thermal conductivity, coals 23
thermal dissociation, water-bearing structures 92 ff.
thermal hydrolysis 333
thermobalance, calcite decomposition 11
thermocouple 10
thermodilatometry 11 f.
thermoluminescence 4
thermomagnetometry 63, 73
thermomechanical analysis 63
time scale, kinetics 172
TiO_2, cross-linking in montmorillonite 323
titanium slags 205 f.
todorokite, DSC 29 f.
topaz, synthesis 215 ff.
topaz formation 196 f., 202
topochemical reactions 121 f.
topotaxy 121
transformation rate calculation, melt crystallization 180

Undercooling effects, phase diagram study 163
undercooling rate, melt crystallization 178
uranyle nitrate, CRTA 139 f.
urea, montmorillonite 347

V-chlorite 323
variable atmosphere control 62 ff.
vermiculite, use 226
vernadite, DSC 29 f.
viscosity, melt crystallization 175

Subject Index

Vitruvius 4
volatilisation reactions, oil sands 46 f.
volkonskoite 323
volume increase, kaolinite
   dehydroxylation 126
volume scale, kinetics 173

Wasiliew apparatus, swelling /
   shrinkage measurement 258
waste disposals, clay barriers 355
water detection method, errors 237 f.
-, rapid in raw materials 235 ff.
water release, organo-clay complexes
   330 f.
water retention, bentonites 247 f.

water sorption, bentonite 248 ff.
water species, minerals 84 ff.
weathering, sulphate formation 103
Wedgwood pyrometer 6
weighing error, ΔH determination in
   DSC 22
weight loss, bentonites 249 f.
-, hectorite 306
-, illite 276
welding flux, rapid water detection 234
Wyoming type, montmorillonite 293 f.

Zeolitic water 86
-, dehydration parameters 99
zircon, desilication 201

Lecture Notes in Earth Sciences

Vol. 30: E.G. Kauffman, O.H. Walliser (Eds.), Extinction Events in Earth History. VI, 432 pages. 1990.

Vol. 31: K.-R. Koch, Bayesian Inference with Geodetic Applications. IX, 198 pages. 1990.

Vol. 32: B. Lehmann, Metallogeny of Tin. VIII, 211 pages. 1990.

Vol. 33: B. Allard, H. Borén, A. Grimvall (Eds.), Humic Substances in the Aquatic and Terrestrial Environment. VIII, 514 pages. 1991.

Vol. 34: R. Stein, Accumulation of Organic Carbon in Marine Sediments. XIII, 217 pages. 1991.

Vol. 35: L. Håkanson, Ecometric and Dynamic Modelling. VI, 158 pages. 1991.

Vol. 36: D. Shangguan, Cellular Growth of Crystals. XV, 209 pages. 1991.

Vol. 38: W. Smykatz-Kloss, S. St. J. Warne (Eds.), Thermal Analysis in the Geosciences. XII, 379 pages. 1991.

Lecture Notes in Earth Sciences aims to report new developments in research and teaching in the entire field of earth sciences – quickly, informally and at high level. The type of material considered for publication includes:

1. Single author or co-authored monographs which may later be published in full book form
2. Class lecture notes in their original form
3. Seminar work-outs
4. Reports of meetings, provided they are
   a) of exceptional interest and
   b) devoted to a single topic.

The timeliness of a manuscript is more important than its form, which may be unfinished or tentative. If possible, a subject index should be included. Publication of Lecture Notes is intended as a service to the international community of earth scientists, in that a commercial publisher, Springer-Verlag, can offer a wider distribution for documents which would otherwise have a restricted readership. Once published and copyrighted, they can be documented in the scientific literature.

Manuscripts

Manuscripts should be no less than 100 and preferably no more than 500 pages in length.
They are reproduced by a photographic process and therefore must be typed with extreme care. Symbols not on the typewriter should be inserted by hand in indelible black ink. Corrections to the typescript should be made by pasting in the new text or painting out errors with white correction fluid. Authors receive 75 free copies and are free to use the material in other publications. The typescript is reduced slightly in size during reproduction; best results will not be obtained unless the text on any one page is kept within the overall limit of 18 x 26.5 cm (7 x 10½ inches). On request, the publisher will supply special paper with the typing area outlined. More detailed typing instructions are also available on request.

Manuscripts should be sent directly to Springer-Verlag Heidelberg.

Springer-Verlag, Heidelberger Platz 3, D-1000 Berlin 33
Springer-Verlag, Tiergartenstraße 17, D-6900 Heidelberg 1
Springer-Verlag, 175 Fifth Avenue, New York, NY 10010/USA
Springer-Verlag, 37-3, Hongo 3-chome, Bunkyo-ku, Tokyo 113, Japan

ISBN 3-540-54520-4
ISBN 0-387-54520-4

LOOK FOR BARCODE →